Demco, Inc. 38-293

DATE DUE

RENEWALS 691-4574

			AUG 12
			AUG 09

Fundamentals of Experimental Design

Fundamentals of Experimental Design

SECOND EDITION

Jerome L. Myers

University of Massachusetts

Allyn and Bacon, Inc., Boston

Contents

Preface to the First Edition

To be useful a book on experimental design must clearly present designs and appropriate data analyses, supplemented by specific instances of application and by numerical examples. One purpose of this volume is such a presentation. However, recipes for designs and analyses are in themselves not sufficient for the training of independent researchers. It is hoped that the development of this book will provide a basis for selecting among designs and analyses and for extending the principles presented to designs and analyses which are not discussed here. In short, this book is concerned not merely with what is done but also with why it is done.

The reader should be familiar with material usually covered in a one-semester introductory statistics course—such topics as the binomial, normal, chi-square, and t distributions, and bivariate regression and correlation. No previous exposure to the analysis of variance is assumed. Ideally, the reader should have some foundation in such general inferential topics as estimation, Type I and Type II errors, power, and confidence intervals. A brief review of these topics is contained in Chapter 2. Notational usage, which is also considered to be part of a well-taught introductory course, is reviewed in Chapter 3. No mathematics beyond college algebra is required.

In organizing a book of this sort, the writer may exhaustively discuss a design within a single chapter, at that point treating the wide variety of data manipulations which he deems pertinent. A somewhat different strategy is followed here: Each design chapter covers the model and calculations related to the overall F tests which dominate our experimental literature. Changes in the model and in the computations are followed and the pertinent inferential problems noted as new design principles are added—matching, repeated measurements, nesting of variables, and counterbalancing. With a firm grasp (hopefully) of how the statistical model is developed and of the computations which it leads to, the reader should then be ready to turn to other data manipulations which may be performed on the designs which have been previously treated. Therefore, the later chapters consider the concepts and calculations relevant to such diverse topics as confidence intervals, covariance adjustment, range tests, and trend analyses.

Since this is meant to be a basic experimental design book rather than a handbook of designs and analyses, a number of topics have been excluded. The criteria for inclusion were whether or not the material was considered fundamental for the development of subsequent topics (as well as topics which have not been included), and the likelihood that the material might be of use to the researcher. Those topics that have been included in the book have generally received extensive consideration. It is hoped that the resulting book will provide a reasonably sound foundation in experimental design and analysis.

The author is indebted to the literary executor of the late Sir Ronald A. Fisher, F.R.S., Cambridge; to Dr. Frank Yates, F.R.S., Rothamsted; and to Messrs. Oliver & Boyd Ltd., Edinburgh, for permission to reproduce Tables A-3, A-4, and parts of A-5 from their book *Statistical Tables for Biological, Agricultural and Medical Research.*

There are many people who have made a contribution to the development of this volume. I particularly wish to express my gratitude to my secretary, Mrs. Dorothy Thayer, whose contribution in preparing the manuscript is incalculable; to Miss Virginia Kochanowski for drawing the figures; to several anonymous reviewers for their comments; and to Dr. Mary M. Suydam, who painstakingly read the manuscript. A multilithed edition was used as a text for my graduate course in experimental design; I am indebted to the many students who detected errors and who suggested improvements in exposition. A special note of gratitude is due my wife and co-worker, Dr. Nancy A. Myers, not only for reading and commenting on the manuscript, but also for encouraging me throughout the writing process.

J. L. M.

Preface to the Second Edition

This revision was undertaken for several reasons. First, a few errors—both typographical and substantive—required correction; I hope that I have not introduced too many new errors in the process. Second, several years of experience with the first edition indicated that there were several points at which a different or expanded statement might enhance clarity. Third, additional material seemed to warrant inclusion on the grounds that such material involved aspects of experimental design that I now judge to be useful. Fourth, and finally, this revision is a response to many instructors who requested answers to the exercises at the end of each chapter. I have revised the exercises, discarding those I now find ambiguous and adding others that seem to test the knowledge I have sought to impart; answers—hopefully correct—are now available.

Some indication of the major revisions in material may be helpful to the instructor who is familiar with the first edition and contemplates adopting this revision. Chapter 2 now includes a discussion of random variables, both continuous and discrete, and of the relationships among certain specific distributions relevant to subsequent developments. The discussion of robustness in Chapter 4 is considerably longer than before and includes an extensive statement of the consequences of violations of assumptions for both Type I and Type II errors. Chapter 10 is approximately twice as long as before; I have included some variations of the Latin square design that were previously omitted, but most of the increase is due to coverage of applications in which each cell contains different subjects. Chapter 11 now contains a discussion of the use of expected mean squares to estimate proportions of population variance due to various factors. Chapter 13 now provides what I hope is a clear distinction between planned and unplanned comparisons, orthogonal and nonorthogonal, with a detailed discussion of the accompanying analyses. In Chapter 14, I have provided a more extensive discussion of the rationale of trend analysis, with additional examples of the sorts of experiments in which such analyses might prove profitable. There are many other additions, but these are the major ones. Furthermore, as I remarked above, I have attempted to improve upon the style of presentation at several points.

I have resisted the temptation to increase the number of chapters—for example, to tackle nonparametric statistics, quasi-experimental designs, and incomplete block designs. I have sacrificed additional topics for the sake of depth in covering the topics presented, with emphasis upon rationale, particularly upon models of design and their implications.

The purpose of this revision, as of its predecessor, is to serve as a text for a one-semester course in experimental design, rather than as an all-purpose handbook. Hopefully the book will provide a solid base to which the student can subsequently add further statistical knowledge.

J. L. M.

Planning the Experiment

1.1 INTRODUCTION

A psychological experiment is undertaken in order to determine the factors which influence a certain behavior and the extent and direction of their influence. The experimenter seeks answers to such questions as, What are the relative effects of these three drugs upon the number of errors made in learning a maze? Which of these training methods is more effective? What changes in auditory acuity occur as a function of these changes in sound intensity? If an experiment is to answer such questions adequately, the investigator must first specify those factors whose effects are to be studied (*independent variables*); minimize the operation of factors which are not of interest at the time (*irrelevant variables*); carefully select a measure, or measures, of the behavior which he is investigating (*dependent variables*); and choose those whose behavior is to be measured (*subjects*). Planning these four basic aspects of an experiment is the first and most critical step in obtaining answers about behavior. Therefore, this first chapter presents a general discussion of these considerations. Although much of this discussion may appear obvious to the well-trained and experienced researcher, it is hoped that the student of experimental design will profit from this review of the many things to be considered in planning an experiment.

1.2 THE INDEPENDENT VARIABLE

Once the experimenter has decided upon the independent variable or variables that are to be studied, he must choose the actual treatments: the levels—specific types or amounts—of the independent variable which will be tested in the experiment. He must decide which drugs, which training methods, which sound intensities will be compared. In considering this class of decisions, it is helpful to distinguish between two types of independent variables, quantitative and qualitative, and to discuss these separately.

1.2.1 Quantitative variables. A quantitative independent variable is a variable whose levels differ in amount. Examples of such a variable are amount of reward, intensity of shock, and number of practice trials. Generally, the experimenter is not interested in the specific numerical levels chosen for inclusion in the experiment. For example, in a study of the effects of inter-trial interval upon the speed of learning lists of words, the experimenter may choose 2, 4, and 6 seconds as his level for interval length. Probably 1.8, 3.8, and 5.8 seconds would be just as adequate for his purpose, but he tends to think in terms of whole numbers. The levels of a quantitative independent variable are usually of interest only to the extent that they permit the experimenter to determine whether any change in the quantity manipulated results in a change in behavior, and, if so, what are the characteristics (e.g., the shape, slope, and position) of the function relating the independent and dependent variables. This being the case, the levels of the independent variable should be chosen to cover a wide enough range to detect any behavioral change which might result, and in sufficient number and close enough together so that the shape of the function will be clearly defined.

In any single experiment it may be difficult to achieve the ideal of broadly covering the continuum of the independent variable with many levels, close together. It is not always possible to decide without some pilot experimentation how many and how close the levels should be. Further, the limitations of time, money, and subjects may make the ideal difficult to realize. Therefore, it will often be best in initial experiments to determine generally whether any behavioral change occurs as the independent variable is manipulated and to attain a rough description of the shape of the function relating the independent and dependent variables. If desirable, subsequent experiments can be designed to yield a more precise definition of the function.

For example, suppose that an experimenter is interested in the relationship between x and y of Figure 1-1. Assuming that there are other independent variables which he also wishes to investigate, he may not be able to include as many levels of x as would seem ideal. In his first experiment, he might include the levels x_1, x_3, and x_5, thus learning that variations in x do result in variations in y and obtaining some idea of the slope of the function and the minimum range of x over which y varies. In subsequent experiments, levels might be chosen between x_2 and x_4, giving a more specific picture of the function.

Note that the decision about the selection of levels depends upon the results of previous research. However, such decisions may also be influenced by theoretical considerations. For example, suppose that the experimenter is trying to decide which of two theories is correct, one of which predicts gradual changes in y (in Fig. 1-1) with changes in x, while the other predicts a "staircase" effect, stepwise changes in the function. In this case, an examination of a broad range of levels of x might be sacrificed in order to concentrate more levels within a narrow range.

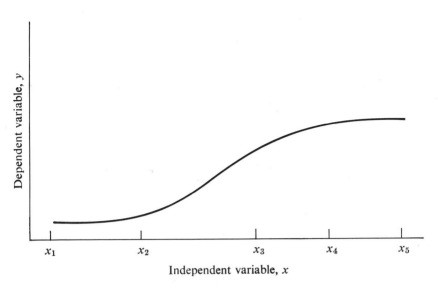

Independent variable, x

FIGURE 1–1 *A function relating dependent and independent variables*

While on the topic of quantitative independent variables, one might note some of the inferential pitfalls that occur, particularly when only a few levels of the variable are included in an experiment. Suppose that the variable x of Figure 1-1 is under investigation and that the relationship between independent and dependent variable is that depicted in the figure. If no other information were available, and if only x_1 and x_2 or x_4 and x_5 were selected for investigation, the data might lead the experimenter to conclude that the independent variable, x, does not influence behavior. If the experimenter happened to choose x_1, x_3, and x_4, he might conclude that performance and x were related by a linear function. The first instance is an example of extrapolation beyond the range of levels selected, and the second instance is an example of interpolation between the levels selected. It is wise to recognize the tentative status of inferences which go beyond the levels that have been included in the experiment.

1.2.2 Qualitative variables. A qualitative independent variable is one whose "levels"* differ in type. Examples of such a variable are type of punishment, method of training, and type of instructions. The particular "levels"—the specific types—of the variable which are included in the experiment are usually of direct interest to the experimenter, in contrast to the numerical levels of quantitative variables, which, as previously mentioned, are often

---

* The word "levels" is actually not applicable to the qualitative variable inasmuch as all definitions of the word suggest an ordering or ranking. However, since it is a useful concept for later chapters, it has been employed for both the quantitative and qualitative variables.

less important in themselves than for the information they provide about some function relating the independent and dependent variables. Qualitative treatments are chosen because previous research, theoretical predictions, or practical considerations dictate their choice. For example, two particular methods of teaching reading may be chosen for experimental comparison because (*a*) previous experimentation suggests that these are the two most efficient procedures available, (*b*) the results of the comparison should differentiate between two educational theories in which the experimenter is interested, or (*c*) other methods require so much time and money that they are impractical to experiment with and would not be used in practice even if effective.

1.2.3 Control levels. A consideration common to both qualitative and quantitative variables is the selection of *control* levels. These are treatments which may not be of interest in themselves but which provide additional information about the effects of one or more other treatments. Suppose that one wished to compare the effects of teaching, and of not teaching, reading in kindergarten upon reading scores obtained at the end of first grade. The experiment demands two groups of first graders, both with kindergarten experience, but only one of which has had reading experience in kindergarten. Adding a control group, children who had no kindergarten experience at all, would permit evaluation of the possibility that organized activity alone is helpful preparation for grade-school performance.

1.2.4 Fixed and random variables. Implied in the discussion of this section has been the premise that we are dealing with *fixed* variables in our research, variables whose levels are arbitrarily chosen by the experimenter. A second, less frequently occurring, but equally important possibility is that the variable is *random*, i.e., that its levels were chosen from some larger population of levels on the principle that all members of the population have an equal opportunity to be chosen. The experimenter chooses his levels in this way when his major consideration is to obtain a reliable estimate of the variability in the population. For example, in order to determine whether newly manufactured calculating machines perform similarly, a sample, representative of the entire population of such machines, would be tested. Another random variable which is of major concern in psychological research is the subject. Individual subjects are rarely chosen for unique personal attributes. Generally, subjects in psychological experimentation are a random sample from some population, perhaps all the students taking basic psychology at some college.

The distinction between fixed and random independent variables is important to our inferential processes. When the levels of the variable have been arbitrarily chosen, any inferences about differences among the effects of the levels are limited to the particular levels chosen. Having compared the effects of three arbitrarily chosen methods of teaching reading, it is possible to draw inferences about these three methods, not about any broader population of methods. On the other hand, the observed variability among performances on five calculators chosen randomly from a factory's output leads to conclusions about the variability in the population of machines (the factory's output). Our inferential statements extend beyond the actual levels sampled (the five machines in the experiment) to the broader population from which they were randomly sampled. This distinction also has important implications for the analysis of data, but this aspect will be considered at a later point in the text.

1.3 IRRELEVANT VARIABLES

Drawing conclusions about a certain behavior would present no problems if the only variables affecting the behavior were those selected for study by the experimenter. Unfortunately, behavior is seldom so simply caused: it may be a function of such variables as the intelligence, prior experience, attitude, and age of the subject, the time of day at which the data are collected, and even the tone of voice in which instructions are read. If one is not interested in the effects of these variables in some particular experiment, they are irrelevant in the limited sense that the experiment has not been performed to investigate them. However, since such "irrelevant" variables do influence behavior, care must be taken to ensure that their effects are minimized for the duration of the experiment in which they are not being studied. In this section, ways of accomplishing this will be considered.

It is difficult to ensure that different treatment groups will be perfectly matched with regard to all irrelevant variables which might influence the data. For example, the assignment of subjects to two problem-solving groups might, by chance, result in one treatment being applied to a group of subjects with a higher average intelligence than the second group. This bias is called a *random bias* if each subject had an equal chance of being assigned to either group. In this case, our statistical techniques, which assume that biases are random, will take the bias into account.

We can assess the likelihood that differences in the abilities of our subjects were sufficient to account for differences in the average performances of the two experimental groups. If this likelihood is very small, we will conclude that the two levels of the independent variable differ in their effects upon problem-solving performance. However, if the assignment of the brighter subjects to one treatment occurred because of a lack of careful randomization, we have a *systematic bias*, which makes it exceedingly difficult to draw valid inferences from the data.

If the brighter students, under treatment *A*, perform better, is intelligence or the treatment responsible? If the two groups perform equally well, are the treatments equally effective, or is it possible that the more effective treatment was handicapped by being applied to the less able problem solvers? Clearly, such systematic biases are one possible danger resulting from the presence of irrelevant variables.

Even when irrelevant variables do not result in systematic biases, their presence is reflected in error variance, that variability among scores which cannot be attributed to the effects of the independent variables. There will always be some error variability, even among scores that have been obtained under the same experimental treatment, for these scores will come either from different individuals who differ in such variables as intelligence, attitudes, and age or from the same individuals at different points in time, who will show change in such variables as attentiveness, practice, and fatigue. The greater such error variability, the more difficult it is to determine the effects of independent variables. The following example may illustrate why this is so.

Assume that the two sets of data below have been obtained under two treatments, *A* and *B*, each applied to three subjects. The mean performance under treatment *B* is better, as reflected by a mean of 6 as against one of 5 for treatment *A*. However, the individuals within a group differ by at least as much as do the group means, and we cannot be sure whether the difference between the *A* and *B* means is due to individual or treatment differences.

A	B
4	5
5	6
6	7

Now suppose that all the scores for *A* were 5 and all those for *B* were 6. The treatment means are the same as before, but if people treated alike do not differ and people treated differently do, it seems reasonable to conclude that the treatments do have different effects.

In the remainder of this section we will consider the control of irrelevant variables; that is, the elimination of systematic biases and the reduction of error variance. The approaches to this control problem may be grouped in the following categories.

1.3.1 Uniform application of the irrelevant variable. If only one level of the irrelevant variable is present in the experiment, it contributes no variability at all and therefore cannot give advantage to any one level of the independent variable, nor can it contribute to error variance. A major aid in uniformly applying the irrelevant variable is automation. Thus, electronic timers allow us to keep inter-trial intervals constant over trials, subjects, and experimental conditions; tape-recorded instructions are presented in the same words and tone to every subject; and automated animal test cages (which also serve to house the animals) eliminate handling and provide a uniform environment for all animals. However, while automation is helpful, the careful experimenter can do much to minimize variability without it. Care can be taken to read the same instructions to all subjects or to provide similar living conditions and a minimal amount of handling for all animal subjects. There will also be some variables which cannot be uniformly applied through the use of automated equipment. For example, if the experimenter expects the age of his subjects to influence their behavior, and if he is not interested in this variable, he may choose subjects who fall within certain narrow age limits. Every experiment will have its own potential sources of irrelevant variation, but a careful analysis of the situation can result in the elimination or minimization of many of these.

1.3.2 Randomization. Randomization guards against the danger of systematic biases in the data. Suppose one is interested in comparing the effects of two sets of instructions upon problem-solving performance. One obvious potential source of bias is problem-solving ability; it is necessary to guard against a systematic bias due to the application of one set of instructions to the better problem solvers. One way to do this is to assign the subjects randomly to treatments; that is, by some method that ensures each subject is equally likely to be assigned to either set of instructions. Each subject could draw a number from a hat, then the odd-numbered subjects could be assigned to one treatment and the even-numbered subjects to the second. Note that randomization does not ensure that the two experimental groups are perfectly matched on those variables which might influence problem solving. Randomization does ensure that over many replications of the experiment neither treatment will have an advantage. In any one experiment one group could have an advantage (the odd-numbered subjects might, by chance, have higher intelligence), but statistical procedures that assume randomization take these biases into account.

Randomization does not only apply to the assignment of subjects to the levels of independent variables. It can also be applied, for example, to the selection of orders of presentation of treatments when each subject is tested at a number of levels of the independent variable. In fact, there is a vast array of different schemes for collecting data. Some of these will be considered next, with particular emphasis on the ways in which error variance may be further reduced.

1.3.3 Experimental design and analysis. An experimental design is a plan for running the experiment. One such plan is complete randomization, in which each subject is randomly assigned to only one combination of levels of the independent variables. One might extend such a design by including irrelevant variables as independent variables. For example, in the experiment on instructions and problem solving, the subject pool might be divided into three levels of intelligence—low (I.Q. under 85), medium (86–115), and high (above 115). Low I.Q. subjects would then be randomly assigned to the two instructional sets, and similarly for medium and high subjects, giving six combinations of I.Q. and instructions with an equal number of subjects in each group. The advantage of this plan over the original completely random assignment of subjects to the two treatments is that it permits a more accurate assessment of the effects of instructions. One may now remove, through statistical analysis, variability in the data due to differences in problem-solving ability among the three levels of intelligence. This design, in which the levels of the independent variable are matched on some irrelevant variable, is said to be more *efficient*, i.e., to result in less error variability, than the completely randomized design first considered. A more extensive discussion of the matching design appears in Chapter 6.

Matching of treatments may be made on the basis of variables other than the subjects' personal attributes. For example, if it were necessary to divide the subject testing between two experimenters, subjects could be randomly assigned to experimenters regardless of treatment group, or one could ensure that half of each treatment group was run by each experimenter. The second method is similar to the matching of intelligence previously suggested and would be recommended if there were any reason to suspect that experimenter differences might be a source of variability in the data.

There are advantages to running each subject through all levels of the independent variable. For example, a subject might do one problem-solving task under one set of instructions and a second equivalent task under the second set of instructions. Since all subjects experience both treatments, there can be no systematic bias due to personal attributes such as intelligence. Furthermore, computations exist for removing variability due to individual differences, permitting evaluation of the effects of instructions against a smaller error variability. The major drawback of this design is the possibility of a systematic bias due to temporal effects. If one set of instructions is always given first, the second set might profit from practice or be handicapped because of fatigue. For this reason the order of presentation should be random; each treatment should have an equal chance of being assigned to each position in the sequence of treatment presentations. Repeated measurement designs of this sort are treated at length in Chapter 7.

The randomization just described might be restricted to ensure that each treatment appeared equally often in each position in the sequence of presentations. In the problem-solving experiment, each of 20 subjects could be assigned a number from 1 to 20; each number would be assigned exactly once. If the number is even, one instructional set is given first; otherwise the other set comes first. Such a design further reduces error variance by permitting removal of variability due to temporal effects in addition to that due to individual differences. This type of design is generally referred to as a Latin square design, and is more fully discussed in Chapter 10.

With any design, further control of the effects of irrelevant variables is possible if one can somehow analyze out that part of each score which is due to the irrelevant variable. For example, rather than matching subjects on intelligence test scores, one could use the design described earlier, in which subjects were randomly assigned to treatments regardless of intelligence level. Each subject's intelligence score could then be used as a basis for adjusting his problem-solving scores. The method of adjustment, called analysis of covariance, is discussed in Chapter 12.

If control of irrelevant variables were the only concern of the experimenter, there would be greater uniformity in the selection of experimental designs and possibly more extensive application of the analysis of covariance. However, there are at least three other factors to consider in selecting a design—the information desired, the model for the data analysis, and the practical requirements of the situation. These factors are considered next.

1.4 FACTORS IN SELECTING EXPERIMENTAL DESIGNS

1.4.1 Information. Different types of designs yield different sorts of information. If information about the effects of time is wanted, a repeated measurements design is required. If there is interest in the effects of the order of presentation of treatments, a Latin square design is indicated. If the experimenter hypothesizes that the effect of the independent variable will change as a function of certain characteristics of his subjects, some sort of matching on these characteristics might be undertaken. If certain single or joint effects of variables are of more interest than others, there are designs which permit a more efficient evaluation of these effects at the expense of a loss of information on others. These are but a few examples of the ways in which designs may differ with respect to the information they provide.

1.4.2 The model. The validity of any inference drawn from a statistical analysis rests upon the validity of an underlying model, a set of assumptions about the data. The violation of any one of the assumptions may result in an incorrect inference about the effects of the independent variable. Since the model is a function of the design employed, selection of the experimental design must involve consideration of which assumptions are implied, whether they are likely to be met, and how failure to meet them will affect the validity of the inferences drawn from the statistical analysis. For example, the use of the Latin square design assumes that the size of the treatment effects does not change as a function of the order of presenting the treatments. This assumption will often be false, with the usual result being an increased probability of concluding that the treatment has no effect when in fact it has.

1.4.3 Practical requirements. The selection of designs must often be dictated, or at least narrowed, by such considerations as the number of available subjects or the time available per subject. For example, in animal research it may be more convenient, and certainly less expensive, to run a few subjects through many conditions than to run many groups of subjects each under a different treatment. On the other hand, in research with children, where teachers may object to any one child losing too much class time, it may be more reasonable to use a completely randomized design, with fewer measurements on more children.

1.5 THE DEPENDENT VARIABLE

1.5.1 Choice of the dependent variable. The choice of an appropriate dependent variable, or measure, may appear to be a trivial problem—after all, one is interested in leadership, or aggressive behavior, or learning, or visual acuity, and this is what will be measured. Unfortunately, there are always a number of measures which can be reasonably interpreted as indices of the behavioral process under investigation. For example, consider the learning of a list of words. Shall one measure the number of trials to attain some predetermined criterion? Shall the speed of each response be measured? Shall the basic datum be the number of errors in each block of five trials? ten trials? all trials? Should errors of commission (incorrect responses) and errors of omission (failures to respond) be analyzed separately? Certainly these alternatives are not mutually exclusive, but there are practical limits to the number of measures which can be considered in the analysis of any single experiment. In this section we will consider some of the factors involved in selecting the dependent variable or variables.

There are several characteristics of dependent variables upon which the accuracy of inferences depends. Ideally, the dependent variable should be reliable, sensitive, and distributed in a way which conforms to the assumptions of the data analysis model. Reliability will be a factor insofar as measures which are equivalent in all other respects differ in their variability. That measure which is least variable under constant experimental conditions is preferred.

Sensitivity refers to the fact that certain measures show greater differential effects than do other measures as a function of changes in the independent variable. For example, in a study of escape from conflict, animals under conflict might not differ from control subjects who were not under conflict in the number of escape responses (presses of a platform that result in removal of the conflictual stimuli). However, the two groups might differ in mean duration of the escape responses.

Statistical analysis is often complicated because the distribution of the measure does not conform to the assumptions of available statistical models. If all other considerations are equal, we want that measure whose distribution is consistent with the model associated with the statistical analysis to be applied. This implies a thorough knowledge of previous research and the type of results obtained with different measures, as well as a consideration, prior to data collection, of the statistical model.

As in every other phase of experimental planning, there are practical considerations involved in the choice of measures. All else being equal, we want measures that can be easily obtained. For example, in research on personality, if a paper and pencil test and a projective test are equally reliable and sensitive and conform similarly to the statistical model, the paper and pencil test is preferable; it is administered and scored much more rapidly. Of course, all other things are rarely equal; the experimenter may consider the projective test to be more sensitive, while the paper and pencil test is probably more reliable and possibly more likely to result in a normal distribution of data. The moral is that in all phases of experimental planning the ideal is rarely attained; the experiment is a compromise among the factors that have been indicated.

Theory may also be an issue in the selection of measures. For example, consider an experiment in which the subject guesses which of two events will occur on each trial. The measure generally taken in such experiments is the percent of each type of guess, partly because it is easily obtained and partly because pertinent theories of the behavior under investigation yield predictions of this measure. With the recent advent of theories which generate exact quantitative predictions of response latency for choice behavior, latency will probably be more frequently used as a measure in the future.

1.5.2 Choice of the measuring technique. In many instances the choice of a measure still leaves unanswered the question of how the measure is to be recorded. In deciding this point, the experimenter should consider the ease of obtaining the data as well as the probable degree of reliability of the recording technique. Again, consider the situation in which the subject must guess which of two events will occur on each trial. There are at least three methods of recording which response occurred on each trial: (*a*) the experimenter could manually record the subject's choice on each trial, (*b*) responses could be automatically recorded by some type of event-pen system, and (*c*) responses could be automatically punched out onto IBM cards. The manual technique is the least reliable, since it is subject to recording errors by the experimenter. However, the manually recorded data are generally easier to score and tabulate than ink records, and this nonautomated recording system is less subject to breakdowns. The event-pen method frees the experimenter's time during experimental sessions, involves an initial expenditure of several hundred dollars, is generally reliable in recording the data, but involves record-reading labor and possible error at that time. The data punch-out method is efficient in both the collection and analysis of data, is highly reliable, but may involve an initial expenditure of several thousands of dollars. The automatic system may also be less reliable from day to day in the sense that breakdowns may be more frequent than with the alternative procedures. Clearly, the choice of a system for recording the data is, like all other decisions made during the planning of an experiment, a compromise among numerous considerations.

1.6 SUBJECTS

There are basically two classes of decisions which the experimenter must make with regard to subjects: How many should be run? What population should they be drawn from? In this section we will consider the factors which should be involved in such decisions.

1.6.1 How many subjects? A primary consideration in deciding upon the number of subjects is the *power* desired. Roughly, power is the probability of correctly concluding that differences among the effects of treatments exist. Power depends upon the direction and size of the effect to be detected, how large a risk of wrongly concluding that the treatments have different effects one is willing to take, and the error variance expected. The error variance, as noted earlier in this chapter, in turn depends upon the experimental design employed. For the time being, we merely note that decisions about the factors influencing power will influence the number of subjects required to attain a desired degree of power. In Chapters 3 and 4 we will consider the relationships among all these factors in much greater detail.

The second major determiner of the number of subjects selected is the number of subjects available. This depends largely on the subject population from which we draw—rats, college sophomores, first-grade children, and so forth. We will next present some of the factors which dictate choice of subject population.

1.6.2 The subject population. The major consideration in deciding which type of subject to employ in any experiment is *availability*. Since much psychological research is done by college professors and by graduate students in pursuit of a degree, the college sophomore is an extremely popular subject, perhaps the most popular subject in studies of human performance. The rat, small, easily housed and fed, is another convenient and therefore popular subject for research. However, there are other considerations, and some of these should influence the choice of subjects more than they usually do.

One can expect less *error variance* with some subjects than with others. Suppose an experiment on problem solving with children involved instructions which could be expected to be clear to all five-year-olds but which might prove too complex for some four-year-olds; one might then expect more error variability with the four-year-olds as subjects. In short, subjects should be drawn from a population which is as homogeneous as possible with regard to irrelevant variables that might influence scores.

Control of the subject's previous history is a factor in choosing the subject population and is an oft-stated reason for the abundance of data on animals. With such subjects, who are often born and reared in the laboratory, the experimenter can control eating habits, genetic history, and environmental influences. As Tolman has remarked,* ". . . rats live in cages; they do not go on binges the night before one has planned the experiment. . . ."

The *purpose of the research* is often a factor in selecting subjects. Certainly there are basic processes which can be studied in many different types of subjects. On the other hand, if the experimenter is specifically interested in human development, or schizophrenic performance, his choice of subject populations is immediately narrowed.

Theoretical considerations may also play a role in the choice of subjects. For example, adult human subjects, when asked to guess which of two events will occur, tend to guess one event if the other event has had a long run, that is, has come up several times in succession. This behavior conflicts with the predictions of some prominent theories of choice behavior. It has been hypothesized that the behavior results from the fact that the subject has experienced only short event runs in his pre-experimental history and therefore expects short runs in the laboratory. A test of this hypothesis might involve the use of subjects with a limited previous exposure to event sequences, for example, young children or rats.

1.7 CONCLUDING REMARKS

Several times in this chapter, it has been implied that experimental plans are in part related to the data analysis. We have noted that the choice of designs and of dependent variables is a function of the demands of the statistical model. This point is now made explicit: the statistical analysis should be planned *in detail* before a single subject is run. The alternative is a post-experiment search for an analysis that is consistent with the design and the distribution of the measure, and that may not exist. One can have reasonable assurance that an appropriate analysis exists only if one considers design and analysis together, before the data are collected. It is worth noting that when alternative analyses exist, the choice depends upon factors similar to those which were previously cited in Section 1.4 in discussing the choice among experimental designs. Thus one should take into account the relative *efficiencies* of analyses, the resulting *information*, the *computational labor* involved, as well as *assumptions* and whether or not they are likely to be met.

In this chapter we have discussed those considerations involved in

---

* E. C. Tolman, "A Stimulus-Expectancy Need-Cathexis Psychology," *Science*, 101:160–166 (1945).

planning experiments. Sometimes these considerations will point to a single decision; more often they will be in conflict. Thus, the simplest measure to obtain may be the least reliable, and the most efficient design may imply a statistical model to which our data will not conform. Experimental planning is a weighing of such considerations, a compromise among them. It is impossible to state a single set of rules for weighing these considerations since one would need different rules for each experiment. However, we have attempted to state what the important factors influencing our decisions should be, and why they are important.

2

Notation

If the reader is to follow the derivations and computational formulas presented in this text, we must develop a common language that will be both explicit and relatively brief. Such a language, a notational system, is presented in this chapter. Whatever effort is required to attain a complete command of this material will be amply repaid in subsequent chapters. Our general approach will be to present a few simple rules and then to show their application to some elementary statistical quantities.

2.1 A SINGLE GROUP OF SCORES

2.1.1 Some basic rules. Consider a group of scores: Y_1, Y_2, Y_3, Y_4, ..., Y_n. The subscript has no purpose other than to distinguish among the individual scores. The quantity n is the total number of scores in the group. Suppose $n = 5$, and we wish to indicate that all five scores are to be added together. We could write

$$Y_1 + Y_2 + Y_3 + Y_4 + Y_5$$

or more briefly,

$$Y_1 + Y_2 + \cdots + Y_5$$

Still more briefly, we write

$$\sum_{i=1}^{5} Y_i$$

This expression is read as "sum the values of Y for all i from 1 to 5." In general, $i = 1, 2, \ldots, n$ (i.e., i takes on the values from 1 to n), and the summation of a group of n scores is indicated by

$$\sum_{i=1}^{n} Y_i$$

15

The quantity i is the *index*, and 1 and n are the *limits* of summation. Where the context of the presentation permits no confusion, the index and limits are often dropped. Thus we may often indicate that a group of scores are to be summed by $\sum Y$.

We will next consider three rules for summation and then some illustrative examples.

RULE 1. *The sum of a constant times a variable equals the constant times the sum of the variable.* That is,

$$(2.1) \qquad\qquad \sum CY = C\sum Y$$

C is a constant in the sense that its value does not change as a function of i; the value of Y depends upon i, and Y is therefore a variable with respect to i. Equation (2.1) is easily proven.

$$\sum CY = CY_1 + CY_2 + CY_3 + \cdots + CY_n$$
$$= C(Y_1 + Y_2 + Y_3 + \cdots + Y_n)$$
$$= C\sum Y$$

RULE 2. *The sum of a constant is equal to n times the constant, where n equals the number of quantities summed.* That is,

$$(2.2) \qquad\qquad \sum C = C + C + \cdots + C = nC$$

RULE 3. *The summation sign operates like a multiplier on quantities within parentheses.* Two examples follow:

EXAMPLE 1.

$$\sum_{i}^{n} (X_i - Y_i) = \sum_{i}^{n} X_i - \sum_{i}^{n} Y_i$$

PROOF.

$$\sum(X - Y) = (X_1 - Y_1) + (X_2 - Y_2) + \cdots + (X_n - Y_n)$$
$$= (X_1 + X_2 + \cdots + X_n) - (Y_1 + Y_2 + \cdots + Y_n)$$
$$= \sum X - \sum Y$$

EXAMPLE 2.

$$\sum(X - Y)^2 = \sum X^2 + \sum Y^2 - 2\sum XY$$

PROOF.

$$\sum(X - Y)^2 = (X_1 - Y_1)^2 + \cdots + (X_n - Y_n)^2$$
$$= (X_1^2 + Y_1^2 - 2X_1Y_1) + (X_2^2 + Y_2^2 - 2X_2Y_2) + \cdots$$
$$\quad + (X_n^2 + Y_n^2 - 2X_nY_n)$$
$$= (X_1^2 + X_2^2 + \cdots + X_n^2) + (Y_1^2 + Y_2^2 + \cdots + Y_n^2)$$
$$\quad - 2(X_1Y_1 + X_2Y_2 + \cdots + X_nY_n)$$
$$= \sum X^2 + \sum Y^2 - 2\sum XY$$

Next, consider the application of these three rules to some commonly computed statistics. The mean of a group of n scores is given by

$$(2.3) \qquad \bar{Y} = \frac{\sum Y}{n}$$

The operation of subtracting each score from the mean and adding the deviations may be represented by $\sum(Y - \bar{Y})$. Applying Rule 3, we have

$$(2.4) \qquad \sum(Y - \bar{Y}) = \sum Y - \sum \bar{Y}$$

However, $\bar{Y}$ is a constant; its value remains the same regardless of the value of the index, i. It should be clear that we are summing over i from 1 to n throughout this presentation, even though the index and limits are not explicitly presented in each expression. Applying Rule 2, we rewrite Equation (2.4) as

$$(2.5) \qquad \sum(Y - \bar{Y}) = \sum Y - n\bar{Y}$$

At this point we substitute Equation (2.3) into Equation (2.5), giving

$$(2.6) \qquad \sum(Y - \bar{Y}) = \sum Y - n\left(\frac{\sum Y}{n}\right) = \sum Y - \sum Y = 0$$

It has been proven that the sum of deviations of scores about their mean is zero, a rather elementary but, nevertheless, important result. Of course, our purpose has been primarily to illustrate how the application of the summation rules may result in proofs of basic statistical properties.

2.1.2 Raw Score Formulas. Next we will show how the summation rules may be applied to simplify computations. Two examples will be used, the sample variance and the sample correlation. The sample variance, when used as an estimator (as it will be throughout this book) is defined as follows:

$$(2.7) \qquad S^2 = \frac{\sum(Y - \bar{Y})^2}{n - 1}$$

Computations, particularly with a desk calculator, are greatly simplified by obtaining a *raw score formula* for S^2, so called because the mean is eliminated and only individual, or raw, scores are manipulated. To obtain such a formula, only the numerator of the right side of Equation (2.7) will be manipulated. First, expand the quantity within the summation sign. Thus,

$$(2.8) \qquad \sum(Y - \bar{Y})^2 = \sum(Y^2 + \bar{Y}^2 - 2Y\bar{Y})$$

Rule 3 is applied, permitting elimination of the parentheses:

$$(2.9) \qquad \sum(Y - \bar{Y})^2 = \sum Y^2 + \sum \bar{Y}^2 - \sum 2Y\bar{Y}$$

Application of Rule 2 leads to a further change, when it is noted that $\bar{Y}^2$ is a constant:

$$(2.10) \qquad \sum(Y - \bar{Y})^2 = \sum Y^2 + n\bar{Y}^2 - \sum 2Y\bar{Y}$$

The quantity $2\bar{Y}$ is a constant and can, by Rule 1, be placed in front of the summation sign. Thus,

$$(2.11) \qquad \sum(Y - \bar{Y})^2 = \sum Y^2 + n\bar{Y}^2 - 2\bar{Y}\sum Y$$

Now substitute for $\bar{Y}$, using Equation (2.3).

$$(2.12) \qquad \sum(Y - \bar{Y})^2 = \sum Y^2 + n\frac{(\sum Y)^2}{n^2} - 2\left(\frac{\sum Y}{n}\right)\sum Y$$

The final step is to obtain the simplest possible form of the above expression, which is

$$(2.13) \qquad \sum(Y - \bar{Y})^2 = \sum Y^2 - \frac{(\sum Y)^2}{n}$$

Dividing the right-hand side of Equation (2.13) by $n - 1$ gives the raw score formula for S^2 that was sought.

A similar approach to the one just illustrated may be used to obtain a raw score formula for the Pearson product-moment correlation coefficient. The equation is generally given as

$$(2.14) \qquad r = \frac{\sum(X - \bar{X})(Y - \bar{Y})}{\sqrt{\sum(X - \bar{X})^2}\sqrt{\sum(Y - \bar{Y})^2}}$$

The raw score formula for the denominator is already known from Equation (2.13). We will therefore concern ourselves with the transformation of the numerator. First expand, obtaining

$$(2.15) \qquad \sum(X - \bar{X})(Y - \bar{Y}) = \sum(XY - X\bar{Y} - \bar{X}Y + \bar{X}\bar{Y})$$

Parentheses are then removed according to Rule 3, yielding

$$(2.16) \qquad \sum(X - \bar{X})(Y - \bar{Y}) = \sum XY - \sum X\bar{Y} - \sum \bar{X}Y + \sum \bar{X}\bar{Y}$$

Application of Rule 2 permits removal of one of the summation signs, and

$$(2.17) \qquad \sum(X - \bar{X})(Y - \bar{Y}) = \sum XY - \sum X\bar{Y} - \sum \bar{X}Y + n\bar{X}\bar{Y}$$

Application of Rule 1 results in a further change:

$$(2.18) \qquad \sum(X - \bar{X})(Y - \bar{Y}) = \sum XY - \bar{Y}\sum X - \bar{X}\sum Y + n\bar{X}\bar{Y}$$

Substituting for the means results in

$$(2.19) \qquad \sum(X - \bar{X})(Y - \bar{Y}) = \sum XY - \left(\frac{\sum Y}{n}\right)(\sum X)$$

$$- \left(\frac{\sum X}{n}\right)(\sum Y) + n\left(\frac{\sum X}{n}\right)\left(\frac{\sum Y}{n}\right)$$

and algebraic simplification gives the final result,

(2.20) $$\sum(X - \bar{X})(Y - \bar{Y}) = \sum XY - \frac{(\sum X)(\sum Y)}{n}$$

Thus the raw score formula for the correlation coefficient is

(2.21) $$r = \frac{n\sum XY - (\sum X)(\sum Y)}{\sqrt{n\sum X^2 - (\sum X)^2}\sqrt{n\sum Y^2 - (\sum Y)^2}}$$

Note that a little insight provides a short cut. Equation (2.13) could be rewritten as

(2.13′) $$\sum(Y - \bar{Y})^2 = \sum(Y - \bar{Y})(Y - \bar{Y}) = \sum YY - \frac{(\sum Y)(\sum Y)}{n}$$

By analogy, Equation (2.20) follows.

Obtaining raw score formulas from those formulas originally used to define a statistic is a common enough problem to warrant a brief summary of the steps involved in the preceding two examples. They are expansion, application of Rule 3 (removal of parentheses), application of Rule 2 (replacing summation signs that precede constants by appropriate multipliers), application of Rule 1 (placing constants before summation signs), substitution for quantities not presently in raw score form, and algebraic simplifications.

2.1.3 Variance of a sum. The notational rules developed in this chapter are also helpful in deriving many statistical relationships. We will next consider one such derivation, in part as an additional example of the manipulation of statistical symbols, and in part because this particular result is closely related to later developments in the text.

Consider a score composed of several component parts; thus, the total score for the ith individual is the sum of the scores on the p parts:

$$X_{it} = X_{i1} + X_{i2} + \cdots + X_{ij} + \cdots + X_{ip} = \sum_{j=1}^{p} X_{ij}$$

We will derive a relationship between S_t^2, the variance of the total score, and the p values of S_j^2, the variance on part j. By definition,

$$S_t^2 = \frac{1}{n} \sum_{i=1}^{n} (X_{it} - \bar{X}_{it})^2$$

We know that $X_{it} = \sum_{j=1}^{p} X_{ij}$. Furthermore,

$$\frac{\sum_{i=1}^{n} X_{it}}{n} = \frac{\sum_{i=1}^{n} X_{i1}}{n} + \frac{\sum_{i=1}^{n} X_{i2}}{n} + \cdots + \frac{\sum_{i=1}^{n} X_{ip}}{n}$$

or $$\bar{X}_{.t} = \bar{X}_{.1} + \cdots + \bar{X}_{.p}$$

Then,

$$(2.22) \quad S_t^2 = \frac{1}{n} \sum_{i=1}^{n} [(X_{i1} + \cdots + X_{ij} + \cdots + X_{ip})$$

$$- (\overline{X}_{.1} + \cdots + \overline{X}_{.j} + \cdots + \overline{X}_{.p})]^2$$

$$= \frac{1}{n} \sum_{i=1}^{n} [(X_{i1} - \overline{X}_{.1}) + \cdots + (X_{ij} - \overline{X}_{.j}) + \cdots$$

$$+ (X_{ip} - \overline{X}_{.p})]^2$$

Expanding the squared term, we have

$$(2.23) \quad S_t^2 = \frac{1}{n} \sum_{i=1}^{n} [(X_{i1} - \overline{X}_{.1})^2 + \cdots + (X_{ij} - \overline{X}_{.j})^2 + \cdots + (X_{ip} - \overline{X}_{.p})^2$$

$$+ 2(X_{i1} - \overline{X}_{.1})(X_{i2} - \overline{X}_{.2}) + \cdots + 2(X_{ij} - \overline{X}_{.j})$$

$$\times (X_{ij} - \overline{X}_{.j}) + \cdots + 2(X_{i,p-1} - \overline{X}_{.p-1})(X_{ip} - \overline{X}_{.p})]$$

Note that any term of the form $(1/n)\sum_i(X_{ij} - \overline{X}_{.j})^2$ is the variance of the n scores on part j. Furthermore, dividing numerator and denominator of Equation (2.14) by n, the correlation of scores on any two parts, j and j', is

$$r_{jj'} = \frac{\sum_i(X_{ij} - \overline{X}_{.j})(X_{ij'} - X_{.j'})/n}{S_j S_{j'}}$$

Substituting in Equation (2.23), we have

$$(2.24) \quad S_t^2 = \frac{1}{n} \sum_{j=1}^{p} S_j^2 + 2 \sum_{\substack{j,j' \\ j \neq j'}} r_{jj'} S_j S_{j'}$$

In words, the variance of the total equals the sum of the variances of the parts plus twice the sum of all possible covariances; the covariance is the product of the two standard deviations and the correlation coefficient. If, for all parts, the scores are independent of those obtained on the other parts, the covariances are zero and $S_t^2 = \sum_j S_j^2$.

2.2 SEVERAL GROUPS OF SCORES

The simplest possible experimental design is one involving several groups of scores. Thus one might have a groups of n subjects each, which differ in the amount of reward they receive for their performance on some learning task. In setting the data down on paper, there would be a column for each level of amount of reward; i.e., for each experimental group. The scores for a group could be written in order within the appropriate column. In referring to a score, we would designate it by its position in the column (or experimental group) and by the position of the column. Table 2–1 illustrates this procedure.

TABLE 2–1 *A two-dimensional matrix*

	Groups				
	Y_{11}	Y_{12}	$\cdots$ Y_{1j}	$\cdots$	Y_{1a}
	Y_{21}	Y_{22}	$\cdots$ Y_{2j}	$\cdots$	Y_{2a}
Subjects	$\vdots$	$\vdots$			$\vdots$
	Y_{i1}	Y_{i2}	$\cdots$ Y_{ij}	$\cdots$	Y_{ia}
	$\vdots$	$\vdots$			$\vdots$
	Y_{n1}	Y_{n2}	$\cdots$ Y_{nj}	$\cdots$	Y_{na}

Note that the first subscript refers to the position in the group (row), the second to the position of the group (column). Thus Y_{22} is the second score in group 2 and, in general, Y_{ij} is the ith score in the jth group.

Suppose we wish to refer to the mean of a single column. The term used previously, $\overline{Y}$, is obviously misleading since it does not designate the row or column that we want. Even $\overline{Y}_1$ is not sufficient, since it might easily refer to the mean of the first row as of the first column.* The appropriate designation is $\overline{Y}_{.1} = [(1/n)\sum_i^n Y_{i1}]$; the dot represents summation over i, the index that ordinarily appears in that position. Similarly, the mean of row i would be designated by $\overline{Y}_{i.} = [(1/a)\sum_j^a Y_{ij}]$; summation is over the index j. The mean of all an scores would be designated by $\overline{Y}_{..} = [(1/an)\sum\sum Y_{ij}]$, or merely $\overline{Y}$.

Some examples of the application of the double summation ($\sum_i \sum_j$) may be helpful. Suppose we have

$$\sum_{j=1}^{a} \sum_{i=1}^{n} Y_{ij}^2$$

This is an instruction to initially set i and j at 1; the resulting score, Y_{11}, is then squared. Holding j at 1, we step i from 1 to n, squaring each score thus obtained and adding it to those previously squared. When n scores have been squared and summed, we reset the index i at 1 and step j to 2; the squaring and summing is then carried out for all Y_{i2}. The process continues until all an scores have been squared and summed. The process just described may be represented by

$$(Y_{11}^2 + Y_{21}^2 + \cdots + Y_{na}^2)$$

If we have

$$\sum_{j=1}^{a} \left(\sum_{i=1}^{n} Y_{ij} \right)^2$$

the notation indicates that a sum of n scores is to be squared. We again set j at 1 and, after adding together all the Y_{i1}, square the total. The index j is

---

* In the design we used for an example, the mean of the first row would not be a quantity of interest, since we stipulated that the order within each column was arbitrary. However, there are designs giving rise to tables similar to Table 2–1 for which it is as interesting to obtain row means as it is to obtain column means.

then stepped to 2 and i is reset at 1; we obtain another sum of n scores, which is squared and added to the previous squared sum. We again continue until all an scores have been accounted for. The process may be represented by

$$(Y_{11} + Y_{21} + \cdots + Y_{n1})^2 + \cdots + (Y_{1a} + Y_{2a} + \cdots + Y_{na})^2$$

A third possibility is

$$\left(\sum_{j=1}^{a} \sum_{i=1}^{n} Y_{ij} \right)^2$$

which indicates that the squaring operation is carried out once on the total of an scores; we then have

$$[(Y_{11} + Y_{21} + \cdots + Y_{n1}) + \cdots + (Y_{1a} + Y_{2a} + \cdots + Y_{na})]^2$$

Note that the indices within the parentheses indicate how many scores are to be summed prior to squaring while the indices outside the parentheses indicate the number of squared totals to be summed. When no parentheses appear, as in $\sum\sum Y^2$, we treat the notation as if it were $\sum\sum(Y^2)$. When no indices appear outside the parentheses, it is understood that we are dealing with a single squared term, as in $(\sum\sum Y)^2$. When several indices appear together, whether within or outside the parentheses, the product of their upper limits tells us the number of terms involved. Thus, $(\sum_{j=1}^{a} \sum_{i=1}^{n} Y)^2$ indicates than an scores are summed prior to the squaring operation.

Our three illustrations of the double summation may be further clarified if we use some numbers. Let us use the three groups of four scores each shown in Table 2–2.

TABLE 2–2 *Some sample data*

	Group 1	Group 2	Group 3
	4	1	6
	1	7	4
	3	2	5
	2	4	4
$\sum_i Y_{ij} = 10$		14	19
$\sum_i Y_{ij}^2 = 30$		70	93

Now,

$$\sum_j \sum_i Y_{ij}^2 = 30 + 70 + 93 = 193$$

and

$$\sum_j \left(\sum_i Y_{ij} \right)^2 = (10)^2 + (14)^2 + (19)^2 = 657$$

and

$$\left(\sum_j \sum_i Y_{ij} \right)^2 = (10 + 14 + 19)^2 = 1,849$$

As another example of the use of the double summation, we might derive a raw score formula for the average group variance, often referred to as the *within-groups mean square*. This is the sum of the group variances divided by a, the number of groups, or

$$\frac{1}{a}\left[\frac{\sum_{i=1}^{n}(Y_{i1} - \overline{Y}_{.1})^2}{n-1} + \cdots + \frac{\sum_{i=1}^{n}(Y_{ia} - \overline{Y}_{.a})^2}{n-1}\right]$$

More briefly, this average is indicated by

$$\frac{1}{a(n-1)}\sum_{j}^{a}\sum_{i}^{n}(Y_{ij} - \overline{Y}_{.j})^2$$

Now, expanding the numerator (or "sums of squares") of the above quantity, we obtain

$$(2.25) \qquad \sum_{j=1}^{a}\sum_{i=1}^{n}(Y_{ij} - \overline{Y}_{.j})^2 = \sum_{j=1}^{a}\sum_{i=1}^{n}(Y_{ij}^2 + \overline{Y}_{.j}^2 - 2Y_{ij}\overline{Y}_{.j})$$

We "multiply through" by $\sum_i$, noting that $\overline{Y}_{.j}$ varies only with j; it is constant when i is the index of summation. Terms are also rearranged so that sums are premultiplied by constants.

$$(2.26) \qquad \sum_{j}\sum_{i}(Y_{ij} - \overline{Y}_{.j})^2 = \sum_{j}\left(\sum_{i}Y_{ij}^2 + n\overline{Y}_{.j}^2 - 2\overline{Y}_{.j}\sum_{i}Y_{ij}\right)$$

Substituting raw score formulas for the group means gives

$$(2.27) \qquad \sum_{j}\sum_{i}(Y_{ij} - \overline{Y}_{.j})^2 = \sum_{j}\left[\sum_{i}Y_{ij}^2 + n\frac{(\sum_i Y_{ij})^2}{n^2} - 2\left(\frac{\sum_i Y_{ij}}{n}\right)\sum_{i}Y_{ij}\right]$$

Simplifying results in

$$(2.28) \qquad \sum_{j}\sum_{i}(Y_{ij} - \overline{Y}_{.j})^2 = \sum_{j}\left[\sum_{i}Y_{ij}^2 - \frac{(\sum_i Y_{ij})^2}{n}\right]$$

which may also be written as

$$\sum_{j}\sum_{i}Y_{ij}^2 - \frac{\sum_j(\sum_i Y_{ij})^2}{n}$$

2.3 MORE THAN TWO INDICES

The principles thus far developed may be extended to any number of summations and variables. For example, we might wish to represent operations on a data set obtained from children varying with respect to both age and socioeconomic level. Let A_j represent the jth level of age and B_k represent the kth socioeconomic level. If we let i take on values from 1 to n indicating children within each AB classification, the typical score is Y_{ijk}

representing the value obtained for the ith of n children at the jth age level and the kth socioeconomic level. Table 2–3 presents a sample data matrix.

TABLE 2–3 *Some sample data involving three indices*

	B_1	B_2
	64	70
A_1	78	67
	67	83
	70	81
	$\Sigma_i Y_{i11} = 279$	$\Sigma_i Y_{i12} = 301$
	68	84
A_2	82	76
	79	81
	74	70
	$\Sigma_i Y_{i21} = 303$	$\Sigma_i Y_{i22} = 311$
	75	91
A_3	84	87
	88	78
	83	86
	$\Sigma_i Y_{i31} = 330$	$\Sigma_i Y_{i32} = 342$

In the course of many of the statistical analyses encountered in this text, a variety of operations will be performed on data matrices similar to the one in Table 2–3. For example, we might encounter the notation

$$\sum_{j=1}^{a} \left(\sum_{k=1}^{b} \sum_{i=1}^{n} Y_{ijk} \right)^2$$

This implies a squared total for each level of A, each of which is based on bn scores; in the present example we would let j equal 1, vary i and k to obtain all eight (bn) values of Y_{ijk}, sum these values and square the total, and then proceed to the next level of A ($j = 2$), proceeding in this manner until we have accounted for all abn scores. In terms of the data of Table 2–3, we would have

$$(279 + 301)^2 + (303 + 311)^2 + (330 + 342)^2$$

Another commonly encountered term is

$$\sum_{j=1}^{a} \sum_{h=1}^{b} \left(\sum_{i=1}^{n} Y_{ijk} \right)^2$$

This indicates six (*ab*) squared totals, each total based on four (*n*) scores. In our example, we have

$$(279)^2 + (303)^2 + \cdots + (342)^2$$

It is equally simple to go from verbal instructions to symbolic representation. For example, the instruction, "Add all *an* scores at each level of *B*, square the totals, and sum the squared terms," implies:

(a) There are *b* squared quantities to be summed, so we have $\sum_{k=1}^{b}(\quad)^2$;

(b) Since the design involves *abn* scores, each of the squared totals must contain *an* scores; then we have $\sum_{k=1}^{b}(\sum_{j=1}^{a}\sum_{i=1}^{n} Y_{ijk})^2$.

EXERCISES

Answers are provided at the end of the book. It is extremely important to try to do each problem before turning to the answers for corroboration or help.

2.1 Write the summation of the *Y*s encircled below, indicating the limits.

$$Y_1 + Y_2 + \boxed{Y_3 + Y_4 + Y_5 + Y_6} + Y_7 + Y_8$$

2.2 Simplify the following expression (*k* is a constant):

(a) $\sum_{i=1}^{n} (Y_i - \bar{Y})$

(b) $\sum_{i=1}^{n} (Y_i + k - n)$

2.3 Show that

$$\frac{1}{k} \sum_{i=1}^{k} (k + kY_i) = k + \sum_{i=1}^{k} Y_i$$

indicating which summation rules have been applied.

2.4 Prove that the variance remains the same if a constant is added to all scores.

2.5 Let the variance of a set of *n* scores be S_y^2. Prove that if each score is multiplied by *C*, the new variance is $C^2 S_y^2$.

2.6 Let $Z_i = (Y_i - \bar{Y})/S_y$. Prove that
(a) $\bar{Z} = 0$
(b) $S_z^2 = 1$

2.7 Let $D_i = X_i - Y_i$. Prove that
(a) $\bar{D} = \bar{X} - \bar{Y}$
(b) $S_D^2 = S_X^2 + S_Y^2 - 2\pi_{XY}S_X S_Y$

2.8 In an experiment on the effects of several variables upon the running times of rats, there are *a* levels of amount of reward (*A*), *d* levels of delay of reward (*D*), and *t* levels of inter-trial interval (*T*). There are *n* subjects in each of the *adt* combinations of treatments, yielding a total of *nadt* subjects. Write the following verbal instructions in a complete and specific notational form, first clearly specifying your indices and limits of summation:

(a) Sum all scores obtained under A_1. Square the sum. Do the same for all other levels of *A*. Add the squared quantities together.

(b) Add all the scores obtained under $D_1 T_1$. Square this sum. Do the same for each DT combination. Add the squared quantities.

2.9 Assume that each of n subjects is tested under each of a conditions. Prove that

$$\sum_{j=1}^{a} \sum_{i=1}^{n} (\bar{Y}_{i.} - \bar{Y}_{..})^2 = \frac{\sum_{i=1}^{n} (\sum_{j=1}^{a} Y_{ij})^2}{a} - \frac{(\sum_{i=1}^{n} \sum_{j=1}^{a} Y_{ij})^2}{an}$$

2.10 There are a levels of A, b levels of B, and n subjects in each of the ab combinations.

Let $i = 1, 2, \ldots, n$
$\quad j = 1, 2, \ldots, a$
$\quad k = 1, 2, \ldots, b$

Prove that

$$na \sum_{k=1}^{b} (\bar{Y}_{.k.} - \bar{Y}_{...})^2 = \frac{\sum_k (\sum_i \sum_j Y_{ijk})^2}{na} - \frac{(\sum_i \sum_j \sum_k Y_{ijk})^2}{nab}$$

2.11

	C_1			C_2		
	A_1	A_2	A_3	A_1	A_2	A_3
B_1	14	3	4	2	1	12
	12	8	11	8	9	5
B_2	3	4	5	4	1	7
	6	7	3	2	8	6
B_3	4	⑦	2	1	2	6
	5	1	3	④	7	3

Y_{1231} Y_{2132}

Let $i = 1, 2$ (scores in cells)
$\quad j = 1, 2, 3$ (levels of A)
$\quad k = 1, 2, 3$ (levels of B)
$\quad m = 1, 2$ (levels of C)

Thus $Y_{2132} = 4$ and $Y_{1231} = 7$. Find

(a) $\sum_j \sum_k \left(\sum_m \sum_i Y_{ijkm} \right)^2$

(b) $\sum_j \sum_k \sum_i \left(\sum_m Y_{ijkm} \right)^2$

(c) $\bar{Y}_{...2}$

(d) $\bar{Y}_{..2.}$

2.12 We have response time (X) and error frequencies (Y) for each of 3 Ss in each of 12 experimental groups. The data are:

X Data

	C_1				C_2			
	B_1	B_2	B_3		B_1	B_2	B_3	
	4	1	8	13	6	4	1	11
A_1	3	9	6	18	8	9	12	29
	4	5	2	11	7	3	10	20
	11	15	16	42	21	16	23	60

	12	8	9	29	16	9	14	39
A_2	9	14	11	34	8	6	7	21
	13	10	7	30	15	13	11	39
	34	32	27	93	39	28	32	99

Y Data

	C_1				C_2			
	B_1	B_2	B_3		B_1	B_2	B_3	
	6	4	8	18	3	7	14	24
A_1	12	3	6	21	11	6	7	24
	9	12	10	31	8	9	12	29
	27	19	24	70	22	22	33	77
	7	6	14	27	16	10	5	31
A_2	14	2	6	22	11	12	13	36
	9	12	10	31	8	7	9	24
	30	20	30	80	35	29	27	91

Let $i = 1, 2, 3$—the level of subject
$j = 1, 2$—the level of A
$k = 1, 2, 3$—the level of B
$m = 1, 2$—the level of C

Set up the calculations appropriate for the following formulas:

(a) $\sum_m \left(\sum_i \sum_j \sum_k X_{ijkm} \right) \left(\sum_i \sum_j \sum_k Y_{ijkm} \right)$

(b) $\sum_k \sum_m \left(\sum_i \sum_j Y_{ijkm} \right)^2$

(c) $\sum_j \sum_k \left(\sum_i \sum_m X_{ijkm} \right) \left(\sum_i \sum_m Y_{ijkm} \right)$

(d) $\sum_i \left(\sum_j \sum_k \sum_m X_{ijkm} \right)^2$

3

Statistical Inference

3.1 INTRODUCTION

Since it is generally impractical to measure the behavior of all individuals in a population, the experimenter usually confines his investigation to a small sample of subjects. For example, an experimenter interested in the effects of amount of reward upon the running times of rats might randomly assign 40 rats to two groups of 20 subjects each; one group receiving one food pellet at the end of a six-foot runway and the other receiving a reward of two food pellets. The experimenter is not usually concerned with whether these two groups of rats differ in their performances. However, he is interested in drawing conclusions about the relative performances of two populations of rats, systematically differing only in the amount of reward for running. If the subjects can be considered random samples from the populations in which the experimenter is interested, it is reasonable to use the sample data as a basis for conclusions about the populations. Conclusions from sample data about such things as the mean and variance of the running times of large populations of rats are statistical inferences. The primary purpose of this chapter is to review briefly the processes by which such inferences are drawn.

The basic problem in statistical inference is that samples are not miniature replicas of the populations from which they have been drawn. Individuals tested under the same experimental conditions will perform differently due to differences in ability, personality, or motivation. The same individual, tested at different moments in time, will exhibit variability in his scores because of changes in set or motivation. The consequence of such error, or chance, variability is that the measures computed from the sample —for example, the mean, the variance, the proportion of latencies longer than five seconds—will vary among samples and no single sample will accurately specify the characteristics of the population in which we are interested.

Consider the example in which the effects of two magnitudes of reward are compared upon running speeds of rats. Because of error variability, it is possible that the experimental results will misrepresent the relative performances of two populations of rats differing in reward magnitude. While for large numbers of animals, the smaller reward may, on the average, result in slower running, we could have by chance placed the faster rats in the one-pellet sample. Thus our experiment might detect no difference between the reward magnitudes, or even lead us to the conclusion that the smaller reward results in faster running times.

Fortunately, there are patterns to error variability and knowledge of these patterns provides the basic tool of statistical analysis. If we have the correct statistical model—that is, if we know how scores are distributed in the population—we can state the probabilities of various sampling results under various hypotheses about the populations. Before we consider these inferential processes in more detail, it will be helpful to consider certain basic concepts.

3.2 RANDOM VARIABLES

Our data analysis begins with sets of numbers; we might have trials to criterion or latency scores for each subject in an experiment, or number of subjects falling into each of several categories such as personality type or political preference. We designate any particular one of these numbers by the symbol Y. Since the various values of Y occur with different frequencies, we can assign a probability to each value of Y. Then Y is called a random variable.

3.2.1 Discrete random variables. A random variable is discrete if it assumes only a finite, or denumerable, number of values. Suppose we have 20 individuals classified as to whether they solve a particular problem; then Y, the number of solvers, can take on only the values from zero to 20. The probability that Y is one of some subset of values—for example, that Y is greater than 15—is the sum of the probabilities of the values comprising the subset; in other terms

$$P(Y > 15) = P(Y = 16) + \cdots + P(Y = 20)$$

In general, we will designate the probability that the random variable Y takes on some specific numerical value y as $p(y)$; this function is a *discrete density function* if for all y, $\sum p(y) = 1$.

If we know the density function of Y, we know its probability distribution, the theoretical probability of each possible value of Y. For example, consider a set of four coins that are repeatedly tossed. If the tosses are independent (the probabilities of heads and tails do not change as a function of the outcome of any other toss), and if each coin is equally likely to come up heads or tails on any one toss, then the probability of y heads is

$$P(y) = \frac{4!}{y!\,(4 - y)!} \left(\frac{1}{2}\right)^4$$

and the tabled probability distribution is

y	$p(y)$
0	1/16
1	1/4
2	3/8
3	1/4
4	1/16

These values of $p(y)$ are the proportions of sets of four tosses having y heads *in the long run* if our assumptions (independence, equal likelihood of heads and tails on a single toss) are correct. If the four coins have been tossed relatively few times, the observed distribution of Y will deviate from the theoretical distribution tabled above; I would be surprised if 16 sets of tosses yielded exactly one set with zero heads, four sets with one head, and so on. On the other hand, 1600 sets of tosses should yield observed proportions of heads within one or two percent of the theoretical values of $p(y)$.

3.2.2 Continuous random variables. A continuous random variable is one that can take on any value within a given interval. A common example in psychological research is response time, which theoretically takes on any value between zero and infinity. Of course, observed response times usually fall between some boundaries such as 200 milliseconds and 10 seconds. Even within such boundaries, continuity is more theoretical than real since the best of laboratory timing devices rarely record in units smaller than thousandths of a second. In general, we will not be able to record all the values that a continuous random variable will take on. Nevertheless, many of the statistics that figure prominently in our inferential processes are continuous variables, so the concept is therefore important.

    A logical problem arises when we attempt to deal with the probabilities of values of Y. This can be illustrated by considering a relatively crude clock capable of registering response times to within .1 seconds. Times longer than .95 seconds but less than 1.05 seconds will be recorded as one second. Suppose we now substitute a more accurate clock capable of registering time to the nearest hundredth of a second. Latencies in the interval .995–1.005 now will be recorded as one second. The proportion of such latencies will clearly be smaller than the proportion in the .95–1.05 interval. Extending the argument, it should become apparent that the probability of a response time of exactly one second duration is essentially zero.

    In general, when Y is a continuous random variable, the concept of the probability of some exact value of Y is useless; if we are to distinguish among continuous random variables having different distributions, we require some function of Y, other than its probability, which determines its distribution. Such a function is $f(y)$, the *probability density function* of Y at some specific value, y. In order to define this function, we first consider the probability that $Y < y$, which we denote by $F(y)$. Similarly, $F(y + \Delta y)$ is the probability that $Y < y + \Delta y$, where Δy is an increment added to y. Then, $F(y + \Delta y) - F(y)$ is the probability that Y lies in the interval Δy. As we noted in the preceding paragraph, this probability approaches zero as Δy becomes very small; however, the ratio $(F(y + \Delta y) - F(y))/\Delta y$ approaches a limiting value that varies as a function of the way Y is distributed and of the particular value of Y under consideration. This limiting value is $f(y)$, the continuous density function. It measures not the probability of y, but rather the rate of change in the probability relative to some small change in y. By plotting $f(y)$ for a wide range of values of y, we can obtain a picture of the probability distribution.

In our discussion of discrete random variables, we noted that the probability of any subset of values of Y was given by the sum of the probabilities of the members of the subset; furthermore, the sum of all $p(y)$ had to be 1 if $p(y)$ was truly a density function. When the random variable is continuous, the probability that it lies within some subset of values, say between $Y = a$ and $Y = b$, is obtained by integration:

$$P(a \leq Y \leq b) = \int_a^b f(y)dy$$

For those unfamiliar with calculus, imagine dividing the area under the curve determined by $f(y)$, between the points $Y = a$ and $Y = b$, into narrow rectangular strips of height $f(y)$ and width Δy. The product of $f(y)$ and Δy roughly gives the area in the strip above Δy—that is, the probability that Y lies in the interval Δy. Integrating between the limits a and b is essentially obtaining the limiting sum of these probabilities as the interval widths become very small. To carry the analogy to the discrete case a step further, the integral over the entire range of values of Y must equal 1 if $f(y)$ is a probability density function.

3.3 EXPECTED VALUES

Frequently, in this and subsequent chapters, we will have occasion to refer to the *expected value* of some random variable. The expected value of Y, denoted by $E(Y)$, is an average theoretically computed over all possible values of Y. To put $E(Y)$ in perspective, consider the more familiar arithmetic mean, $\overline{Y}$. Suppose that $\overline{Y}$ is a discrete random variable which, in a sample of nine scores, takes on the values 3, 5, 5, 8, 9, 9, 9, 12, and 12. Then,

$\overline{Y}$ is computed as the sum of scores, 72, divided by the number of scores, 9. We could obtain the same result by multiplying each distinct value by the proportion of its occurrences, and then summing these terms; that is

$$\overline{Y} = (3)(1/9) + (5)(2/9) + (8)(1/9) + (9)(3/9) + (12)(2/9)$$

It is this representation of the mean which the concept of expected value embodies; in general, we have

(3.1) $$E(Y) = \sum yp(y)$$

where $\sum p(y)$ must equal 1. If the random variable is continuous,

(3.2) $$E(Y) = \int yf(y)dy$$

where the summation sign of Equation (3.1) is replaced by the integral sign and $p(y)$ is replaced by $f(y)dy$.

Why do we require this concept of an average? The usual arithmetic mean is appropriate when we wish to describe the average of an observed finite set of numerical values. However, in drawing statistical inferences, we are often concerned not with the average of a set of values that have been available for observation but with what the average would be in the long run if we were to collect a large number of observations. The game of roulette provides a simple example of this application of the expected value concept. The roulette wheel contains 16 black numbers, 16 red numbers, and "0" and "00", which are neither black nor red. Thus, if we bet on either color on each spin of the wheel, the probability is 16/34 of winning and 18/34 of losing. If each bet is one dollar, the value of the bet is either plus or minus one dollar and the average value per trial in the long run will be

$$E(Y) = (1)(16/34) + (-1)(18/34) = -1/17$$

Due to the presence of "0" and "00", the gambling house has a slight edge and, in the long run, we will lose a little less than six cents (one seventeenth of a dollar) on each spin of the wheel.

Other aspects of the expected distribution are also of interest. In particular, the variance over the long run can be expressed in terms of expected values. Again, it helps to begin with the variance for an observed sample of values. As we showed in Chapter 2, the raw score formula for the observed variance is $\sum Y^2/n - (\sum Y/n)^2$. Note that $\sum Y^2/n$ is the arithmetic mean of Y^2 and $(\sum Y/n)^2$ the square of the arithmetic mean of Y. Replacing arithmetic means by expected values, the variance in the long run would be

(3.3) $$\sigma^2 = E(Y^2) - [E(Y)]^2$$

The quantity, $E(Y^2)$, is generally referred to as the second raw moment and, for a discrete random variable, is $\sum y^2 p(y)$. For a continuous random variable the expression is $\int y^2 f(y)dy$.

In Chapter 4 we will consider some derivations involving expected values. These will be straightforward if we note that the expectation operator, E, follows the same rules that the summation operator, $\sum$, does. For example,

$$E(X + Y) = E(X) + E(Y)$$

and

$$E(cY) = cE(Y)$$

and

$$E(X + Y)^2 = E(X^2) + E(Y^2) + 2E(XY)$$

Furthermore, E and $\sum$ are manipulated in the same manner that two summation signs are. For example,

$$E(\textstyle\sum X) = \sum E(X)$$

and

$$E[\textstyle\sum(X + Y)^2] = E(\textstyle\sum X^2) + E(\textstyle\sum Y^2) + 2E(\textstyle\sum XY)$$
$$= \textstyle\sum E(X^2) + \sum E(Y^2) + 2\sum E(XY)$$

3.4 SOME IMPORTANT PROBABILITY DISTRIBUTIONS

One particular distribution, the F distribution, plays an exceedingly large role in the remainder of this text, as it does in most experimental design texts. In this section, we will briefly consider the nature of this distribution. However, the F is closely related to two other distributions, the normal and chi-square (χ^2); in fact, the normal distribution logically precedes the χ^2 and the χ^2 logically precedes the F. We will therefore consider the normal and χ^2 distributions first.

3.4.1 The normal distribution. A theoretical continuous distribution that plays a major role in statistical inference is the normal, characterized by the density function

(3.4)
$$f(y) = \frac{1}{\sigma\sqrt{2\pi}} e^{-(y-\mu)^2/2\sigma^2}$$

where μ and σ are the mean and standard deviation of the population and π and e are mathematical constants. The random variable Y may take on any value between $-\infty$ and $+\infty$ and the curve is symmetric about the mean. It should be realized, however, that not all symmetric distributions over the range of all real numbers are normal distributions, although the normal distribution may provide an excellent approximation to such symmetric distributions.

The location of the distribution is determined by μ and its spread by σ; thus there are infinitely many possible normal distributions. Fortunately, all of these can be reduced to a single function. Given a random variable, Y, we can transform it by $Z = (Y - \mu)/\sigma$; if Y is normally distributed, Z will be also. However, the mean of the Z distribution will be zero and its standard deviation will be 1 (see Exercise 2.6). The density function is now

$$(3.5) \qquad f(z) = \frac{1}{\sqrt{2\pi}} e^{-z^2/2}$$

This transformation entails no loss of information and is extremely helpful. We can plot or table a single normal distribution, that for Z, and use it to provide information about any normally distributed random variable. Table A–2 presents values of $F(z)$ (labelled $1 - \alpha$ in the table), the probability that a normally distributed random variable with mean zero and standard deviation equal to 1 will be exceeded by a given value of Z. Because the distribution is symmetric, only positive values of Z have been tabled. Thus, if we desire the probability that Z is less than -1.0, we must note that this is equal to the probability that Z is greater than 1.0, or approximately $1.0–.84$. Typically, we want information about the distribution of Y; for example, we might want to know what proportion of individuals have I.Q. scores greater than 130. If we can assume that such scores are normally distributed in the population, and if we know the value of the population mean and variance, Table A–2 solves our problem. If the mean is 100 and the standard deviation is 15, a score of 130 can be transformed into a z-score of 2.00. Less than 2.5 percent of the populations have z-scores larger than this or, returning to the original random variable, I.Q. scores larger than the corresponding value of 130.

The normal distribution is prominently involved in testing hypotheses about population parameters and in estimating these parameters since many random variables, although far from all, are at least approximately normally distributed. Furthermore, regardless of the distribution of Y, the distribution of the sample mean, $\overline{Y}$, will rapidly approach normality as the sample size increases if the observations are independently sampled. The normal also provides a good approximation to several other distributions when the number of cases is large; the binomial, which is a discrete distribution, provides one example. There are several other reasons for the central role played by the normal distribution in statistical inference; of primary concern to us, however, is the fact that the assumption that some random variable, Y, is normally distributed permits us to derive the distribution of other random variables that are functionally related to Y. The chi-square distribution is a case in point, one that we will consider next.

3.4.2 The chi-square distribution. Suppose that we randomly sample n values from a normally distributed population. Each value of the random variable, Y, is then transformed into a z-score which is then squared; the set of n

squared z-scores are then summed. This sum will be referred to as χ^2; that is

$$(3.6) \qquad \chi_n^2 = \sum_{i=1}^{n} \left(\frac{Y_i - \mu}{\sigma} \right)^2$$

If we independently draw many such samples of size n from a normal population, each time calculating χ_n^2, we will have a distribution of values of χ_n^2
with a characteristic density function. The formula for the density function,
$f(\chi_n^2)$, is not particularly revealing and will be omitted; the important point
is that the density depends not only upon the value of χ_n^2, but also upon the
degrees of freedom (df), the number of independent observations that were
summed in obtaining χ_n^2. We actually have a differently shaped distribution
for each value of df.

Suppose that we have one df; that is

$$\chi_1^2 = \left(\frac{y - \mu}{\sigma} \right)^2$$

The relationship between the chi-square and normal distributions is, in this
instance, quite direct. In general, if some proportion, P, of the z-scores lie
between $-z$ and $+z$, then P of the values of χ_1^2, lie between zero and z^2.
For example, approximately 50 percent of the z-scores lie between $-.67$ and
$.67$; then approximately 50 percent of the values of χ^2 lie between zero and $.45$.
The first row of Table A–4, which contains values of χ^2 that will be exceeded
with certain selected probabilities, verifies this observation. If the χ^2 is based
on more observations, it stands to reason that larger values are more likely
to occur. Thus a chi-square of $.45$ is exceeded more than 70 percent of the
time when there are two df, and more than 90 percent of the time when there
are three df. These values are also based on Table A–4.

The χ^2 distribution is of interest for several reasons. A discrete random
variable, generally referred to as χ^2 but actually distributed only approximately
as χ^2 (which is itself continuously distributed), is useful in testing hypotheses
about frequency distributions; for example, whether the proportions of cases
in certain categories conform to some theory held by the investigator. The
χ^2 distribution also forms the basis for testing hypotheses about the population variance; this application will be considered in Chapter 11. It is based
on the fact that the ratio of the numerator of the sample variance to the
population variance is distributed as χ^2 on $n - 1$ df; i.e.,

$$(3.7) \qquad \frac{\sum_{i=1}^{n}(Y_i - \bar{Y})^2}{\sigma^2} = \chi_{n-1}^2$$

This is readily proven. We begin with the identity,

$$(Y_i - \mu) = (Y_i - \bar{Y}) + (\bar{Y} - \mu)$$

Squaring both sides and summing over the n values, we have

$$\sum_i (Y_i - \mu)^2 = \sum_i (Y_i - \overline{Y})^2 + n(\overline{Y} - \mu)^2 - 2(\overline{Y} - \mu)\sum_i (Y_i - \overline{Y})$$

Note the application of the rules of Chapter 2. Since we proved in that chapter that $\sum_i (Y_i - \overline{Y}) = 0$, we can proceed with

$$\frac{\sum(Y - \mu)^2}{\sigma^2} = \frac{\sum(Y - \overline{Y})^2}{\sigma^2} + \frac{n(\overline{Y} - \mu)^2}{\sigma^2}$$

Rewriting the last term as $(\overline{Y} - \mu)^2/(\sigma^2/n)$, and noting that σ^2/n is the sampling variance of $\overline{Y}$, we find that this term is a squared z-score; $(Y - \mu)/(\sigma/\sqrt{n})$ is the deviation of a quantity, $\overline{Y}$, from its expected value, divided by the standard deviation of the quantity, which meets the general definition of a z-score. Furthermore, since the mean of scores that come from a normal distribution is itself normally distributed, the squared z-score is a χ^2 variable on one df. Of course, the term to the left of the "$=$" sign is a χ^2 variable on n df. Then

$$\frac{\sum(Y - \overline{Y})^2}{\sigma^2} = \chi_n^2 - \chi_1^2$$
$$= \chi_{n-1}^2$$

because of the additive property of χ^2 variables. Equation (3.7) is thus verified. This result is critical in defining the next distribution we will consider.

3.4.3 The F distribution. We typically use the sample statistic S^2 to estimate the value of the population variance σ^2; S^2 is defined as $\sum(Y - \overline{Y})^2/(n - 1)$. In Section 3.5 we will consider the properties of this estimator and will justify the use of $n - 1$ in the denominator. For the present, note that $(n - 1)S^2 = \sum(Y - \overline{Y})^2$ and, according to the results of the preceding section,

$$\frac{(n - 1)S^2}{\sigma^2} = \chi_{n-1}^2$$

or, transposing terms,

(3.8) $$S^2/\sigma^2 = \chi_{n-1}^2/(n - 1)$$

Thus, the ratio of the sample variance to the population variance is equivalent to a χ^2 variable divided by its df.

Suppose that we independently drew two samples from the population giving two independent estimates of σ^2, S_1^2, and S_2^2. F, the ratio of these two sample variances is then related to χ^2; that is, using Equation (3.8),

(3.9) $$\frac{S_1^2}{S_2^2} = \frac{S_1^2/\sigma^2}{S_2^2/\sigma^2} = \frac{\chi_{n_1-1}^2/(n_1 - 1)}{\chi_{n_2-1}^2/(n_2 - 1)} = F$$

where n_1 and n_2 are the two sample sizes. If we repeatedly draw pairs of independent samples, we can obtain a frequency distribution of the ratio of two χ^2s divided by their *df*. Such a distribution is called the *F* distribution.

Note the conditions under which we may conclude that the ratio of sample variances is distributed as *F*. First, we assume that the random variable *Y* is distributed normally, in which case we can prove that $\sum(Y - \overline{Y})^2/\sigma^2$ is distributed as χ^2_{n-1}, or that (S^2/σ^2) is distributed as $\chi^2_{n-1}/(n - 1)$. Second, we assume that we have two independent estimates of the variance of the normal population. Then the ratio of the two sample variances is an *F* statistic, distributed on $n_1 - 1$ and $n_2 - 1$ *df*.

Since we are dealing with ratios of variances, *F* can take values between zero and plus infinity. The actual density function is rather complicated and will not be presented. However, the distribution is generally asymmetric with low probabilities of large values, and depends upon the values of df_1 and df_2, the numerator and denominator *df*s. Table A–5 presents values of *F* that will be exceeded with certain selected probabilities. For example, if df_1 equals four and df_2 equals eight, one percent of the values of *F* will exceed 7.01 and ten percent will exceed 2.81.

There is little point in discussing the applications of the *F* statistic in this chapter. Most of the remainder of this book will do exactly that.

3.5 POINT ESTIMATION

A quantity that can be computed from the population data, such as the population mean or variance, is generally referred to as a *population parameter*. Throughout this book, Greek letters will be used to denote parameters. For example, the population mean will be indicated by μ (mu) and the population standard deviation by σ (sigma). Similar quantities computed from the sample data are referred to as sample *statistics*, and Latin letters will be used to denote these. Thus the sample mean and standard deviation are represented by $\overline{Y}$ and *S*, respectively. One purpose of experimentation is to obtain estimates of the magnitude of population parameters on the basis of sample statistics. This type of inference is often referred to as *point estimation* to distinguish it from the estimation of an interval containing the parameter.

3.5.1 Properties of estimators. There are an infinite number of possible estimators of any single population parameter. For example, the population mean might be estimated by the sample mean, the sample median, or even the first score drawn from the sample. The question of which quantity best estimates the parameter can be answered by establishing criteria for good estimators and then examining how closely various estimators meet these criteria. The criteria that are generally agreed upon are based on the knowledge that an estimate will fluctuate from sample to sample. For example, if the value of the sample mean is computed for each of 20 samples independently drawn from the same population, the result will be a distribution of values. The same would be true of the median or mode or of any other sample statistic. It would seem desirable to choose an estimator such that the estimates obtained from different samples would be distributed about the estimated parameter with little variability. Then any one estimate will have high probability of being close to the parameter value. It also seems desirable that this requirement should have greater probability of being met as the size of the sample is increased; increased information should result in increased reliability. The following criteria embody the properties just described.

(a) *Lack of bias.* If the mean of the distribution of a statistic is equal to the parameter being estimated, the statistic is said to be an *unbiased estimator* of the parameter. A familiar example of such an estimator is the sample mean, $\bar{Y}$. If the distribution is plotted for the means of a large number of samples from the same population, the mean of this distribution of sample statistics will equal the population mean. The mean of a distribution is often referred to as the *expected value* of the statistic. Accordingly, the statement that the mean of the distribution of sample means equals the population mean may be more simply expressed by

(3.10) $$E(\bar{Y}) = \mu$$

which is read as "the expected value of the sample mean equals μ." In general, the statement that $\hat{\theta}$, an estimator of some parameter, θ, is unbiased may be expressed by

(3.11) $$E(\hat{\theta}) = \theta$$

For a finite population, the population variance, σ^2, is defined as $\sum(Y - \mu)^2/N$, the sum of the squared deviations of scores about the population mean divided by the number of scores in the population. A logical estimator of σ^2 would seem to be $\sum(Y - \bar{Y})^2/n$, the sum of the squared deviations about the sample mean divided by the number of scores in the sample. However, this estimate is biased since it can be proven that

(3.12) $$E\left[\frac{\sum(Y - \bar{Y})^2}{n}\right] = \left(\frac{n-1}{n}\right)\left[\frac{\sum(Y - \mu)^2}{N}\right] = \left(\frac{n-1}{n}\right)\sigma^2$$

Multiplying both sides of Equation (3.12) by $n/(n - 1)$ yields an unbiased estimator since

(3.13) $$E\left[\frac{\sum(Y - \bar{Y})^2}{n-1}\right] = \sigma^2$$

meeting the condition set forth in Equation (3.11). Because of the absence of bias, the estimator with $n - 1$ in the denominator is most frequently recommended.

(b) *Consistency*. An estimator is said to be consistent if

(3.14) $$P(\hat{\theta} \to \theta) \to 1 \qquad \text{as} \qquad n \to \infty$$

The above expression is read as "the probability approaches 1 that the estimator approaches the parameter as the sample size approaches infinity." An alternative statement is that the probability of obtaining an estimate closer to the estimated parameter increases as the sample size increases. An example of a consistent estimator is the sample mean, $\overline{Y}$. However, it does not follow that an estimator must be unbiased in order to be consistent. If an estimator is to approach the population parameter as n increases, it need be unbiased only for very large values of n. The quantity $\sum(Y - \overline{Y})^2/n$ is an example of a biased estimator which is unbiased when the sample size is very large, and which is consistent. It follows from Equation (3.12) that this estimator is unbiased for large n, since $(n - 1)/n \to 1$ as $n \to \infty$; for such values of n, the expression states that the expected value of the estimator equals the parameter. Nor does it follow that all unbiased estimators are consistent. Mood (1950) points out that $E(Y_1) = \mu$ where Y_1 is the first score drawn from the sample. However, this estimator is not consistent. Note that it is not necessarily equal to μ even when the entire population is sampled.

(c) *Efficiency*. In choosing between two estimators of a parameter, that estimator whose distribution exhibits less spread about the parameter is preferred. For example, assume that a large number of samples are obtained from a normally distributed population, a mean and a median are computed for each sample, and sums of squared deviations about μ are then computed, one for the sample means and one for the sample medians. The variability of the sample means about μ will be 64 percent of the variability of the sample medians about μ. This is expressed by saying that the *relative efficiency* of the median to the mean (as estimators of μ) is 64 percent. Conversely, the relative efficiency of the mean to the median is 1/.64 or 157 percent. The relative efficiency of $\hat{\theta}_1$ to $\hat{\theta}_2$ when these are two different estimators of θ is expressed by

$$\frac{E(\hat{\theta}_2 - \theta)^2}{E(\hat{\theta}_1 - \theta)^2}$$

Thus, relative efficiency is defined as the ratio of two averages of squared deviations of estimators about the same population parameter. Note that this is a measure of the efficiency of the estimator in the denominator relative to that in the numerator.

Since all the properties that have been described are not always present in one estimator, some conclusion regarding the relative importance of these properties is required. Although lack of bias is intuitively appealing, a biased estimator may be very satisfactory if its variability about the parameter is slight relative to that of other estimators and decreases with increasing sample size. The relative importance of efficiency and bias are illustrated in Figure

3–1; although the frequency distribution of $\hat{\theta}_1$ centers about θ, the probability of obtaining an estimate closer to θ is greater when $\hat{\theta}_2$ is the estimator, despite the fact that $\hat{\theta}_2$ is somewhat biased.

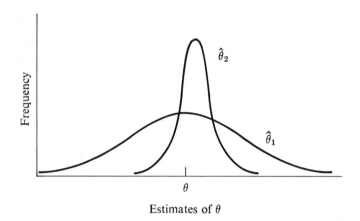

Estimates of θ

FIGURE 3–1 *Sampling distributions of two estimators of θ*

3.5.2 Principles of estimation. Once it is agreed that the properties just discussed are desirable, the problem of deriving estimators having such properties must be faced. The following discussion deals with two principles that lead to such derivations.

(a) *Maximum likelihood estimation.* Assume that we are concerned with the problem of estimating the probability of obtaining heads on the toss of a coin. This is basically the problem of estimating the parameter π, the proportion of heads in an infinite population of tosses. Now suppose that a sample of 10 independent tosses has resulted in seven heads and three tails. The maximum likelihood principle states that the best estimate is the one giving the highest probability, the maximum likelihood, of obtaining the observed data. Clearly $\pi = 1$ or $\pi = 0$ is an exceedingly poor estimate according to this principle, for if $\pi = 1$, the sample should have consisted of all heads, and if $\pi = 0$, all tails would have resulted. Somewhere between zero and one is an estimate that makes the occurrence of the observed sequence of results most likely. The probability of the obtained sequence of outcomes, P, is related to π by

(3.15) $$P = \pi^7(1 - \pi)^3$$

Table 3–1 is the result of substituting different estimates of π into Equation (3.15). When $\pi = .7$, the likelihood of obtaining the observed data is maximal, since P then takes on its largest value. The maximum likelihood estimate is therefore .7.

TABLE 3-1 *The probability, P, of obtaining seven heads and three tails in the observed sequence for various values of* $\hat{\pi}$

$\hat{\pi}$	P
0	0
.1	.0000000729
.2	.0000065535
.3	.0000750133
.4	.00035390
.5	.00097656
.6	.0017886
.7	.0022236
.8	.0016778
.9	.0004783
1.0	0

It is not necessary to work with an actual set of data or to enumerate likelihoods for various estimates. Differential calculus is applied to the equation describing the distribution of the data, and a general formula for the maximum likelihood estimator is obtained. In our example, Y/n would be the general solution, where Y is the observed number of heads and n is the number of tosses.

Maximum likelihood estimators have the important property of being as efficient as, or more efficient than, any other estimator and consistent as well. Although a maximum likelihood estimator is not necessarily unbiased, the center of its distribution is generally close to the value of the parameter being estimated.

(b) *Least-squares estimation.* Assume that any score in a sample is related to μ, the mean of the population, by

(3.16) $$Y_i = \mu + e_i$$

where Y_i is the score of subject i, and e_i is subject i's error component, the deviation of the score from μ. A measure of variability can be obtained by computing $\hat{e}_i (= Y_i - \hat{\mu})$ for all individuals in the sample, squaring these quantities, and summing them. The least-squares principle states that the appropriate estimate of μ is the value that makes the sum of squared deviations as small as possible; i.e., which minimizes the variability about $\hat{\mu}$. For example, the least-squares estimate is $\overline{Y}$ and is arrived at by applying differential calculus to find the minimum point of the function relating the sum of the $\hat{e}_i^2$ to various values of $\hat{\mu}$. The same approach can be used to find estimators for other parameters. If the e_i are uncorrelated and their frequency distribution is normal, the method yields results essentially similar to those of maximum likelihood estimation. However, the least-squares computations are usually much simpler.

Figure 3–2 illustrates the two estimation procedures just discussed. On the top is plotted the probability of obtaining the observed set of data as a function of various estimates of the population parameter. The estimation problem reduces to finding the maxima of the function, the value of $\hat{\theta}$ for which P is greatest. On the bottom is plotted the error variance, the variance

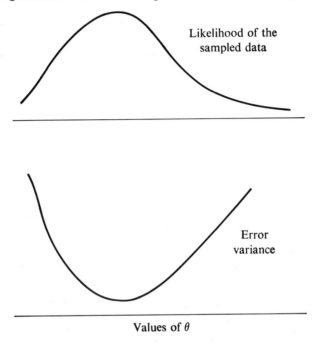

Likelihood of the sampled data

Error variance

Values of θ

FIGURE 3–2 *Two approaches to parameter estimation*

of the data points about the estimated parameter, as a function of various estimates. Now the estimation problem reduces to finding the minima of the function, the value of $\hat{\theta}$ for which the error variance is least. The functions drawn are hypothetical and will, of course, vary depending on the parameter to be estimated and the distribution of the data. However, the principles implied are general ones.

3.6 INTERVAL ESTIMATION

A point estimate will almost never be correct in the sense of equaling the parameter. Nor can the extent of error of estimate be judged on the basis of the estimate itself, since this reflects no information about its sampling distribution. In order to evaluate the adequacy of a point estimate, it is desirable to have an estimate of certain limits within which the parameter falls and a quantitative statement of confidence that the parameter does fall within these limits. Such confidence intervals can be established for many different parameters, but at present the process of establishing such limits will be exemplified by obtaining limits of μ.

The determination of a confidence interval for μ requires the following information:

(a) The value of the sample mean
(b) the value of the population standard deviation or some estimate of it
(c) the equation for the frequency distribution of scores in the population
(d) the size of the sample
(e) the degree of confidence required.

This last item is decided upon by the experimenter and will always be a number between zero and 1, generally .90 or above. For our example we will assume that

(a) $\bar{Y} = 25$
(b) $\sigma = 5$
(c) the distribution is known to be normal
(d) $n = 100$
(e) 95 percent confidence is desirable.

In accordance with the discussion of Section 3.4.1, if Y is normally distributed, the quantity

$$(3.17) \qquad\qquad Z = \frac{\bar{Y} - \mu}{\sigma/\sqrt{n}}$$

is also normally distributed, with mean equal to zero and standard deviation equal to 1. The distribution of z-scores, with percentages lying between the mean and several selected values, is presented in Figure 3–3. The fact that

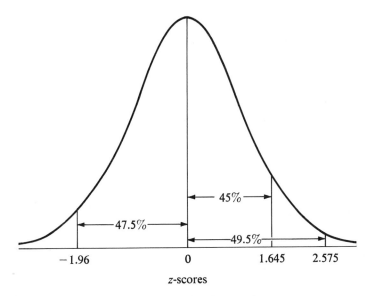

FIGURE 3–3 *Normal distribution of z-scores*

95 percent of the z-scores lie between -1.96 and 1.96 may be interpreted as follows: if 100 samples of scores are collected from a normal distribution, and if the means of these samples are transformed into z-scores according to Equation (3.17), 95 of them would be expected to fall between -1.96 and 1.96. Equation (3.18) expresses this more succinctly.

$$(3.18) \qquad P\left(-1.96 < \frac{\overline{Y} - \mu}{\sigma/\sqrt{n}} < 1.96\right) = .95$$

which is read as "the probability that the z transform of the sample mean lies between -1.96 and 1.96 is .95." Algebraic manipulation of Equation (3.18) results in Equation (3.19), which states the probability that μ lies within the designated limits.

$$(3.19) \qquad P\left(\overline{Y} - \frac{1.96\sigma}{\sqrt{n}} < \mu < \overline{Y} + \frac{1.96\sigma}{\sqrt{n}}\right) = .95$$

Plugging in the appropriate values from our example results in the 95-percent confidence interval, $24.02 < \mu < 25.98$. We have 95-percent confidence in the sense that, while 100 samples might give rise to 100 different sets of limits, 95 percent of such intervals are expected to contain μ. We therefore have high confidence that the one interval actually derived does contain μ.

Three variables that affect the width of the confidence interval and, consequently, the precision of the estimate, are σ, n, and the level of confidence. It can be seen in Equation (3.19) that as variability (σ) decreases, the upper limit of the interval is lowered and the lower limit is raised (becomes less negative). This reduced interval width reflects the increased precision accompanying decreased variability. As a concrete illustration of the relationship between interval size and σ, suppose that σ is 2.5, rather than 5 as in the original example. The limits are now $25 \pm (1.96)(.25)$ or 24.51 and 25.49.

The width of the interval containing μ also decreases as n increases. If n were 200 (σ, $\overline{Y}$, and the confidence level are again 5, 25, and .95, respectively), the limits would be $25 \pm (1.96)(.354)$ or 24.31 and 25.69.

Increased confidence is paid for by wider interval widths. If the 99 percent confidence interval were desired, 2.575 would be substituted for 1.96 in the previous calculations. (Note in Figure 3–3 that 99 percent of the z-scores fall between 2.575 and -2.575). Using the original values of σ, $\overline{Y}$, and n, the interval is now 25 ± 1.29 or 23.71 and 26.29. We have greater assurance than originally that μ falls within the specified interval, but the interval is wider than before and the interval estimate is therefore less precise.

3.7 HYPOTHESIS TESTING

Psychologists have generally been more concerned with the evaluation of specific hypotheses about the value of the parameter than with estimation (point or interval) of the parameter. Thus, they have usually asked questions of the form, Does μ equal 100? rather than, What does μ equal? or, Do these treatments differ in effect? rather than What is the difference in effect among these treatments? This section presents an example of how a specific hypothesis would be tested. We are concerned with those steps common to various statistical tests (such as t, χ^2, or F), although we have had to confine ourselves to one particular test to exemplify the inferential process.

Twenty rats are trained to run to a box containing food; the food box is white for half the rats, black for the other half. On the day following the last training day, all rats are tested for their preference between a white and a black box, neither of which contains food. The purpose is to determine whether a preference has been established for the box previously associated with food. In order to reach a conclusion, it is first necessary to state explicitly two hypotheses, the *null hypothesis*, H_0, and the *alternative hypothesis*, H_1. An appropriate null hypothesis is that, in the population from which these subjects are a sample, the percentage of correct responders (subjects who prefer the box that previously contained food) equals the percentage of incorrect responders. This might be expressed by

$$H_0: \pi = .5$$

where π is the proportion of correct responders in the population. The alternative hypothesis might be that in the population sampled, the pro-portion of correct responders exceeds the proportion of incorrect responders. This may be stated as

$$H_1: \pi > .5$$

This statement of H_1 ignores the possibility that there is a majority of incorrect responders in the population. Such a statement implies that the experimenter is willing to attribute the occurrence of more than 50 percent incorrect responders in the sample to chance, that such an occurrence will be viewed as support for H_0. This situation is described as a test of H_0 against a one-tailed alternative, or simply as a one-tailed test.

Suppose the experimenter considered it possible that a majority of incorrect responders might exist in the population. Possibly, the box that did not contain food is reinforcing because it is a novel stimulus; the subject has not experienced it before. In this situation, H_0 should be tested against a two-tailed alternative, namely

$$H_1: \pi \neq .5$$

In discussing various aspects of hypothesis testing in relation to this example, a one-tailed procedure will first be assumed. Changes in the testing procedure that arise when H_1 is two-tailed will then be noted.

The choice between H_0 and H_1 requires that a test statistic be computed from the data. Such a statistic should have the following properties:

(a) The statistic should reflect the relative merits of the two hypotheses,

(b) its distribution should be obtainable under the assumption that H_0 is true, and

(c) the assumptions necessary to derive the distribution should be reasonably valid for the data at hand.

Of the various statistics that meet these criteria for our example, the number of correct responders seems the least complicated and therefore the best for illustrative purposes. Let us consider how the number of correct responders meets our three requirements.

(a) If the null hypothesis is true, then the average number of correct responders (over many replications of the experiment) will be 10. If the true value of π is greater than .5, the average number of correct responders will be greater than 10. This monotone increasing relationship between π and the expected value of the test statistic meets requirement (a) above.

(b) If the performances of the 20 rats are independent, the probability that the number of correct responders equals x is

$$(3.20) \qquad P = \frac{n!}{x!\,(n-x)!}\,\pi^x(1-\pi)^{n-x}$$

where n is the number of subjects, x is the number of correct responders, and π is the proportion of correct responders in the population. In our

TABLE 3–2 *Probability of the number of correct responders when $n = 20$, assuming $\pi = .5$ and $\pi = .75$*

Number Correct	$\pi = .5$	$\pi = .75$
0	.000	.000
1	.000	.000
2	.000	.000
3	.001	.000
4	.005	.000
5	.015	.000
6	.037	.000
7	.074	.000
8	.120	.001
9	.160	.003
10	.176	.010
11	.160	.027
12	.120	.061
13	.074	.112
14	.037	.169
15	.015	.202
16	.005	.190
17	.001	.134
18	.000	.067
19	.000	.021
20	.000	.003

example $n = 20$, and under the null hypothesis $\pi = .5$. The probability distribution may now be easily obtained by inserting these values of n and π into Equation (3.20) and letting x take on the values $0, 1, \ldots, 20$. The resulting probabilities of various values of x may be found in the column headed "$\pi = .5$" in Table 3–2 and in the solid-line histogram of Figure 3–4.

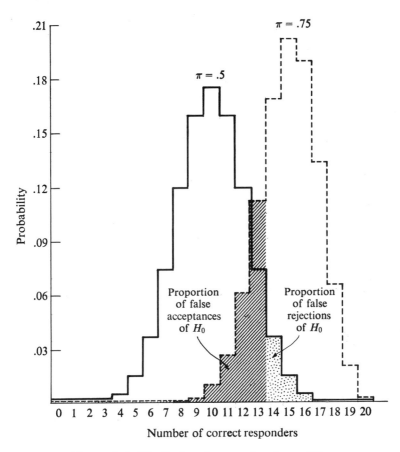

FIGURE 3–4 *Distributions for $\pi = .5$ and $\pi = .75$*

(c) Since each subject may be considered an independent replication of the same experiment, yielding only two classes of responses (correct, incorrect), the data seem appropriate to the assumptions underlying the application of Equation (3.20).

3.7.1 Type I errors. The null and alternative hypotheses have been stated, and a test statistic has been selected to provide a basis for choosing between them. The next step is the determination of those values of the statistic that will result in rejection of H_0 in favor of H_1. Such values constitute a critical region, a set of possible values of the test statistic, consistent with H_1, which are so improbable if H_0 is assumed true, that their occurrence leads us to reject H_0. An arbitrarily chosen value, alpha (α), defines exactly how improbable "so improbable" is. More precisely, α, the *level of significance*, is the probability of obtaining a test statistic that falls within the critical region when H_0 is true. For example, if α equals .05, the test statistic can be expected to fall within the critical region in five of 100 replications of an experiment for which the null hypothesis is true, thus resulting in rejections of true null hypotheses in 5 percent of the replications. The rejection of a true null hypothesis is generally referred to as a *Type I, or α, error*.

The size of the critical region will clearly depend upon the magnitude of α. We have also stated that this region is a set of values consistent with H_1. This means that the critical region is so chosen that its location is consistent with H_1. For example, if the alternative hypothesis is that π is greater than .5, the critical region should include only numbers of correct responders greater than half the sample size. If the alternative is two-tailed, the critical region should include numbers of responders both greater than and less than half the sample size.

In the example of the 20 rats faced with a black-white preference test, assume that α has been set equal to .06. Given the null hypothesis that the two choices are equally likely and the alternative that the correct side is preferred, the critical region can now be determined for the distribution of Table 3–2 and Figure 3–4 ($\pi = .5$). Since the probability of 14 or more responders is almost 6 percent ($.037 + .015 + .005 + .001 = .058$) in the right-hand tail of the distribution, this is the critical region. If 14 or more subjects respond correctly, H_0 will be rejected. This region of rejection is indicated in Figure 3–4.

What if the alternative were two-tailed; i.e., that the proportion of correct responders in the population might be greater than or less than .5? If deviations in both directions are equally important to detect, and if it is still required that the proportion of cases in the critical region does not exceed 6 percent, five or less or 15 or more correct responders would result in rejection of H_0.

To summarize briefly, tests of the null hypothesis require explicit statements of the null and alternative hypotheses, a choice of a statistic, and knowledge of the distribution of the statistic assuming H_0 to be true. These things, together with a stated significance level, result in the selection of a critical region. All these decisions must be made prior to the collection of data; the availability of experimental data could influence the statement of hypotheses or the selection of a critical region. Furthermore, if the test statistic and the assumptions underlying its distribution are not considered before the experiment is carried out, an experiment may be performed resulting in data that are difficult to analyze properly. Once the critical region has been decided upon, the data may be collected, the appropriate statistic computed, and its value compared with those falling within the critical region.

3.7.2 Type II errors. Since α is the probability of rejecting true null hypotheses, why not set it extremely low? The answer is that decreases in α decrease the probability of rejection of both true and false null hypotheses. In the extreme case where α is zero, Type I errors would never be made. However, false null hypotheses would also never be rejected. Obviously, the failure to reject a false null hypothesis is also an error. This failure to establish a treatment effect when it actually exists is generally referred to as a *Type II, or beta (β), error.* The example of the 20 rats in the preference test will be used to illustrate the principles involved in computing β, the probability of a Type II error, as well as to examine the relationships between β and such variables as α and n.

Once again, assume that the null hypothesis $\pi = .5$ is being tested against the alternative that $\pi > .5$. In addition, assume that, unknown to the experimenter, H_0 is false and the proportion of correct responders in the population is actually .75, i.e., $\pi = .75$. The probability distribution for $\pi = .75$ and $n = 20$ is presented in Table 3–2 and in Figure 3–4 (dashed-line histogram). If $\alpha = .06$, as previously stipulated, the experimenter will accept H_0 whenever 13 or fewer rats respond correctly. If π is really equal to .75, the probability of 13 or fewer rats responding correctly is .001 + .003 + .010 + .027 + .061 + .112 = .214. Thus β, the probability of accepting a false null hypothesis (false because $\pi = .75$ is the true state of affairs in the population), is .214. To summarize, the procedure for determining β against a specific alternative value of the parameter consists of dividing the baseline under the H_0 function into a region of acceptance and a region of rejection, and then determining the probability of the statistic falling in the acceptance region, given that the alternative is true.

An interpretation of β in terms of Figure 3–4 may also be helpful. Beta equals that proportion of the total area under the $\pi = .75$ function labeled "proportion of false acceptances of H_0." This region is that part of the $\pi = .75$ distribution lying above abscissa values also underlying the $\pi = .5$ function, but which excludes the critical region of the $\pi = .5$ distribution.

Several sources (e.g., Cohen, 1970) contain tables or graphs for *operating characteristics*, (OC) or *power* functions for some of the commonly used statistical tests. The OC function consists of values of β for the range of possible alternative values of the population parameter. The quantity $1 - \beta$ often appears rather than β. This is commonly referred to as the power of the test and is the probability of rejecting H_0 when it is false. Power or OC functions are important because several tests of the same H_0 may exist but have different power for the same critical region and sample size. If all other factors (e.g., validity of the model for the data, computational ease, availability of tables) are equal, the more powerful test is preferred. Furthermore, power and OC functions yield information on the relationship of power to α, H_1, and sample size, thus facilitating decisions about these variables.

Consider first the relationship between α and β. If the critical region of Figure 3–4 were reduced in size, the region of false acceptances of H_0 would consequently increase. Conversely, if α were increased, enlarging the critical region, there would be a corresponding decrease in the size of the region of false acceptances of H_0, and therefore in β. Because of this inverse relationship between α and β, the choice of α should always reflect a compromise between the relative importance of Type I and Type II errors. In some situations, Type I errors will be more undesirable. When the consequences of a significant finding (and such consequences may involve applications of the findings and/or future experiments) will be costly in time, effort, and money, the experimenter will want to be fairly certain that an effect really does exist before exploring it further. In such an instance, α may be set extremely low, guarding against Type I errors even at the increased risk of Type II errors. The relationship between one's own experimental findings and those of previous studies is also a consideration. If a significant result will conflict with an established body of knowledge, more stringent significance levels (e.g., .01) might be required. On the other hand, in research areas where the variables influencing behavior are less well understood, the experimenter might be willing to take a greater risk of a Type I error, reducing β in an attempt to avoid missing some promising lead.

The above comments are, at best, guides rather than rules for the setting of α. There are no nice, neat formulas for arriving at the appropriate level of α. As a consequence of this arbitrary quality of our inferential process, it follows that statistical significance should not be the sole basis for judgments of experimental effects but merely one important piece of information. Inferences should not be ground out by a computer, but rather should be thought out by an experimenter. We will return to this problem in the last part of this chapter.

The choice between one and two-tailed tests also has implications for β. If α is kept constant, the critical region of Figure 3–4 must be reduced to establish a similar region in the left-hand tail of the $\pi = .5$ distribution. The shift from a one-tailed to a two-tailed alternative permits the detection of π values less than .5, but only at the cost of reduced power to detect π values greater than .5.

Figure 3–5 presents distributions of correct responders for the null hypothesis $\pi = .5$ and for the alternative $\pi = .75$ with α equal to .06 and n equal to 15. Comparing these distributions with those of Figure 3–4, one notes that the decrease in n from 20 to 15 has resulted in an increase in the proportion of area under the $\pi = .75$ function, which consists of false acceptances of H_0. Beta has shifted from 21.4 percent to 31.4 percent. A reduction in the amount of data increases the overlap between the two distributions, consequently reducing the ability to detect false null hypotheses.

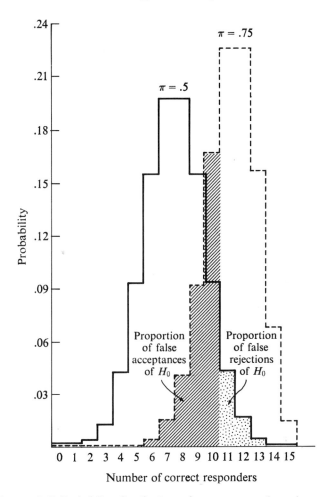

FIGURE 3–5 *Probability distributions of correct responders when n = 15*

The converse is also true; increases in *n* result in decreases in β and increases in power.

Some concept of how β varies as a function of the true value of the parameter may be obtained by looking at Figure 3–6, which shows probability distributions for $\pi = .5$ and for $\pi = .9$. In both cases, α and *n* are again equal to .06 and 20, respectively. Note that the proportion of false acceptances of H_0 constitutes only .2 percent of the area under the $\pi = .9$ function as compared with 21.4 percent for the $\pi = .75$ function in Figure 3–4. This result makes good sense; the further apart the null-hypothesized and true parameter values are, the easier it should be to detect false null hypotheses.

To summarize, β is reduced (and power is consequently increased) as the difference between the value of the parameter under H_0 and the true

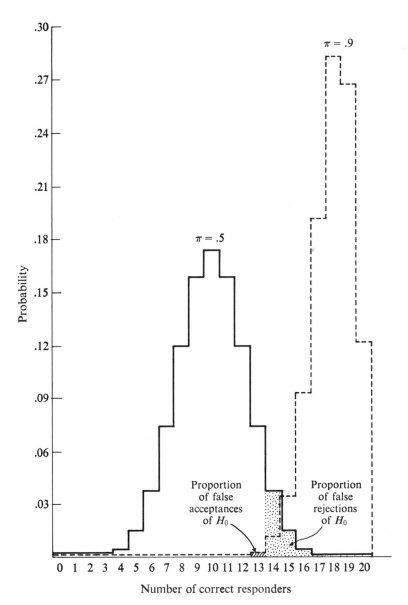

FIGURE 3–6 *Distributions of correct responders for $\pi = .5$ and $\pi = .9$*

value of the parameter increases, as α increases, and as n increases. These relationships are negatively accelerated; i.e., the reduction in β becomes less marked as the value of the parameter increases. Thus, for example, the reduction in β is much less when n is increased from 90 to 100 than when n is increased from 10 to 20. In addition to the effects just cited, a one-tailed test will always be more powerful than a two-tailed test against alternatives in that tail, but less powerful against alternatives in the second direction.

Beta differs from α in the sense that β can only be selected with reference to some minimal effect. To say that one will set β equal to .10 (as we might set α equal to .10) is meaningless; one can only require that β equal .10 for some stated minimal degree of falsity of the null hypothesis. It is not sufficient to select a level of β (or power); one must also decide what deviations from the parameter value assumed under H_0 are important to detect. In the example of the preference experiment with rats, it might be asked how much greater than .5 must π be before it becomes important to reject the null hypothesis. It may seem that any deviation from the value of the parameter assumed under H_0 is important to detect. However, it is rare that the null hypothesis is precisely true. This being the case, we wish to design the experiment so that the rejection of H_0 will not be a trivial result, merely reflecting the collection of a very large amount of data. Our interest lies not in whether H_0 is false (it usually is), but in whether it is so false that the fact is important. For example, suppose that one desires to compare the effect of traditional classroom procedures with the effect of using teaching machines on the rate of learning arithmetic. The null hypothesis might be that the two methods are equally effective, and the alternative hypothesis might be that the use of teaching machines is more effective. It might be considered important to reject H_0 only if the teaching machine was so much more effective that the added expense of developing teaching machine programs and of building machines was judged to be outweighed.

An important consequence of the preceding discussion of β is that there is now a rationale for deciding on the number of subjects to be run. We require an n sufficient to achieve a desired degree of power for rejecting H_0 against some specified alternative, given that the probability of a false rejection is α. The specific steps in deciding upon n are the following:

(a) State the null and alternative hypotheses. Continuing with our example of the rat experiment, we have

$$H_0: \pi = .5$$
$$H_1: \pi > .5$$

(b) Decide on the level of α. Assume that α is set equal to .06.

(c) Decide on the minimal degree of falsity of H_0 that is important to detect. In our example, assume that if $\pi \geq .75$, it is important to reject H_0.

(d) Decide on the level of power, the probability of rejecting H_0 if the true parameter value deviates to the extent specified under (c). Assume that if $\pi \geq .75$, we wish at least an 80-percent chance of rejecting H_0. Consequently, power $= .80$ and $\beta = .20$.

Given the outcomes of the above decisions, the required n can now be determined. Turning to Figure 3–4, we note that if $\alpha = .06$, $\pi = .5$ under

H_0, $\pi = .75$ under H_1, and $n = 20$, then $\beta = .214$. To have $\beta = .20$, approximately 21 subjects would be required. The steps indicated above are essentially the same for all statistical tests. The major difference in the steps between the simple binomial test developed in this chapter and others such as the F and t tests is that the last two require some estimate of the population variance.*

3.8 CONFIDENCE INTERVALS AND SIGNIFICANCE TESTS

In the preceding two sections confidence intervals and significance tests have been separately considered. In this section, some similarities and differences in the two procedures will be briefly noted. Both techniques involve the same assumptions, manipulations of the same quantities, and the same definitions of probability; i.e., the probability of an event is the relative frequency of occurrence of the event. Despite these similarities, the two procedures do result in somewhat different presentations of the information obtained from the data. In this section, a more detailed comparison of the two procedures will be made, using the following example.

Twenty-five rats are run for 500 trials in a T-maze. One goal box contains food on 60 percent of the trials, the other goal box contains food on the remaining 40 percent. One theory of maze behavior would lead the experimenter to predict that the rats will run to each goal box with the same relative frequency that the goal box contains reward: 60 percent of the responses will be made to the more frequently rewarding goal box and 40 percent to the less frequently rewarding box. In order to assess this "matching" hypothesis, the experimenter decides to test the null hypothesis that for the last block of 20 trials there are an average number of 12 runs (i.e., 60 percent) to the more frequently rewarding goal box in a population of rats from which the 25 subjects have been sampled. The t test is the statistic chosen, the alternative hypothesis is that $\mu \neq 12$, and α is chosen as .05. The corresponding critical region consists of values of t less than -2.06 and greater than $+2.06$.

The mean number of runs to the more frequently rewarding goal box is 11, and the standard deviation is 3. Inserting these values, together with the sample size, 25, into the following equation

(3.21)
$$t = \frac{\overline{Y} - \mu}{s/\sqrt{n}}$$

results in $(11 - 12)/(3/5) = 1.67$, which does not fall within the critical

---

*In order to deal with a power function that could easily be computed by the reader, the foregoing discussion has omitted one factor, variability. Power for many tests (e.g., z, t, F) is plotted as a function of a ratio of the parameter value to the population variance. As might be expected, power varies inversely with variability. (A specific example is presented for the F test in Chapter 4.)

region. The experimenter concludes that the mean percentage of responses to a goal box does match the percentage of trials on which it contains food.

The experimenter could also approach this problem by establishing a confidence interval for μ. Starting with Equation (3.22)

$$(3.22) \qquad P\left(-2.06 < \frac{\overline{Y} - \mu}{s/\sqrt{n}} < 2.06\right) = .95,$$

algebraic manipulation and insertion of the values for $\overline{Y}$, s, and n result in the 95-percent confidence interval, $9.76 < \mu < 12.24$. There is 95-percent confidence that the population mean lies between these values, and again there is support for the belief that the true mean is 12.

With the two analyses completed, certain aspects may be compared. In both techniques the same information has been used: the values of the mean, standard deviation, and the size of the sample. Both techniques rest upon the properties of the t distribution and consequently upon the assumption that the scores constitute a random sample from a normally distributed population. The t test of $H_0 : \mu = 12$ provides evidence for the matching hypothesis. The confidence interval does also, since the sample mean, 11, will not differ significantly (at the 5-percent level, if confidence is 95 percent) from any value that falls in the obtained interval, $9.76 < \mu < 12.24$. The confidence interval goes further in permitting the experimenter to note that certain other hypotheses are also tenable; e.g., the hypothesis that the rats develop no preference is consistent with the fact that the value 10 also falls within the interval. In brief, the confidence interval approach permits the simultaneous consideration of all possible null hypotheses.

Suppose the confidence interval had been $4 < \mu < 18$. While it is true that the value predicted by the matching hypothesis falls within this interval, the interval is so wide that it also contains support for numerous alternative theories. Since no theory that predicts means from 5 through 12 can be rejected, this interval would result in a less firm statement that the matching hypothesis was correct. The width of the confidence interval indicates how firmly an experimenter can state an inference about any hypothesized population mean.

Tests of the null hypothesis can also provide the experimenter with an index of assurance when he accepts the null hypothesis. If there is high power of detecting even small deviations from the value assumed under H_0, an experimenter can be quite firm in his support of the matching hypothesis when the statistical test does not yield significant results. If he had little probability of detecting differences of the size observed, the experimenter might be more hesitant about drawing conclusions on the basis of the test. Thus the width of the confidence interval and the power of the statistical test provide similar types of information. However, there is a major difference, which is to the advantage of the confidence interval approach. The computed confidence interval must of necessity provide information about its width, while null hypothesis tests can be, and usually are, made without reference to the appropriate power or OC functions.

To summarize, the confidence interval permits the experimenter to simultaneously consider the possible range of null hypotheses and immediately yields an index of the strength of an inference about any one null hypothesis. In addition, the confidence interval consists of a set of numbers on the same scale as the original data. In the case of the null hypothesis test, the experimenter is further removed from his original measures, since the test statistic is on a different scale. It may therefore be easier to digest the information provided by the confidence interval.

Tests of the null hypothesis are firmly entrenched in the statistical methodology of psychology. The intent of our comparison of null hypothesis tests and confidence intervals has not been to promote one at the expense of the other, but rather to make the reader aware of one useful approach to data analysis that has often been neglected by researchers. This book will generally reflect the overwhelming prevalence of hypothesis testing in the psychological literature. However, in subsequent chapters techniques will be presented for establishing confidence intervals for several parameters. Hopefully, the preceding discussion pointing to the availability of alternative techniques will at least facilitate an intelligent choice between estimation and hypothesis testing. Too often, psychologists have tested in ignorance of the alternative.

References to additional discussions of confidence intervals and null hypothesis tests may be found at the end of the chapter. There are also several references, which present objections to both interval estimation and null hypothesis tests and in which alternative inferential procedures are proposed. Although a discussion of this material is beyond the scope of the present text, the reader should find the source material stimulating.

In concluding this chapter it is important to warn against a too literal translation of statistical results into scientific conclusions. Certainly the significance or nonsignificance of a test statistic should be a major factor in drawing conclusions about treatment effects, particularly when decisions about such things as α, β, and n have been made prior to the experiment. But, assuming nonsignificance, how sure can one be that an important effect does not exist when the probability of the test statistic falls .1 percent above α, or when one finds that the observed variability is greater than the estimate used to decide n? What should be concluded about results that are barely significant or just fall short of significance when the assumptions underlying the test are not met by our data? Should one draw the same inference about two nonsignificant results when the qualitative trends in one set of data are consistent with expectations based on available data and theory, yet no recognizable pattern exists in the second data set? How should published results of others be interpreted, when it is apparent that most experimenters give no thought to β in planning their experiment, that few even pre-select α?

Under such circumstances, the test statistic can be at best a rough indicator of population effects rather than a sharp inferential tool. There are no simple answers, but we reject any one-to-one relationship between the significance or nonsignificance of a test statistic and the existence or nonexistence of treatment effects, or the tenability or nontenability of a theory under investigation. In drawing inferences, the scientist has the responsibility of adding to the test statistic his a priori expectations, his knowledge of the literature, of the particular experimental conditions (e.g., is there reason to suspect that some variable whose effects are not analyzable obscured the effects of independent variables?), and of the size and direction of effects, and to subjectively weight these factors. When the results of the data analysis (regardless of the type) conflict with those factors not built into the test, then the experimenter should reserve judgment. The ultimate criterion of the credability of experimental conclusions is whether or not these conclusions are supported by subsequent replications of the experiment or by differently designed investigations of the hypotheses in question.

EXERCISES

3.1 Define the following terms:

(a) parameter

(b) consistent

(c) efficient

(d) unbiased

(e) expected value

(f) power

(g) Type I error

(h) Type II error

(i) alpha

(j) beta

(k) critical region

(l) one-tailed test

(m) confidence interval

3.2 One hundred random samples, each consisting of 225 scores, are drawn from a population of normally distributed scores, with $\mu = 0$, $\sigma = 1$. The 95 percent confidence interval, $\bar{Y} - 1.96(1/15) < \mu < \bar{Y} + 1.96(1/15)$, is computed for each sample.

(a) Verify that the formula used is correct.

(b) Only 92 of the 100 computed confidence intervals contain the value of the true mean, 0. Does this fact conflict with the statement that these are 95 percent confidence intervals? If a 95 percent confidence interval does not mean that 95 percent of the obtained intervals will include the parameter value, what does it mean?

3.3 I desire a 95 percent confidence interval for μ, the mean of a normally distributed population whose standard deviation is 15. I would like the interval to be no wider than 10 units. How large should my sample be to achieve this criterion?

3.4 In our discussion of hypothesis testing, the critical region was always selected to be in one or both tails of the distribution. For example, against the alternative hypothesis, $\mu_1 > \mu_0$, using the z test with $\alpha = .05$, the critical region would consist of z-values equal to or greater than 1.645. Consider this rule for rejecting H_0: reject H_0 if the test statistic falls between $z = -.065$ and $z = .065$, this area also being approximately 5 percent of the area under the normal curve. What would happen to the power of the test? Present diagrams similar to those in Figures 3–4 to 3–6 for the usual z test and for the suggested alternative to illustrate your argument.

3.5 Consider

$$H_0 : \mu = 100$$

$$H_1 : \mu = 110$$

(a) Assume $n = 25$, $\sigma = 20$, $\alpha = .05$. What is the power of a one-tailed test of H_0 against H_1? Use the normal probability (z) test.
(b) What is the power if the test is two-tailed?

3.6 Prove

$$E\left(\frac{\sum(Y - \bar{Y})^2}{n}\right) = \left(\frac{n - 1}{n}\right)\sigma^2$$

(*Hint:* Begin with $Y - \mu = (Y - \bar{Y}) + (\bar{Y} - \mu)$).

REFERENCES

Cohen, J. *Statistical power analysis for the behavioral sciences.* New York: Academic Press, 1969.
Mood, A. M. *Introduction to the theory of statistics.* New York: McGraw-Hill, 1950.

SUPPLEMENTARY READINGS

For the student with a background in calculus, a more mathematical treatment of the material in this chapter may be found in the following:

Mood, A. M. and Grayhill, F. A. *Introduction to the theory of statistics* (second edition). New York: McGraw-Hill, 1963.

Some articles written by psychologists treating the problems of null hypothesis testing are:

Baken, D., "The test of significance in psychological research." *Psychological Bulletin,* 66: 423–437 (1966).
Binder, A., "Further considerations on testing the null hypothesis and the strategy and tactics of investigating theoretical models," *Psychological Review,* 70: 101–109 (1963).
Edwards, W., "Tactical note on the relation between scientific and statistical hypothesis," *Psychological Bulletin,* 63: 400–402 (1965).
Edwards, W., Lindman, H. and Savage, L. J., "Bayesian statistical inference for psychological research," *Psychological Review,* 70: 193–242 (1963).
Grant, D. A., "Testing the null hypothesis and the strategy and tactics of investigating theoretical models," *Psychological Review,* 69: 54–61 (1962).

LaForge, Rolfe, "Confidence intervals or tests of significance in scientific research," *Psychological Bulletin*, 6: 446–447, (1967).

Lykken, David T., "Statistical significance in psychological research," *Psychological Bulletin*, 3: 151–159 (1970).

Rozeboom, W. W., "The fallacy of the null hypothesis significance test," *Psychological Bulletin*, 57: 416–428 (1960).

Wilson, W., Miller, H. L. and Lower, J. S., "Much ado about the null hypothesis," *Psychological Bulletin*, 3: 188–196 (1967).

Wilson, W. R. and Miller, H., "A note on the inconclusiveness of accepting the null hypothesis," *Psychological Review*, 71: 238–242 (1964).

Many of the above include some discussion of significance tests and confidence intervals. A more detailed comparison is presented in the following.

Natrella, M. G., "The relation between confidence intervals and tests of significance—a teaching aid," *American Statistics*, 14: 20–22, 38 (1960).

An excellent statement of the considerations involved in drawing inferences from data is to be found in

Tukey, J. W., "Conclusions versus decisions," *Technometrics*, 2: 423–433 (1960).

4

Completely Randomized One-Factor Designs

4.1 INTRODUCTION

In this and the following chapter we will consider *completely randomized designs*. These designs are characterized by the random assignment of each subject to only one level of the independent variable, or to only one combination of levels if more than one independent variable is under investigation. The term "random" represents the requirement that each subject has an equal probability of assignment to any level or combination of levels.

As an example of a completely randomized design involving one independent variable, consider the following experiment. Each subject is required to learn the appropriate response to each of 12 stimulus words; the dependent variable is the number of trials (times through the list of 12 stimulus-response pairs) required to attain a criterion of two errorless trials. The independent variable is level of noise intensity, and it is planned to have 40 subjects, 10 tested at each of four levels of noise. The subjects are college students, volunteers from a basic psychology course, who indicate on a sign-up sheet their willingness to participate.

One way to assign these subjects randomly to noise levels would be to write the numbers from 1 to 40 on separate slips of paper, place these in a hat, mix well, and then note the order in which the slips are selected out of the hat. If the slip of paper numbered 17 were selected first, then the 17th subject to volunteer would be placed in the first treatment group; for example, the one corresponding to the lowest level of noise. The first 10 selected pieces of paper designate the subjects who go into the first group, the second 10, those who are placed in the second treatment group, and so on.

A simpler way to accomplish this would be to employ a table of random numbers such as Table A–1 in the Appendix. Two adjacent columns of digits are chosen at random within a five-column set, and a row within the columns is then selected randomly. Beginning with this row, one proceeds down the columns, noting the order of appearance of the numbers from 01 to 40 and ignoring all other numbers. This order provides the basis for assignment to groups in the same way as did the order of drawing numbered slips. Numbers other than 01 to 40 might be used by making 41 equivalent to 01, 42 equivalent to 02, and so on. The numbers 81 to 00 would still be omitted in order to give all numbers an equal chance of being selected.

The procedures described for assigning subjects to experimental treatments ensure that all subjects have an equal probability of assignment to each treatment and, consequently, that there exists no systematic source of error. In contrast with such procedures, note the possible consequences when the first 10 subjects to volunteer are assigned to one treatment, the second 10 volunteers to a second treatment, and so on. Under this procedure it is difficult to defend any inference based upon the data analysis. For example, if the low noise group learns most quickly, is this a function of differences in the effects of noise upon performance, or were the earlier volunteers more interested in the experiment, more motivated to do well?

The completely randomized design has one major advantage over other designs—simplicity. This characteristic of such designs extends to the experimental layout, the model underlying the data analysis, and the computations involved in the data analysis. Let us briefly consider these advantages.

(a) *The experimental layout.* The experimental layout is a statement of who is tested under what conditions. If the design requires that each subject be tested under a number of experimental conditions, the order of presentation of conditions for each subject is part of the layout. For the completely randomized design, only the random assignment described previously is required to lay out the experiment. In contrast, stratified designs (see Chapter 6) require the division of subjects into strata or levels on the basis of some additional measure; then random assignment to the levels of the independent variable is carried out within these strata. Other designs involve still other complications in obtaining the experimental layout.

(b) *The model.* The model underlying the completely randomized design will be considered shortly in some detail. Its important feature is that it involves fewer assumptions than those made for any other experimental design. Consequently, the derivations of parameter estimates are simpler than in any other instance. An even more important consequence of the parsimony of assumptions is that there is less that can go wrong with the inferential machinery than in the case of other, more complex, models. Each additional assumption underlying the derivation of the test statistic is one more assumption that may be violated, undermining the validity of the statistical inference.

(c) *The computations.* The analysis of variance for the completely randomized design involves fewer terms than are required for other designs, and the terms are often easier to compute. Furthermore, related analyses such as the analysis of covariance and the estimation of missing data will also generally be simpler for the completely randomized design.

The major disadvantage of the completely randomized design is its relative inefficiency. The error variance will usually be large compared to that resulting from the use of other designs. This is in part offset by the fact that no design yields as many degrees of freedom for the error variance as does the completely randomized design, assuming some fixed amount of data. The degrees of freedom will be discussed at greater length later in this chapter; for now it is enough to note only that the power of a test increases monotonically with degrees of freedom.

In order to facilitate the learning of the subject matter of this book, we shall first consider only limited aspects of a topic, then gradually develop additional concepts and computations in later chapters. Thus, in the present chapter, the following limitations have been placed on the presentation:

(a) We consider only that subset of completely randomized designs that involve only one independent variable; we refer to these as one-factor designs.

(b) We consider only independent variables whose levels are fixed; i.e., we define the population of levels as consisting only of those that have been selected for the experiment.

(c) We consider only the test of the general null hypothesis that $\mu_1 = \mu_2 = \cdots = \mu_j = \cdots = \mu_a$, where μ_j is the mean of a population of individuals tested under A_j, the jth level of the independent variable A.

The above restrictions are loosened in subsequent chapters. In Chapter 5, designs involving several independent variables are discussed; in Chapter 7, the implications of randomly sampling a subset of the possible levels of an independent variable are examined; and in Chapters 13 and 14 the rationale and computations for several additional tests on the data are discussed.

4.2 A MODEL FOR THE COMPLETELY RANDOMIZED ONE-FACTOR DESIGN

Consider an infinitely large population of individuals. Assume that each individual in this parent population is randomly assigned to exactly one of a possible treatments (levels of the independent variable). There are now a very large *treatment populations*, systematically differing from each other only with respect to the levels of the independent variable A. Next consider a completely randomized one-factor experiment in which *an* subjects are randomly distributed among a treatments in the manner described earlier or by some other procedure that equally assures randomness of assignment. The a experimental groups of n subjects may be considered as a random samples, one from each of the treatment populations hypothesized previously. In general, we are interested in drawing inferences about $\mu_1, \mu_2, \ldots, \mu_j, \ldots,$ μ_a, the means of the treatment populations. At present, our specific concern is with developing a test of the null hypothesis that the μ_j are equal. To develop such a test and to answer other questions about the population parameters, we require some statement about the relationship between the data and the parameters of the population. A simple possibility is the following:

(4.1) $$Y_{ij} = \mu + \alpha_j + \varepsilon_{ij}$$

where Y_{ij} is the score i of treatment group j; μ is the mean of the μ_j or, equivalently, the mean of the parent population prior to the establishment of treatment populations; $\alpha_j = \mu_j - \mu$, the *effect* of treatment A_j; and $\varepsilon_{ij} = Y_{ij} - \mu - \alpha_j = Y_{ij} - \mu_j$.

The quantity ε_{ij} is the error associated with the ith score in the jth group and is a unique contribution of the individual, a deviation of the total score from μ, which cannot be accounted for by the treatment effect. The variability in ε_{ij} may be due to differences among subjects in such factors as ability, set, motivation, the tone in which instructions are read, or the temperature of the room.

Before developing the consequences of Equation (4.1), it may be helpful to consider the rationale underlying the equation. If our subjects were identical individuals, identically treated, we would have

$$Y_{11} = Y_{21} = \cdots = Y_{ij} = \cdots = Y_{na} = \mu$$

But each group of n individuals has been treated differently. Assuming n identical individuals in each treatment population, but now allowing for the possibility that the treatments do not have identical effects, we obtain

$$Y_{11} = Y_{21} = \cdots = Y_{i1} = \cdots = Y_{n1} = \mu_1$$
$$Y_{1j} = Y_{2j} = \cdots = Y_{ij} = \cdots = Y_{nj} = \mu_j$$
$$Y_{1a} = Y_{2a} = \cdots = Y_{ia} = \cdots = Y_{na} = \mu_a$$

Finally, we take into account the fact that even individuals treated alike with respect to the independent variable will rarely perform in an identical manner. Due to individual differences, the score of subject i in group j will deviate from μ_j by an amount ε_{ij}. Consequently

$$Y_{11} = \mu_1 + \varepsilon_{11}$$

(4.2) $$Y_{ij} = \mu_j + \varepsilon_{ij}$$

$$Y_{na} = \mu_a + \varepsilon_{na}$$

Equation (4.2) is unchanged if we add and subtract the constant μ, resulting in

$$Y_{ij} = \mu + \mu_j - \mu + \varepsilon_{ij}$$

Since $\mu_j - \mu = \alpha_j$ by definition, Equations (4.2) and (4.1) are equivalent.

Thus the model underlying the data analysis asserts that the score of the ith individual in the jth group is the sum of the following three components:

(a) μ *the parent population mean.* This quantity is a constant component of all scores in the data matrix.

(b) α_j, *the effect of treatment* A_j. This effect is a constant component of all scores obtained under A_j, but may vary over treatments (levels of j). If the a levels of A are arbitrarily selected, as assumed, they exhaust the population of levels of A, and therefore $\sum_j \alpha_j = 0$, since the sum of all deviations of scores (μ_j) about their mean (μ) is zero. The null hypothesis asserts that $\alpha_1 = \alpha_2 = \cdots = \alpha_j = \cdots = \alpha_a = 0$.

(c) ε_{ij}, *the deviation, due to uncontrolled variability, of the ith score in group j from the jth treatment population mean.* This component of Y_{ij} is the only source of variance among scores in the jth group and, if the null hypothesis is true, the only source of variance in the data matrix.

Equation (4.1) is not a sufficient basis for the derivation of parameter estimates and statistical tests. In addition, the following assumptions about the distribution of the ε_{ij} are required:

(a) The ε_{ij} are independently distributed. This means that the probability of sampling some value of ε_{ij} does not depend on any other values of ε_{ij} in the sample. An important consequence of this is that the ε_{ij} are uncorrelated.

(b) The distribution of the ε_{ij} is normal, with zero mean, in each of the a treatment populations.

(c) The distribution of the ε_{ij} has variance of σ_e^2 in each of the a treatment populations; i.e., $\sigma_1^2 = \sigma_2^2 = \cdots = \sigma_j^2 = \cdots = \sigma_a^2 = \sigma_e^2$.

Since the ε_{ij} are solely responsible for the variability in the jth treatment population, it follows that the Y_{ij} should also be normally and independently distributed with mean μ_j and variance equal to σ_e^2. This fact provides a basis for testing the above assumptions. A more detailed analysis of the assumptions will be presented later; for the present some consequences of the model are shown.

4.2.1 Estimates of the population parameters. Under the normal distribution assumption, the least-squares and maximum likelihood procedures described in Chapter 2 result in identical values of $\hat{\mu}$ and $\hat{\alpha}$, the estimates of μ and α. Both procedures ordinarily require the application of differential calculus to obtain the estimates; we will therefore omit the derivation but will offer an algebraic proof that the least-squares estimates of μ and α_j are $\overline{Y}_{..}$ and $\overline{Y}_{.j} - \overline{Y}_{..}$ respectively.

The simplest approach is to first derive an expression for $\hat{\mu}_j$, the least-square estimate of μ_j. Since μ is an average of the μ_j, $\hat{\mu}$ can then readily be obtained by averaging the $\hat{\mu}_j$. Furthermore, since $\alpha_j = \mu_j - \mu$, then $\hat{\alpha}_j = \hat{\mu}_j - \hat{\mu}$. A simple relationship between μ_j and the error component, ε_{ij}, is provided by Equation (4.2); transposing terms and replacing the parameters by their estimators, we have

(4.2')
$$\hat{\varepsilon}_{ij} = Y_{ij} - \hat{\mu}_j$$

We desire a value of $\hat{\mu}_j$ that minimizes the quantity $\sum_i \sum_j \varepsilon_{ij}^2$, the sum of squared errors. We therefore square both sides of Equation (4.2') and, after summing, we have

(4.3)
$$\sum_j \sum_i \varepsilon_{ij}^2 = \sum_j \sum_i (Y_{ij} - \hat{\mu}_j)^2$$

The least-squares estimate, $\hat{\mu}_j$, may be represented as a value some distance from the group mean, $\overline{Y}_{.j}$; that is, $\hat{\mu}_j = \overline{Y}_{.j} + C$. Substituting in Equation (4.3) gives us

$$\sum_j \sum_i \varepsilon_{ij}^2 = \sum_j \sum_i (Y_{ij} - \overline{Y}_{.j} - C)^2$$

(4.4)
$$= \sum_j \sum_i (Y_{ij}^2 + \overline{Y}_{.j}^2 + C^2 - 2Y_{ij}\overline{Y}_{.j} - 2Y_{ij}C + 2\overline{Y}_{.j}C)$$

$$= \sum_j \sum_i (Y_{ij}^2 + \overline{Y}_{.j}^2 - 2Y_{ij}\overline{Y}_{.j}) + anC^2 - 2C \sum_j \sum_i Y_{ij} + 2Cn \sum_j \overline{Y}_{.j}$$

Note the application of the rules of Chapter 2. Equation (4.4) may be simplified if we remember that $\overline{Y}_{.j} = \sum_i Y_{ij}/n$; then $2Cn \sum_j \overline{Y}_{.j} = 2Cn \sum_j \sum_i Y_{ij}/n$, which cancels the preceding term. We now have

(4.5)
$$\sum_j \sum_i \varepsilon_{ij}^2 = \sum_j \sum_i (Y_{ij}^2 + \overline{Y}_{.j}^2 - 2Y_{ij}\overline{Y}_{.j}) + anC^2$$

If C is not zero, anC^2 is a positive term. Therefore, the error variability is smallest when C equals zero, or when $\hat{\mu}_j = \overline{Y}_{.j}$. Estimates for the remaining parameters follow readily. We have

(4.6)
$$\hat{\mu}_j = \overline{Y}_{.j}$$
$$\hat{\mu} = \overline{Y}_{..}$$
$$\hat{\alpha}_j = \overline{Y}_{.j} - \overline{Y}_{..}$$
$$\hat{\varepsilon}_{ij} = Y_{ij} - \overline{Y}_{.j}$$

If the ε_{ij} are, as assumed, normally distributed, then the above estimates are efficient and consistent.

4.2.2 The F ratio. If the assumptions of the model are met, an appropriate test of the null hypothesis that the μ_j are equal is provided by

(4.7)
$$F = \frac{n \sum_j \hat{\alpha}_j^2/(a - 1)}{\sum_i \sum_j \hat{\varepsilon}_{ij}^2/a(n - 1)} = \frac{n \sum_j (\overline{Y}_{.j} - \overline{Y}_{..})^2/(a - 1)}{\sum_i \sum_j (Y_{ij} - \overline{Y}_{.j})^2/a(n - 1)}$$

The numerator of F is generally referred to as the *between-groups mean square*. We will use the notation MS_A, where the subscript designates the independent variable. The denominator of the F ratio is generally referred to as the *within-groups mean square* and we will use the notation $MS_{S/A}$, where the subscript is read as "subjects within levels of A." Thus $F = MS_A/MS_{S/A}$. Each MS is a ratio of a *sum of squares* (SS) to its *degrees of freedom* (df). Thus, $MS_A = SS_A/df_A$ and $MS_{S/A} = SS_{S/A}/df_{S/A}$. The origins of these components of the F ratio will be discussed later.

Let us now consider why the ratio of mean squares should be sensitive to violations of the null hypothesis. The numerator of the F ratio is n times the variance of the treatment group means. These means will differ simply because each is based on a different set of n individuals: error variance is contributing to the variance of group means. There is another *possible* source of the variance of group means. If the treatments really differ in their effects, the μ_j are not all equal; one would expect this to be reflected in the spread among the group means. Thus, there are two possible sources of the variability among the group means and therefore of the magnitude of the MS_A individual differences and treatment effects.

Next consider the denominator of the F ratio, $MS_{S/A}$. The variance among individuals in the jth treatment group is $\sum_{i=1}^{n}(Y_{ij} - \overline{Y}_{.j})^2/(n - 1)$. Summing over groups and dividing by a yields the variance of individuals averaged over groups. Treatment effects do not contribute to this mean square, for if a constant (e.g., $\hat{\alpha}_j$) is added to all the scores in a group, the variance of the scores is unchanged. The $MS_{S/A}$ is a function only of σ_e^2, the error variance.

In view of the preceding line of reasoning, we would expect that the average values of MS_A and $MS_{S/A}$ over many replications of the experiment would equal each other, *if the μ_j were equal*. In other words, if the null hypothesis is true, both numerator and denominator of the F ratio reflect only error variance and $E(MS_A) = E(MS_{S/A}) = \sigma_e^2$. In any single experiment, we would not expect the numerator and denominator mean squares to be identical since, by chance, there might be greater (or smaller) differences among individuals in different groups than among individuals in the same group. However, if MS_A were considerably larger than $MS_{S/A}$, there would be grounds for suspecting that treatment effects were being added to error variance in the numerator of the F ratio.

If these intuitive arguments can be proved, we will have met one important criterion for a test statistic: it should reflect the relative merits of the null and alternative (that the μ_j are not equal) hypotheses. Such a proof follows. Specifically, it will be proved that if H_0 is true, and if the assumptions that the ε_{ij} are independently distributed with variance σ_e^2 are true, on the average over many replications of the experiment the two mean squares are equal; if H_0 is false, the average value of MS_A is greater than that of $MS_{S/A}$.

Our approach is to derive separately the expectations of $n \sum_j (\overline{Y}_{.j} - \overline{Y}_{..})^2$ and $\sum_i \sum_j (Y_{ij} - \overline{Y}_{.j})^2$. These quantities are the numerators of the mean squares and are respectively referred to as the *sum of squares for A* (SS_A) and the *sum of squares within groups* ($SS_{S/A}$). First the SS_A is redefined in terms of the analysis of variance model. From Equation (4.1),

$$\overline{Y}_{.j} = \frac{\Sigma_i(\mu + \alpha_j + \varepsilon_{ij})}{n}$$

(4.8)
$$= \frac{n\mu + n\alpha_j + \Sigma_i \varepsilon_{ij}}{n}$$

$$= \mu + \alpha_j + \frac{\Sigma_i \varepsilon_{ij}}{n}$$

and

$$\overline{Y}_{..} = \frac{\Sigma_i \Sigma_j(\mu + \alpha_j + \varepsilon_{ij}}{an}$$

(4.9)
$$= \frac{an\mu + n\Sigma_j \alpha_j + \Sigma_i \Sigma_j \varepsilon_{ij}}{an}$$

$$= \mu + \frac{\Sigma_i \Sigma_j \varepsilon_{ij}}{an}$$

Since $\Sigma_j \alpha_j = 0$ and, since $SS_A = n\Sigma_j(\overline{Y}_{.j} - \overline{Y}_{..})^2$, substituting on the basis of Equations (4.8) and (4.9) and taking the expected value of SS_A over many replications of the experiment, we have

(4.10)
$$E(SS_A) = E\left[n\sum_j\left(\alpha_j + \frac{\Sigma_i \varepsilon_{ij}}{n} - \frac{\Sigma_i \Sigma_j \varepsilon_{ij}}{an}\right)^2\right]$$

Expanding Equation (4.10), we obtain

(4.11) $\quad E(SS_A) = nE\left[\sum_j \alpha_j^2 + \frac{\Sigma_j(\Sigma_i \varepsilon_{ij})^2}{n^2} + \frac{\Sigma_j(\Sigma_i \Sigma_j \varepsilon_{ij})^2}{a^2 n^2} + \frac{2\Sigma_i \Sigma_j \alpha_j \varepsilon_{ij}}{n}\right.$

$$\left. - \frac{2\Sigma_j \alpha_j \Sigma_i \Sigma_j \varepsilon_{ij}}{an} - 2\left(\frac{\Sigma_i \Sigma_j \varepsilon_{ij}}{an}\right)\left(\frac{\Sigma_i \Sigma_j \varepsilon_{ij}}{n}\right)\right]$$

Since the expectation of a sum equals the sum of the expectations, the expectation of each term on the right side of Equation (4.11) may be evaluated separately.

(4.11a)
$$E(n\sum \alpha_j^2) = n\sum_j \alpha_j^2$$

since α_j is assumed to be constant over replications of the experiment, and the expectation of a constant is the constant.

(4.11b)
$$E\left[\frac{n\Sigma_j(\Sigma_i \varepsilon_{ij})^2}{n^2}\right] = \sum_j E\left(\frac{\Sigma_i \varepsilon_{ij}^2}{n}\right)$$

$$= a\sigma_e^2$$

where σ_e^2 is the variance of ε_{ij} in the jth treatment population. We might do well to consider Equation (4.11b) in more detail. Note that

$$E \left(\sum_i \varepsilon_{ij} \right)^2 = E(\varepsilon_{ij}^2 + \cdots + \varepsilon_{nj}^2 + 2\varepsilon_{1j}\varepsilon_{2j} + \cdots + 2\varepsilon_{ij}\varepsilon_{i'j} + \cdots$$
$$+ 2\varepsilon_{n-1,j}\varepsilon_{nj}) \qquad (i \neq i')$$

As we noted earlier, the εs are deviations of scores about the treatment population means. The term, $E(\varepsilon_{ij}\varepsilon_{i-j})$, implies that, in each of the many replications of the experiment, we have obtained the product of the ith and i'th error scores sampled from the jth treatment population, and that this product has then been averaged over the replications of the experiment. The average product of such deviations about means is the numerator of a correlation coefficient and must be zero since we have assumed that the εs are independently distributed. Hence,

$$E \left[\frac{(\sum_i \varepsilon_{ij})^2}{n} \right] = E \left(\frac{\sum_i \varepsilon_{ij}^2}{n} \right) = \frac{\sum_i E(\varepsilon_{ij}^2)}{n}$$

Since the average squared deviation of scores about its mean is the variance of the scores,

$$E(\varepsilon_{ij}^2) = \sigma_{\varepsilon_{ij}}^2$$

The variance of the ith error score sampled from the jth treatment population should, in the long run, be equal to the variance of the jth treatment population, regardless of the value of i. Furthermore, the variances of the a treatment populations have been assumed to be identical. Therefore, for any value of i and j,

$$\sigma_{\varepsilon_{ij}}^2 = \sigma_e^2$$

and

$$\frac{\sum_j \sum_i E(\varepsilon_{ij}^2)}{n} = \frac{\sum_j \sum_i \sigma_{\varepsilon_{ij}}^2}{n} = a\sigma_e^2$$

Moving to the next component of Equation (4.11),

(4.11c)
$$E \left[\frac{n \sum_j (\sum_i \sum_j \varepsilon_{ij})^2}{a^2 n^2} \right] = \sigma_e^2$$

The proof follows that for Equation (4.11b).

The next two terms vanish. Since α_j is a constant independent of ε_{ij},

(4.11d)
$$E \left(\frac{\sum_i \sum_j \alpha_j \varepsilon_{ij}}{n} \right) = 0$$

Since $\sum_j \alpha_j = 0$,

(4.11e)
$$E \left(\frac{\sum_j \alpha_j \sum_i \sum_j \varepsilon_{ij}}{an} \right) = 0$$

Finally

(4.11f)
$$E\left[-2n\left(\frac{\sum_i \sum_j \varepsilon_{ij}}{an}\right)\left(\frac{\sum_i \sum_j \varepsilon_{ij}}{n}\right)\right] = -2\sigma_e^2$$

The proof is similar to those for Equations (4.11b) and (4.11c).
Combining terms we obtain

(4.12)
$$E(SS_A) = (a - 1)\sigma_e^2 + n\sum_j \alpha_j^2$$

and for the expected mean square of A, $E(MS_A)$, we have

(4.13)
$$E(MS_A) = E\left(\frac{SS_A}{a - 1}\right) = \sigma_e^2 + n\theta_A^2$$

θ_A^2 rather than σ_A^2 is used to represent the variability among the μ_j for two reasons: (a) in a strict sense $\sum(\mu_j - \mu)^2/(a - 1)$ is not a population variance because of the divisor, and (b) the use of θ^2 is a reminder that the effects of A are fixed rather than random.

The derivation of $E(MS_{S/A})$ is similar to that for $E(MS_A)$. Applying Equation (4.2),

(4.14)
$$E\left[\sum_i \sum_j (Y_{ij} - \bar{Y}_{.j})^2\right] = E\left[\sum_i \sum_j \left(\varepsilon_{ij} - \frac{\sum_i \varepsilon_{ij}}{n}\right)^2\right]$$

Expansion yields

(4.15)
$$E\left[\sum_i \sum_j (Y_{ij} - \bar{Y}_{.j})^2\right] = E\left[\sum_i \sum_j \varepsilon_{ij}^2 - \frac{\sum_j (\sum_i \varepsilon_{ij})^2}{n}\right]$$
$$= a(n - 1)\sigma_e^2$$

Dividing both sides of Equation (4.15) by $a(n - 1)$ yields

(4.16)
$$E(MS_{S/A}) = \sigma_e^2$$

We now can state that

(4.17)
$$\frac{E(MS_A)}{E(MS_{S/A})} = \frac{\sigma_e^2 + n\theta_A^2}{\sigma_e^2}$$

If H_0 is true, $\theta_A^2 = 0$ and the above ratio equals 1. If H_0 is false, $\theta_A^2 > 0$ and $E(MS_A)/E(MS_{S/A}) > 1$. Thus the ratio of mean squares satisfies one requirement of a test statistic; its magnitude can be expected to reflect the validity of H_0. The second requirement of the test statistic is that its distribution under H_0 be known. The distribution of F must be known in order to determine the critical region, those values whose probability of occurrence is less than α, given that H_0 is true. If the F obtained from an experiment is so large that it falls within the critical region, we reject H_0 in favor of the alternative hypothesis that $\theta_A^2 > 0$.

As we pointed out in Section 3.4.2, the distribution of F is known to be the distribution of the ratio of two independently distributed chi-squares (χ^2) divided by their df; i.e.,

(4.18)
$$F = \frac{\chi_1^2/df_1}{\chi_2^2/df_2}$$

If the conditions of normality, independence, and homogeneity of variance hold, and if H_0 is true, the ratio of mean squares meets the condition described by Equation (4.18). (Note that the normality assumption was not invoked in deriving the $E(MS)$; however, it is necessary if the sums of squares are to have the χ^2 distribution.) Consequently, the probability of obtaining various F values can be computed. Values of F required for significance are presented for several combinations of dfs and α levels in Table A–5 in the Appendix. As an example of the use of Table A–5, assume an experiment with five levels of the independent variable A and six subjects at each level. Then the df of the numerator (df_1) are 4 and the df of the denominator (df_2) are 25. If α equals .05, then values of F greater than 2.76 will result in the rejection of H_0. If the selected α equaled .01, then Fs greater than 4.18 would be required for significance.

4.2.3 Assumptions underlying the F test. The third requirement for a statistic to be an appropriate test of H_0 is that the data conform to the assumptions underlying the test. The ratio of mean squares will be distributed as F if:

    (a) the ε_{ij} are independently distributed,
    (b) the ε_{ij} are normally distributed,
    (c) the variance of the ε_{ij} is the same for all treatment populations, and
    (d) the null hypothesis is true.

If the first three assumptions are valid, then significant Fs may be attributed to the falsity of assumption (d). However, if any of the first three assumptions are false, the distribution of mean squares may be such that the true probability of obtaining a result in the critical region is actually not α; whether the true probability of a Type I error is greater or less than α will depend on the form of the violation, although an excess of Type I errors is the more usual occurrence. Figure 4–1 exemplifies one potential problem. The solid line represents the F distribution, the distribution of the ratio of mean squares, when all four assumptions are valid. Under some departures from assumptions (a) through (c), the distribution of the ratio of mean squares might resemble that described by the dotted line. In this case, the probability of obtaining an F in the critical region is less than α. An alternative result of violations of any of the first three assumptions is represented by the dashed line in Figure 4–1. The distribution of the ratio of mean squares is now such that the probability of obtaining an F in the critical region is greater than α, and the risk of a Type I error is therefore greater than that which the experimenter wishes to assume.

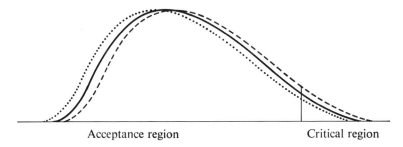

Acceptance region   Critical region

FIGURE 4–1 *An example of the distribution of the ratio of mean squares when assumptions are valid (solid line), when violations result in loss of power (dotted line), and when violations result in increased risk of Type I errors (dashed line)*

The validity of the independence assumption, (a), depends upon the experimenter's manipulations, upon the care he takes to ensure the random assignment of treatments to subjects. For example, suppose that the scores of subjects tested on the same day are positively correlated while those of subjects tested on different days are negatively correlated. This might happen because of slight day-to-day changes in the wording of instructions. If treatment A_1 is given on day 1, treatment A_2 on day 2, etc., differences among treatment means may merely reflect day-to-day differences which, in turn, are a function of the correlation among the scores referred to above. Under such circumstances, inferences based on the statistical test are invalid. Randomization will not eliminate the correlations among measures, but does provide for an equal opportunity for any two treatments to appear in adjacent and in nonadjacent points in time. Randomization provides a mechanism by which the expected correlations (as n increases) tend to cancel, leaving treatment effects independent of day-to-day effects.

The validity of the normality assumption, (b), will depend upon the measure chosen. For example, projective test scores tend to exhibit a skewed distribution, and percentage scores will be binomially distributed. The χ^2 test can be used to evaluate the significance of departures of the obtained distribution from the assumed normal distribution; the test would be applied to each of the a sets of n scores in turn. Generally, such a procedure will be unnecessary, since the distribution of the ratio of mean squares seems little affected by departures from normality. Both mathematical proofs (Scheffé, 1959) and empirical studies (Boneau, 1960; Bradley, 1964; Donaldson, 1968; Lindquist, 1953, pp. 78–90),* attest to this conclusion. In the empirical studies, large treatment populations of scores were created, samples were drawn repeatedly from each population with F ratios computed for each set

---

*An additional, unpublished, computerized sampling experiment has been carried out by J. W. Clinch and the author, producing results consistent with the conclusions presented in this section.

of a samples, and the empirical distribution of the ratio of mean squares was then compared to the theoretical F distribution (Table A–5). When the treatment populations are markedly skewed, but homogeneous in shape and variance, the probabilities of large values of F are slightly deflated; when n is four, the Fs required for significance at the .01 and .05 levels are exceeded in about .003–.009 and .03–.04 of the samples, respectively. Symmetric nonnormal distributions seem to result in slight inflation of the Type I error probabilities; the uniform distribution, in which the density is constant for all values of y, typically yields operative α values of .01–.018 and .05–.065 when the theoretical values are .01 and .05. The worst result I have noted in any of the studies of nonnormality involved three leptokurtic distributions with samples of size three; the operative α levels were .023 and .078. Clearly, even with samples considerably smaller than experimenters typically employ, nonnormality has a limited impact upon Type I error rates. Furthermore, the situation rapidly improves as n increases. This optimistic picture of the role of the normality assumption should be qualified by noting that substantial errors can occur in the estimation of intervals for variance components (e.g., confidence intervals for σ_e^2) if the treatment populations are not normally distributed.

The validity of the homogeneity of variance assumption, (c), will depend upon how careful the experimenter is to administer the experimental treatments in a consistent manner to all subjects; upon the type of measure taken; and upon the treatment levels employed. Certainly, the uniform application of treatments to subjects will tend to stabilize variances from group to group. However, despite proper experimental methodology, heterogeneity of variance can occur. For example, some measures (response frequency is a common instance) are Poisson-distributed, in which case the mean and variance are equal, and either both are homogeneous or both are heterogeneous over treatments. Several methods exist for detecting departures from homogeneity of variance (Bartlett, 1937; Cochran, 1941; Hartley, 1950). They are, however, usually overly sensitive to departures from normality and therefore will not be considered further.

Both mathematical derivations by Box (1954) and the empirical studies cited earlier indicate that the α level is inflated by heterogeneity of variance. If all treatment populations are approximately normally distributed, and if all groups are of the same size, the inflation is slight; with two groups, n equal to five, and a 20:1 variance ratio, the Fs required for significance at the .01 and .05 levels were exceeded in .02 and .07 of the computerized experiments we ran. Similar results were obtained with more groups. Over several investigations, variance ratios of about 4:1 yielded still better results, the operative α levels consistently falling below .02 and .07. Increases in sample size yield slight decreases in the degree of inflation but some distortion of Type I error rates will occur even with very large n. Unequal ns are more troublesome. Boneau's results (1960), obtained with normally distributed populations, are typical; with variances of one and four and ns of five and fifteen, the Fs required for significance at the .01 and .05 levels were exceeded in .001 and .01 of the cases; with the ns reversed so that the group with the smaller variance had the larger n, the operative α levels were .06 and .16. In general, where there is a negative correlation of variance and sample size, or when the ns are equal, the true probability of a Type I error is above that assumed; if the correlation is negative, heterogeneity at variance may deflate the Type I error rate.

Clearly, the best protection against distortion of the Type I error rate is the use of equal ns. However, if experimental groups are not of equal size, or if the error variances differ extremely over treatment levels, it is still possible to carry out the significance test at the desired α level, *provided that the treatment population distributions are approximately normal*. When a equals 2, a reasonable approximation is obtained (Welch, 1937) by computing the quantity

$$(4.19) \qquad t' = \frac{\overline{Y}_{.1} - \overline{Y}_{.2}}{\sqrt{S_1^2/n_1 + S_2^2/n_2}}$$

which is then evaluated for significance in Table A-3 with

$$(4.20) \qquad df = \frac{(n_1 - 1)(n_2 - 1)}{(n_2 - 1)c^2 + (n_1 - 1)(1 - c)^2}$$

where

$$(4.21) \qquad c = \frac{S_1^2/n_1}{S_1^2/n_1 + S_2^2/n_2}$$

and S_j^2 and n_j are the variance and sample size for the jth treatment group. As an example of the use of this approach, suppose that $S_1^2 = 1$, $S_2^2 = 4$, $n_1 = 5$, $n_2 = 15$, $\overline{Y}_{.1} - \overline{Y}_{.2} = 1.8$ and the required α level is .05. The usual t test (which is equivalent to the F test since $t^2 = F$ when two treatment levels are being compared) would fail to reject the null hypothesis; we would use the formula

$$(4.22) \qquad t = \frac{\overline{Y}_{.1} - \overline{Y}_{.2}}{\sqrt{\left[\dfrac{(n_1 - 1)S_1^2 + (n_2 - 1)S_2^2}{n_1 + n_2 - 2}\right]\left[\dfrac{1}{n_1} + \dfrac{1}{n_2}\right]}}$$

which gives

$$t = \frac{1.8}{.943}$$

which is less than 2.101, the value required for significance at the .05 level with 18, $n_1 + n_2 - 2$, df. However, we arrive at a different conclusion using Welch's solution for the unequal variance problem. The adjusted df, using Equations (4.20) and (4.21), is 14.5; thus the critical value of t' lies between 2.131 and 2.145. The value obtained using Equation (4.19) is

$$t' = \frac{1.8}{.683} = 2.64$$

which is clearly significant. Because of the positive correlation of variance and n, the usual significance test, which wrongly assumed homogeneity of variance, failed to detect a result that proved significant when heterogeneity of variance was taken into account. If the variances and sample sizes had been negatively correlated, the usual test might have resulted in significance when Welch's test did not.

When a is greater than 2, other adjustments for heterogeneity of variance are possible. One approximate technique proposed by Box (1954) uses the usual ratio of mean squares as the test statistic. However, the df require adjustment and, if the ns are unequal, the F required for significance must be multiplied by a factor, b. To test for significance at a given level of α, we turn to Table A–5, find the F required for significance on h' and $h\ df$, multiply that value by b if the ns are not equal, and then note whether the ratio of mean squares computed from our data exceeds this value. The appropriate formulas are

(4.23)
$$b = \frac{(N - a) \sum_j (N - n_j) S_j^2}{N(a - 1) \sum_j (n_j - 1) S_j^2}$$

(4.24)
$$h' = \frac{[\sum_j (N - n_j) S_j^2]^2}{(\sum_j n_j S_j^2)^2 + N \sum_j (N - 2n_j) S_j^2)^2}$$

(4.25)
$$h = \frac{[\sum_j (n_j - 1) S_j^2]^2}{\sum_j (n_j - 1)(S_j^2)^2}$$

where $N = \sum_j n_j$.
  Thus, if

$$S_1^2 = 1 \qquad\qquad S_2^2 = 2 \qquad\qquad S_3^2 = 3$$

and

$$n_1 = 7 \qquad\qquad n_2 = 5 \qquad\qquad n_3 = 3$$

then

$$b = 1.28 \qquad\qquad h' = 1.86 \qquad\qquad h = 10.00$$

Assuming $\alpha = .05$, we will reject H_0 if the ratio of mean squares is larger than 1.28 times the F value of F in Table A–5 that is associated with .05 as well as 1.86 and 10 df. Of course, there is some difficulty in finding a column labelled 1.86 df in Table A–5; however, it is clear that the required F lies between 4.10 and 4.96, presumably closer to 4.10. Linear interpolation yields 4.22, a fair approximation. The critical region therefore consists of those values of F equal to or greater than 1.28 × 4.22, or 5.40.

In summary, if the independence and normality assumptions hold, and if the ns are equal, Type I error rates will generally be only slightly inflated above their nominal levels. Unequal ns may result in much larger distortions of the error rates and should generally be avoided; however, if the ns are unequal, procedures exist for obtaining approximately the desired error rate.

The combination of two violations, heterogeneity of variance and non-normality, is not especially worse than heterogeneity of variance alone, provided that all treatment populations have the same distribution function. Increasing n is helpful and n should be constant over groups.

When treatment population distribution functions and variances are both heterogeneous, Type I error rates are again inflated. The degree of inflation will vary with the set of distribution functions as well as with the variances; consequently, it is difficult to provide a definitive description of the effects of these violations. Some feeling for the consequences may be obtained from the results of several computerized investigations. Boneau (1960), using samples from two of these three—normal, exponential, uniform —and variances of 1 and 4, found that the Fs required for significance at the .01 and .05 levels were exceeded in about 1–2 percent and 5–7 percent of his cases, respectively. Norton (Lindquist, 1953), with a 45:1 ratio of largest to smallest variance and four degrees of skewness varying from a normal to a J-shaped distribution, obtained .036 and .10 for the nominal .01 and .05 levels with $n = 3$. The error rates dropped to .029 and .081 with $n = 10$. Bradley (1964) employed a normal and skewed distribution and a 4:1 variance ratio. With $a = 2$ and ns equal, the results were similar to Boneau's. With a equal to 3 and 4, the operative α levels varied greatly as a function of the number of groups sampled from each population and the particular variance-distribution combinations employed; when normal populations with small variances were pitted against skewed populations with larger variances, error rates tended to be more inflated than when the normal populations had the larger variance; in the former case, Fs required for significance at the .01 and .05 levels were sometimes exceeded in .05 and .12 of the computerized "experiments" while, in the latter case, the operative α levels were about .02–.03 and .06–.07. Nor does increased sample size necessarily improve the results.

In summary, large distortions of the Type I error rate are possible. Our knowledge of the consequences of these violations of the analysis of variance model is limited to only a few combinations of variance and distribution function. In addition, we lack adjustment procedures comparable to the Welch and Box solutions for the normal distribution case. Consequently, it is important to attempt to assess before the experiment how the distribution of the dependent variable might be affected by the independent variable. Occasionally, a choice of dependent variables will be available and we can aim for homogeneity of population shape, if not variance. Unequal ns should certainly be avoided wherever possible. If the data do reflect heterogeneity of both shape and variance, the experimenter must recognize that the nominal α level is at best an approximation to the actual probability of a Type I error, probably an underestimate ranging from one percent to seven percent, depending on the nominal α level and the particular combination of violations of assumptions; if the ns are not equal, the distortion may be far greater.

Thus far we have concentrated on the consequences of violations of assumptions upon the Type I error rate. Considerably less is known about the effect of such violations about Type II error rate or its complement, power. Boneau (1962) has concluded that when $a = 2$, the power to detect a particular difference between means is approximately that associated with the operative, as opposed to the nominal, α level. Thus, if violations of assumptions cause the F required for significance at the 5 percent level to be obtained in the 10 percent of experiments in which H_0 is true, the probability of rejecting false null hypotheses will be increased to the level typical when $\alpha = 10$ percent. More recent work suggests that this is an oversimplification as we would expect if we consider that, in violating assumptions, we are not shifting the critical value of the test statistic but rather actually changing the distribution of the ratio of mean squares. Occasionally, the violation of an assumption will reduce power to detect some effects and increase power to detect others.

Some preliminary information is available from computerized sampling experiments carried out by Donaldson (1968) and by Clinch and myself. When the parent populations were skewed, but homogeneous with respect to shape and variance, the power of the F test was actually slightly higher than when the assumptions were met; the discrepancy was most marked for small values of θ_A^2/σ_e^2, the measure of treatment effects. With increases in a or n, the obtained power function approached that derived under the normality assumption.

If the parent distributions are all normal but variances differ, power is not too greatly affected as long as the ns are equal (Scheffé, 1959; Boneau, 1962). My impression is that heterogeneity of variance results in a slight elevation of the power function to a degree related to the magnitude of differences among variances, but rarely by more than 5 percent for any value of $\theta_A^2/\bar{\sigma}_j^2$.

Donaldson considered exponential distributions, which are skewed and have $\sigma_j^2 = \mu_j^2$. In this case, power to detect small effects was slightly reduced but power to detect larger effects was increased. Again, group sizes were equal.

In summary, those violations that have been examined have had little impact on power. However, the scope of the investigations has been limited. In particular, we have little information about the combined impact of a heterogeneity of distribution function and variance. In addition, information about the consequences of heterogeneity of variance with distributions other than the normal and exponential would be welcome.

4.2.4 Transformations of the data. If there is evidence of some systematic relationship between treatment population mean and variance, homogeneity of variance may be obtained through an appropriate transformation of the data. Bartlett (1947) has presented a formula for deriving such transformations provided that the relationship between μ_j and σ_j^2 is known. In many instances where the nature of the relationship is unclear, the experimenter can, through trial and error, find a transformation that will stabilize the within-group variances. Three of the more useful transformations are noted below.

(a) *The square-root transformation.* This transformation is applicable when $\sigma_j^2 = k\mu_j$; i.e., when the means and variances are proportional for each treatment. This situation is not unusual when the data are in the form of frequency counts; for example, when the dependent variable is number correct or number of "yes" responses. In such cases, the analysis of variance is carried out on Y' rather than on Y, where

$$(4.26) \qquad\qquad Y'_{ij} = \sqrt{Y_{ij}}$$

and Y_{ij} is the score originally obtained from the *i*th subject in the *j*th group. If some of the values of Y_{ij} are less than 10, homogeneity of variance is more likely to be produced by the transformation,

$$(4.27) \qquad\qquad Y'_{ij} = \sqrt{Y_{ij} + .5}$$

(b) *The arc sine transformation.* This transformation is applicable when $\sigma_j^2 = \mu_j(1 - \mu_j)$; i.e., when the scores are proportions; for example, percentage correct or percentage predictions of some event. In such cases, the appropriate transformation is

$$(4.28) \qquad\qquad Y'_{ij} = \text{arc sin } \sqrt{Y_{ij}}$$

The transformed score is the angle whose sine is equal to the square root of the original score. If the original score is .50, its square root is approximately .707. Turning to a table of natural trigonometric functions (available in most books of mathematical and/or statistical tables), we find that the sine of $45°$ is .707; Y'_{ij} is 45. The transformation may be made directly, without computation of the square root, if tables of arc sin $\sqrt{Y}$ are available. Such a table may be found in Snedcor and Cochran (1967) or Fisher and Yates (1957).

(c) *The logarithmic transformation.* This transformation is applicable when $\sigma_j^2 = k\mu_j^2$; i.e., when the treatment standard deviation is proportional (k is a constant of proportionality) to the treatment mean. This situation will sometimes arise when the distribution of scores is markedly skewed; thus, reaction time scores may be amenable to this transformation. This transformation is also applicable when the scores are standard deviations. Equation (4.29) characterizes the transformation.

$$(4.29) \qquad\qquad Y'_{ij} = \log Y_{ij}$$

If some of the measures are small, the recommended transformation is

(4.30) $$Y'_{ij} = \log(Y_{ij} + 1)$$

The interpretation of effects upon the transformed data may prove difficult. Most of us would find relatively unintelligible a point estimate of $\mu'_1 - \mu'_2$ where μ'_j is based on the arc sin $\sqrt{Y_{ij}}$. One solution is to retransform $\overline{Y}'_{.j}$, the mean of the transformed data, to the original scale. Thus, assuming an arc sin transformation, if $\overline{Y}'_{.j}$ is 45 we take as our estimate of μ_j, .50, the value whose transform is 45. Note that this value, .50, will generally not be equal to $\overline{Y}_{.j}$, the mean on the original scale. Bartlett (1947) succinctly summarizes the justification for this approach. "If the variability in the data varies with the mean level for different blocks or groups, an unweighted average of the *observed* treatment responses is not necessarily the best estimate of the *true* treatment response, and the average on the transformed scale will often be the better estimate when reconverted to the original scale." It should be clear that the reconverted mean will be a better estimate than the original unconverted mean only if the transformation provided a stability of variance lacking in the original data.

In addition to reducing heterogeneity of variance, transformations sometimes result in a closer approximation to the normal distribution. Furthermore, they are most useful in more complex designs, where it is desirable to eliminate certain effects of combinations of variables that are unaccounted for by the usual analysis of variance models. We reserve discussion of this problem for Chapter 7, when the relevant designs will first be met. For further information on transformations, including additional transformations and estimates of what the variance will be on the new scale, see Bartlett's paper (1947).

The use of nonparametric statistics has been recommended as an alternative approach to data analysis when assumptions have been violated (Siegel, 1956). However, these techniques are also sensitive to differences in parameters other than the mean and are not necessarily more powerful than the F test when the assumptions of that test are violated (Boneau, 1962). We feel that the use of nonparametric statistics should usually be limited to instances in which the data are originally in the form of ranks or frequency counts, as in contingency table problems, or when a quick approximate indication of significance is required. An extensive discussion of the pros and cons of nonparametric statistics and of their advantages and disadvantages relative to those of analysis of variance is beyond the scope of this book; the topic is covered by Siegel (1956), Gaito (1959), and Anderson (1961), among others. The major reason for the author's restraint in utilizing nonparametric statistics is simply that they are not versatile enough, that the person who uses the nonparametric approach is limited in the designs he can use and in the questions he can ask of his data.

4.2.5 *F* ratios less than 1. If the null hypothesis is correct, the ratio of expected mean squares is 1 and, on the basis of sampling variability, one may expect occasional *F* ratios of less than 1. Such occurrences are regarded merely as support for the null hypothesis. However, the occurrence of *F*s so small that their reciprocals are significant or the occurrence of many *F*s less than 1 in a single analysis of variance merits further consideration. Such findings suggest that the model underlying the analysis of variance has in some way been violated. A frequent occurrence is the presence of some systematic effect that is not described by the analysis of variance model and consequently is not accounted for in the analysis of the data. For example, if all experimental groups consist of some subjects run by one experimenter and some subjects run by another experimenter, within-groups variability may be increased without a concomitant increase in between-groups variability. If the systematic factor can be designated, its contribution to the error variance can often be removed with the loss of a few *df*s in the error term.

4.3 THE ANALYSIS OF VARIANCE

Having considered the relevant theory, we now turn to the actual data analysis process. It will be shown that SS_A and $SS_{S/A}$ account for the total variability in the data; then raw score formulas for these quantities will be developed. Finally, these formulas will be applied in the analysis of a sample set of data.

4.3.1 Components of variability. Consider the following identity, which states that the deviation of a score from the grand mean consists of two components: (a) the deviation of a score from the mean of its experimental group, and (b) the deviation of the group mean from the grand mean.

(4.31) $$Y_{ij} - \overline{Y}_{..} = (Y_{ij} - \overline{Y}_{.j}) + (\overline{Y}_{.j} - \overline{Y}_{..})$$

Squaring both sides of Equation (4.31) results in

(4.32) $$(Y_{ij} - \overline{Y}_{..})^2 = (Y_{ij} - \overline{Y}_{.j})^2 + (\overline{Y}_{.j} - \overline{Y}_{..})^2$$
$$+ 2(Y_{ij} - \overline{Y}_{.j})(\overline{Y}_{.j} - \overline{Y}_{..})$$

If we sum over *i* and *j* for both sides of Equation (4.32), remembering the notational rules presented earlier, the result is

(4.33) $$\sum_i^n \sum_j^a (Y_{ij} - \overline{Y}_{..})^2 = \sum_i^n \sum_j^a (Y_{ij} - \overline{Y}_{.j})^2 + n \sum_j^a (\overline{Y}_{.j} - \overline{Y}_{..})^2$$
$$+ 2 \sum_i^n \sum_j^a (Y_{ij} - \overline{Y}_{.j})(\overline{Y}_{.j} - \overline{Y}_{..})$$

The cross-products term, $\sum_i^n \sum_j^a (Y_{ij} - \overline{Y}_{.j})(\overline{Y}_{.j} - \overline{Y}_{..})$, equals zero. This can be proved by rearranging terms.

$$(4.34) \quad \sum_i^n \sum_j^a (Y_{ij} - \bar{Y}_{.j})(\bar{Y}_{.j} - \bar{Y}_{..}) = \sum_j [(\bar{Y}_{.j} - \bar{Y}_{..}) \sum_i (Y_{ij} - \bar{Y}_{.j})]$$

$$= \sum_j (\bar{Y}_{.j} - \bar{Y}_{..})(0)$$

$$= 0$$

Consider Equation (4.33) again, ignoring the cross-products term, which has been proven equal to zero. The term $\sum_i \sum_j (Y_{ij} - \bar{Y}_{..})^2$ is the numerator of the variance of all scores about the grand mean and will henceforth be referred to as the total sum of squares (SS_{tot}). The term $n \sum_j (\bar{Y}_{.j} - \bar{Y}_{..})^2$ is n times the numerator of the variance of the group means about the grand mean. This is usually (and ungrammatically) referred to as the between-groups sum of squares (SS_A). The term $\sum_i \sum_j (Y_{ij} - \bar{Y}_{.j})^2$ is the within-groups sum of squares ($SS_{S/A}$, the subscript representing "subjects within levels of A").

The formulas presented above define the quantities SS_{tot}, SS_A, and $SS_{S/A}$. In practice, they are clumsy to work with and it is helpful to transform these formulas into raw score formulas, simplifying the computations. This was previously done for the $SS_{S/A}$ in Chapter 3. Using the same general procedure, we can obtain raw score formulas for SS_{tot} and SS_A. From the development in Chapter 3,

$$(4.35) \quad SS_{S/A} = \sum_i^n \sum_j^a Y_{ij}^2 - \sum_j^a \frac{(\sum_i^n Y_{ij})^2}{n}$$

In the same manner,

$$(4.36) \quad SS_{tot} = \sum_i^n \sum_j^a (Y_{ij} - \bar{Y}_{..})^2$$

$$= \sum_i^n \sum_j^a (Y_{ij}^2 + \bar{Y}_{..}^2 - 2Y_{ij}\bar{Y}_{..})$$

$$= \sum_j^a \left(\sum_i^n Y_{ij}^2 + n\bar{Y}_{..}^2 - 2\bar{Y}_{..} \sum_i^n Y_{ij} \right)$$

$$= \sum_i^n \sum_j^a Y_{ij}^2 + an\bar{Y}_{..}^2 - 2\bar{Y}_{..} \sum_i^n \sum_j^a Y_{ij}$$

$$= \sum_i^n \sum_j^a Y_{ij}^2 + an \frac{(\sum_i^n \sum_j^a Y_{ij})^2}{a^2 n^2} - 2 \frac{(\sum_i^n \sum_j^a Y_{ij})(\sum_i^n \sum_j^a Y_{ij})}{an}$$

$$= \sum_i^n \sum_j^a Y_{ij}^2 - \frac{(\sum_i^n \sum_j^a Y_{ij})^2}{an}$$

and

$$(4.37) \quad SS_A = n \sum_j^a (\bar{Y}_{.j} - \bar{Y}_{..})^2$$

$$= n \sum_j^a (\bar{Y}_{.j}^2 + \bar{Y}_{..}^2 - 2\bar{Y}_{..}\bar{Y}_{.j})$$

$$= n \sum_j^a \bar{Y}_{.j}^2 + an\bar{Y}_{..}^2 - 2n\bar{Y}_{..} \sum_j^a \bar{Y}_{.j}$$

$$= n \sum_j^a \frac{(\sum_i^n Y_{ij})^2}{n^2} + an \frac{(\sum_i^n \sum_j^a Y_{ij})^2}{a^2 n^2} - 2n \frac{(\sum_i^n \sum_j^a Y_{ij})}{an} \sum_j^a \frac{(\sum_i^n Y_{ij})}{n}$$

$$= \sum_j^a \frac{(\sum_i^n Y_{ij})^2}{n} - \frac{(\sum_i^n \sum_j^a Y_{ij})^2}{an}$$

4.3.2 Summarizing the analysis of variance. Table 4–1 summarizes much of the material presented thus far. The first column on the left contains the *sources of variance* (*SV*). These follow from Equation (4.1), which states that the deviation of a score from the population mean consists of a treatment component (*A*) and a component due to individual differences (*S/A*).

The second column from the left in Table 4–1 contains the *df* associated with each *SV*. In deriving the expected mean squares (*EMS*), we noted that the *df* associated with the *A* effect are $a - 1$, those associated with *S/A* are $a(n - 1)$. Let us consider why this is so.

Suppose that we are asked to choose four numbers that sum to 100. The first three numbers chosen can be any three finite numbers. However, the fourth number must be 100 minus the sum of the first three numbers; only three numbers are chosen freely. For example, if the first three numbers chosen are 41, 3 and -18, the fourth number, k_4, must be $100 - 41 - 3 - (-18) = 74$. We characterize this situation by saying that there is one restriction on the data, causing us to lose one *df*. If it is required that the first two numbers chosen sum to 50 and all four sum to 100, there are two restrictions on the data and two *df* are thus lost. Only two numbers may be freely chosen. Analogous geometrical examples exist. Suppose that one is told to draw a triangle, the only restriction being that the figure must be a closed three-sided one. This restriction causes the loss of one *df*; while the first two sides may be of any length and form any angle, for example,

the third side must be a line connecting the points A and C. From these examples, we draw the generalization:

df = number of independent observations

(4.38)   = total number of observations minus the number
of restrictions on the observations.

Next, this line of reasoning is extended to statistical tests by noting that the estimation of a population parameter places a restriction on the data. The computation of the variance of a group of scores involves the summing of squared deviations about the group mean, imposing the restriction that $\sum_i Y_i = n\overline{Y}$, or $\sum_i (Y_i - \overline{Y}) = 0$. Consequently, the variance of a single group of scores is based on $(n - 1)$ df. In the computation of $MS_{S/A}$, a such quantities are calculated, resulting in $a(n - 1)$ df. In computing the MS_A, we encounter the restriction that $\sum_j \overline{Y}_{.j} = a\overline{Y}_{..}$. We therefore have $(a - 1)$ df.

The role of df may be clearer if the above development for the F test is contrasted with that for the normal deviate, or z test, where $z = (\overline{Y} - \mu)/(\sigma/\sqrt{n})$. Note that the population standard deviation is *known rather than estimated* and consequently there is no restriction on the data. The concept of df is not relevant to this test. An alternative way of viewing the relevance or irrelevance of df to a statistical test is to look at the distribution of the statistic. The shape of the normal curve is not a function of the number of observations, while that of the F distribution is affected by the numerator and denominator df.

The importance of df lies not only in the fact that they are necessary components of the F ratio and determiners of the F distribution. These quantities provide a check on the SV in more complex designs where a source may be overlooked or the variance wrongly analyzed in some other manner. The check assumes that we have the correct df for each listed source, in which case they must sum to the total number of scores minus one. Furthermore, the df provide an alternative basis for arriving at raw score formulas for the SS. This point will be considered next.

The formulas for the SS were previously derived in Section 4.3.1. For more complex designs such a procedure is tedious, and it is therefore desirable to establish some simple rule that makes it possible to set down SS formulas quickly. The rule should generalize to all the terms of the analysis, and to analyses for any design. Such a rule follows if one notes the isomorphism existing between df and SS. There is a df for each squared quantity involved in the computations of SS, and any operation applied to the df is applied to the squared quantities. Consider the $SS_{S/A}$ first:

(a) Expand the df expression.

$$df_{S/A} = a(n - 1) = an - a$$

Table 4-1 Analysis of variance for the completely randomized one-factor design

SV	df	SS	MS	EMS*	F
Total	$an - 1$	$\sum_i^n \sum_j^a Y_{ij}^2 - \dfrac{(\sum_i^n \sum_j^a Y_{ij})^2}{an}$			
A (between groups)	$a - 1$	$\sum_j^a \dfrac{(\sum_i^n Y_{ij})^2}{n} - \dfrac{(\sum_i^n \sum_j^a Y_{ij})^2}{an}$	$\dfrac{SS_A}{a-1}$	$\sigma_e^2 + n\theta_A^2$	$\dfrac{MS_A}{MS_{S/A}}$
S/A (within groups)	$a(n - 1)$	$\sum_i^n \sum_j^a Y_{ij}^2 - \sum_j^a \dfrac{(\sum_i^n Y_{ij})^2}{n}$	$\dfrac{SS_{S/A}}{a(n-1)}$	σ_e^2	

$$* \quad \theta_A^2 = \frac{\sum (\mu_j - \mu)^2}{a - 1}$$

(b) Write an expression in terms of squared quantities that corresponds to the above result.

$$\sum_i^n \sum_j^a (\ )^2 - \sum_j^a (\ )^2$$

(c) Any summation operations not indicated outside the parentheses should be indicated within the parentheses. All indices of summation should appear somewhere in the formula.

$$\sum_i^n \sum_j^a (Y_{ij})^2 - \sum_j^a \left(\sum_i^n Y_{ij} \right)^2$$

(d) Divide each quantity by the number of scores within the parentheses.

$$SS_{S/A} = \sum_i^n \sum_j^a (Y_{ij})^2 - \frac{\sum_j^a (\sum_i^n Y_{ij})^2}{n}$$

For the SS_A, the four steps are:

(a) $df_A = a - 1$

(b) $\sum_j^a (\ )^2 - (\ )^2$

(c) $\sum_j^a \left(\sum_i^n Y_{ij} \right)^2 - \left(\sum_j^a \sum_i^n Y_{ij} \right)^2$

(d) $SS_A = \frac{\sum_j^a (\sum_i^n Y_{ij})^2}{n} - \frac{(\sum_j^a \sum_i^n Y_{ij})^2}{an}$

The SS_{tot} is left as an exercise for the student.

The entries in the *MS* column of Table 4–1 are the ratios of *SS* to *df*. The MS_A is *n* times the variance of the group means about the grand mean, and the $MS_{S/A}$ is the average over groups of the variances of scores about group means. The *EMS* have been derived previously (Section 4.2.1) and have been shown to justify the use of the *F* ratio as a test statistic.

4.3.3 Numerical examples. Table 4–2 contains speeds of traversing a runway, in feet/second, for four groups of eight rats. All rats were allowed access to a solution of sucrose for 20 seconds at the end of each run. The groups differed with respect to the percentage of sucrose in the solution.

The first step in the analysis is to obtain $\sum_i Y_{ij}$ and $\sum_i Y_{ij}^2$ for each group; i.e., for each value of *j*. The two sums can be obtained simultaneously on most desk calculators. They provide the basic ingredients for the *F* test of the null hypothesis that μ_j are equal. They are entered in Table 4–2 together with group means and variances; the variances are computed according to Equation (4.39)

(4.39)　　　　　　　$$S_j^2 = \frac{\sum_i Y_{ij}^2 - (\sum_i Y_{ij})^2/n}{n - 1}$$

TABLE 4–2 *Data for four groups of rats in a runway study*

	Percent Sucrose in Water			
	8	16	32	64
	1.4	3.2	6.2	5.8
	2.0	6.8	3.1	6.6
	3.2	5.0	3.2	6.5
	1.4	2.5	4.0	5.9
	2.3	6.1	4.5	5.9
	4.0	4.8	6.4	3.0
	5.0	4.6	4.4	5.9
	4.7	4.2	4.1	5.6
$\sum_i Y_{ij} =$	24.0	37.2	35.9	45.2
$\bar{Y}_{.j} =$	3.00	4.65	4.50	5.65
$\sum_i Y_{ij}^2 =$	86.54	186.78	171.67	264.24
$S_{y_j}^2 =$	2.08	1.97	1.51	1.27

Now we may proceed to the actual analysis of variance. According to Equation (4.36), $SS_{tot} = \sum_i \sum_j Y^2 - (\sum_i \sum_j Y)^2/an$. Since terms like $(\sum_i \sum_j Y)^2/an$ appear in many components of all analyses of variance, a special notation is used. Such terms are called *correction terms* and are designated by C, where

(4.40) $$C = \frac{(\text{sum of all scores in the data matrix})^2}{\text{total number of scores}}$$

In our example,

$$C = \frac{(24.0 + 37.2 + 35.9 + 45.2)^2}{32} = 632.79$$

and

$$SS_{tot} = 86.54 + 186.78 + 171.67 + 264.24 - 632.79$$
$$= 709.23 - 632.79$$
$$= 76.44$$

According to Equation (4.37)

$$SS_A = \frac{\sum_j (\sum_i Y)^2}{n} - C$$
$$= \frac{(24.0)^2 + (37.2)^2 + (35.9)^2 + (45.2)^2}{8} - 632.79$$
$$= 661.46 - 632.79$$
$$= 28.67$$

The $SS_{S/A}$ may be obtained as the residual variability,

$$SS_{S/A} = SS_{tot} - SS_A$$

(4.41)
$$= 76.44 - 28.67$$

$$= 47.77$$

and, as a check, by pooling the sums of squares for each group,

$$SS_{S/A} = \sum_j \left[\sum_i Y^2 - \frac{(\sum_i Y)^2}{n} \right]$$

(4.42)
$$= \left(86.54 - \frac{(24.0)^2}{8} \right) + \cdots + \left(264.24 - \frac{(45.2)^2}{8} \right)$$

$$= 47.77$$

The first procedure corresponds to $(an - 1) - (a - 1)$ *df*, the second to $(n - 1) + \cdots + (n - 1)$. The result in both instances is $a(n - 1)$; correspondingly, the two computations on SS should yield identical results.

The MS and the F ratio are now easily computed.

$$MS_A = \frac{SS_A}{a - 1} \qquad\qquad MS_{S/A} = \frac{SS_{S/A}}{a(n - 1)}$$

$$= \frac{28.67}{3} \qquad\qquad\qquad = \frac{47.77}{28}$$

$$= 9.56 \qquad\qquad\qquad\quad = 1.71$$

$$F = \frac{MS_A}{MS_{S/A}}$$

$$= \frac{9.56}{1.71}$$

$$= 5.59$$

The results of our analysis are summarized in Table 4–3. Turning to Table A–5, we find that for 3 and 28 *df*, an F of 4.57 is required for significance at

TABLE 4–3 *Analysis of variance for data from a completely randomized one-factor experiment*

SV	df	SS	MS	F
Total	31	76.44		
A	3	28.67	9.56	5.59*
S/A	28	47.77	1.71	
				*p < .01

the .01 level. Since our computed F exceeds this critical value, we feel justified in concluding that the treatment population means differ. The statement "$p < .01$" indicates that if H_0 is true, the observed F of 5.59 will occur in less than 1 percent of the replications of the experiment.

In handling actual experimental data, the analysis would not be concluded at this point. Many interesting questions are still unanswered. Do all four means differ significantly from one another? Or are the 16 percent and 32 percent treatments essentially equivalent, as a quick look at the data suggests? If the treatment means are plotted as a function of concentration, what type of equation best describes the relationship? In order to deal with such questions, a number of additional conceptual and computational factors must be considered. Treatment of these questions is reserved for Chapters 13 and 14.

Throughout this chapter the presentation has been restricted to the case where n is equal for all groups, since both derivations of expectations and computations in the analysis of variance are simpler under this condition. This restriction will now be removed in order to present computations for the unequal n case, in which n_j, the number of subjects in the jth treatment group, varies over levels of j. Before considering the computations, we note that the analysis of variance model is fundamentally the same as previously; the major difference in the theoretical development is that now

$$(4.43) \qquad E(MS_A) = \sigma_e^2 + \frac{\sum_j n_j(\mu_j - \mu)^2}{a - 1}$$

When n_j is constant over j, we have the previously presented EMS. In both cases, the F ratio involves the MS_A and the $MS_{S/A}$.

Table 4–4 contains data for the runway experiment previously analyzed, with some scores randomly discarded from the data matrix of Table 4–2.

Following Equation (4.40), we compute the correction term,

$$C = \frac{(12.1 + 26.1 + 22.4 + 45.2)^2}{(5 + 6 + 5 + 8)}$$

$$= \frac{(105.8)^2}{24}$$

$$= 466.40$$

The SS_{tot} are now computed as previously:

$$SS_{tot} = 38.21 + 123.13 + 105.10 + 264.24 - C$$

$$= 64.28$$

The revised equation for the SS_A for the unequal n case is

$$(4.44) \qquad SS_A = \sum_j \frac{(\sum_i Y)^2}{n_j} - C$$

TABLE 4-4 *Data for the runway experiment with unequal n*

| | Percent Sucrose in Water | | |
	8	16	32	64
	1.4	3.2	6.2	5.8
	1.4	5.0	3.2	6.6
	5.0	2.5	4.5	6.5
	2.0	6.4	4.4	5.9
	2.3	4.8	4.1	5.9
		4.2		3.0
				5.9
				5.6
$\sum_i Y_{ij} =$	12.1	26.1	22.4	45.2
$n_j =$	5	6	5	8
$\bar{Y}_j =$	2.42	4.35	4.48	5.65
$\sum_i Y_{ij}^2 =$	38.21	123.13	105.10	264.24
$S_{y_j}^2 =$	2.23	1.92	1.19	1.27

Therefore, we have

$$SS_A = \frac{(13.1)^2}{5} + \frac{(26.1)^2}{6} + \frac{(22.4)^2}{5} + \frac{(45.2)^2}{8} - C$$

$$= 32.15$$

The $SS_{S/A}$ may again be computed as a residual.

$$SS_{S/A} = SS_{tot} - SS_A$$

$$= 32.13$$

The mean squares are then computed as before, and F is again the ratio of mean squares.

$$MS_A = \frac{32.15}{3} \qquad\qquad MS_{S/A} = \frac{32.13}{(4 + 5 + 4 + 7)}$$

$$= 10.72 \qquad\qquad\qquad = 1.61$$

$$F = 6.66$$

Entering Table A–5 with 3 and 20 *df*, we find that the critical value at the 1-percent level, 4.94, is again exceeded by the obtained F. The analysis is summarized in Table 4–5.

TABLE 4–5 *Analysis of variance for data from a completely randomized one-factor experiment with unequal n*

SV	df	SS	MS	F
Total	23	64.28		
A	3	32.15	10.72	6.66*
S/A	20	32.13	1.61	
				*p < .01

4.4 POWER OF THE F TEST

The power of the F test is usually plotted against an alternative measured by ϕ, where ϕ^2 may be conceptualized as a ratio of mean squares based on the treatment populations; i.e., it is n times the variance among the treatment population means divided by the population error variance. Thus,

$$(4.45) \qquad \phi^2 = \frac{n\sum_j(\mu_j - \mu)^2/a}{\sigma_e^2}$$

Turning to Table A–8 in the Appendix, we see that power is graphed as a function of ϕ for several combinations of df and α levels. Note that power increases with increasing values of ϕ, df_1, df_2 and α. These relationships have been previously discussed in Chapter 2. Note also that for any constant variance among the treatment population means, power will increase as error variance decreases, for ϕ will vary inversely with σ, as is apparent from Equation (4.45).

The power functions of Table A–8 provide a basis for deciding on the number of subjects to be included in the experiment. For appropriate use of the table, we require that the following be selected or estimated:

(a) *The α level.* This reflects our willingness to risk Type I errors.

(b) *The level of power.* This is the probability of rejecting a false null hypothesis. This value is not selected in the same way that α is, for power varies as a function of ϕ and the exact value of this latter quantity is unknown. However, one can select a probability of rejecting H_0, given that ϕ is equal to or greater than some critical value, ϕ'.

(c) *The error variance,* σ_e^2. Previous experimentation or a pilot study with the dependent variable of interest will provide an estimate of σ_e^2. If several estimates are available, the largest should be chosen. The larger the value of σ_e^2, the larger will be the value of n required to achieve a given level of power.

(d) *The critical variance among treatment population means.* This is the minimum value of $\sum(\mu_j - \mu)^2/a$ that it is important to detect. In applied research problems, there may be some magnitude of variance of treatment effects worth considering; below this level any differences in mean performance may be unimportant relative to differences in cost or efficiency in applying the treatments. In other research problems, we may seek to determine whether treatment effects estimated from a previous pilot study are significant. Or in the course of developing a theory, we may wish to, at first, model only those independent variables whose effects reach a certain level. In brief, the choice of a critical value will vary with the purpose of the experiment and our knowledge of the research area.

The following example should illustrate how the above factors combine to permit a selection of n. We assume

$$a = 4$$

$$\alpha = .01$$

$$\text{power} = .80$$

$$\sigma_e^2 = 200$$

$$\frac{\sum(\mu_j - \mu)^2}{4} = 125$$

The last figure was arrived at by assuming that it was important to reject H_0 if the successive μ_j were 10 or more units apart. Then,

$$\mu_1 = \mu - 15$$

$$\mu_2 = \mu - 5$$

$$\mu_3 = \mu + 5$$

$$\mu_4 = \mu + 15$$

and $\sum(\mu_j - \mu)^2/4 = (1/4)[(-15)^2 + (-5)^2 + (5)^2 + (15)^2] = 125$. We may now calculate $\phi' = \sqrt{n(125)/200} = .79\sqrt{n}$. We require a value of n such that the quantities $\phi' = .79\sqrt{n}$, $\alpha = .01$, $df_1 = 3$, and $df_2 = 4(n - 1)$ result in power $= .80$. Turning to the chart for $df_1 = 3$ and $\alpha = .01$, we consider various values of n. Suppose that we first try $n = 5$. Then df_2 are 16 and ϕ' is 1.77. Using the .01 scale, we project upwards from a point on the abscissa corresponding to 1.77 until we intersect with a point slightly above the $df_2 = 15$ curve. A horizontal line drawn from that point would intersect the ordinate somewhere between .4 and .5, clearly less than the required power of .8. We must try a larger value of n. A trial value of nine results in $df_2 = 32$ and $\phi' = 2.37$; the resulting power value on the ordinate is now slightly less than .9 but clearly above .8. We can run groups smaller than size nine. An n of 8 does quite nicely; the error df are now 28, $\phi' = 2.2$, and the resulting power is just about .80. Therefore we decide on eight subjects for each treatment group.

4.5 CONCLUDING REMARKS

Direct applications of the one-dimensional design are infrequent in psychological research. Many studies involve repeated measurements on subjects, necessitating the parceling out of subject effects in addition to treatment effects. Even studies in which there are no repeated measurements usually involve more than one treatment variable. Nevertheless, the material in the present chapter is extremely important. Within the context of a relatively simple design, we have considered notation, derivations, computations, null hypothesis testing, and the model underlying the use of the F test, all of which are involved in subsequent chapters. A thorough comprehension of the material of the present chapter will facilitate understanding of the more complicated designs and analyses of subsequent chapters.

EXERCISES

4.1 In deriving *EMS* we assume that $\sum_j^a \alpha_j = 0$. However, we do not assume that $\sum_i^n \sum_j^a \varepsilon_{ij} = 0$, but rather that

$$E\left(\sum_i^n \sum_j^a \varepsilon_{ij} = 0\right)$$

Why does this distinction between α_j and ε_{ij} exist? How would the situation change if the a levels of the variable A were randomly selected from a population of levels?

4.2 Prove Equations (4.14c) and (4.14e).

4.3 Do analyses of variance on the following sets of data:

(a)

A_1	A_2	A_3
28	38	64
23	39	73
21	57	61
38	36	48
38	38	72
49	48	52
28	52	54
33	40	60
34	39	54
29	45	60

(b)

A_1	A_2	A_3	A_4	A_5
24	48	42	96	73
07	91	82	67	81
46	63	75	88	33
45	69	76	24	44
97	26		92	94
	22		83	77
	45			60
				89
				25

4.4 In an attempt to determine the effect of time in therapy upon schizophrenic patients in a state hospital, the clinical staff agree on the following experiment. n subjects will be individually treated on a daily basis for one year, n subjects will be treated for six months, and n others will receive no special therapy. As a measure of the success of therapy, the clinicians will use a 100-point scale which differentially weights several aspects of personality and which has proven sensitive to experimental manipulations in other studies. They agree that each six months of the individual therapy should produce a mean gain of at least ten points for the technique to be worth the staff's time. They set $\alpha = .05$, $\beta = .10$, and on the basis of previous research estimate σ_e^2 to be 225. How many patients are required for the study?

REFERENCES

Anderson, N. H., "Scales and statistics: parametric and nonparametric," *Psychological Bulletin*, 58: 305–316 (1961).

Bartlett, M. S., "Properties of sufficiency and statistical tests," *Proceedings of the Royal Society of London*, A, 160: 238 (1937).

Bartlett, M. S., "The use of transformations," *Biometrics*, 3: 39–52 (1947).

Boneau, C. A., "The effects of violations of assumptions underlying the *t* test," *Psychological Bulletin*, 57: 49–64 (1960).

Boneau, C. A., "A comparison of the power of the *U* and *t* tests," *Psychological Review*, 69: 246–256 (1962).

Bradley, J. V. *Studies in Research Methodology VI. The Central Limit Effect for a Variety of Populations and the Robustness of Z, T, and F.* AMRL Technical Documentary Report 64–123, 6570th Aerospace Medical Research Laboratories, Wright-Patterson Air Force Base, Ohio, December, 1964.

Box, G. E. P., "Some theories on quadratic forms applied in the study of analysis of variance problems: I. Effect of inequality of variance in the one-way classification." *Annals of Mathematical Statistics*, 25: 290–302 (1954).

Cochran, W. G., "The distribution of the largest set of estimated variances as a fraction of their total," *Annals of Eugenics*, 11: 47–52 (1941).

Donaldson, T. S., "Robustness of the *F*-test to errors of both kinds and the correlation between the numerator and denominator of the *F*-ratio," *Journal of American Statistical Association*, 322: 660–676 (1963).

Fisher, R. A. and Yates, F. *Statistical Tables for Biological, Agricultural and Medical Research*, Edinburgh: Oliver and Boyd, 1955.

Gaito, J., "Nonparametric methods in psychological research," *Psychological Review*, 5: 115–125 (1959).

Hartley, H. O., "The maximum *F*-ratio as a short-cut test for heterogeneity of variance," *Biometrika*, 37: 308–312 (1950).

Lindquist, E. F. *Design and Analysis of Experiments in Psychology and Education* (Boston: Houghton Mifflin, 1953), pp. 78–90.

Scheffé, H. *The Analysis of Variance*, New York: Wiley, 1959.

Seigel, S. *Nonparametric Statistics for the Behavioral Sciences.* New York: McGraw-Hill, 1956.

Snedecon, G. W. and Cochran, W. G. *Statistical Methods*, 6th ed., Ames: Iowa State University Press, 1967.

Welch, B. L., "The significance of the difference between two population means when the population variances are unequal." *Biometrika*, 29: 350–362 (1937).

5

Completely Randomized Multi-Factor Designs

5.1 INTRODUCTION

The modern theory of experimental design permits the experimenter to study a number of independent variables, or factors, within the same experiment. This has two major advantages. It is efficient, saving the time and effort of the experimenter and, equally important, it permits him to investigate the joint effects of variables. This chapter deals with the analysis of data from completely randomized designs involving more than a single factor. In such designs subjects are randomly assigned to combinations of treatment levels. The advantages and disadvantages of the completely randomized one-factor design were cited in the preceding chapter; that discussion is relevant for the designs of this chapter as well. It is again assumed that the levels of each variable have been arbitrarily, rather than randomly, selected.

As an example of the type of design that will be presented, consider the following experiment. We are interested in comparing the effects of four different amounts of reward, of three different delays of reward, and of the 12 combinations of amount and delay upon discrimination learning in rats. Each of 120 rats is randomly assigned to one of the 12 combinations of amount and delay of reward, with the restriction that there be exactly 10 rats exposed to each combination. Ten rats might receive one food pellet 1 second after a correct response, 10 rats might receive two food pellets 3 seconds after a correct response, and so on. In general, such a design involves the random assignment of n (10, in our example) subjects to each of ab ($4 \times 3 = 12$, in our example) combinations of treatment levels. The layout of such a design is presented in Table 5–1. The first subscript, represented by i in the general case, indexes the subjects within each treatment cell and varies from 1 to n. The second subscript, j, indexes the levels of the independent variable A and varies from 1 to a. The third subscript, k, indexes the levels of the variable B and varies from 1 to b. Thus, there are a levels of A, b levels of B, ab cells corresponding to the ab treatment combinations, with abn scores, n in each

TABLE 5–1 *Data matrix for a two-factor design*

		B_1	B_2	$\cdots$	B_k	$\cdots$	B_b
		Y_{111}	Y_{112}		Y_{11k}		Y_{11b}
		Y_{211}	Y_{212}		Y_{21k}		Y_{21b}
A_1		$\vdots$	$\vdots$		$\vdots$		$\vdots$
		Y_{i11}	Y_{i12}		Y_{i1k}		Y_{i1b}
		$\vdots$	$\vdots$		$\vdots$		$\vdots$
		Y_{n11}	Y_{n12}		Y_{n1k}		Y_{n1b}
		Y_{121}	Y_{122}		Y_{12k}		Y_{12b}
		Y_{221}	Y_{222}		Y_{22k}		Y_{22b}
A_2		$\vdots$	$\vdots$		$\vdots$		$\vdots$
		Y_{i21}	Y_{i22}		Y_{i2k}		Y_{i2b}
		$\vdots$	$\vdots$		$\vdots$		$\vdots$
$\vdots$							
		Y_{1j1}	Y_{1j2}		Y_{1jk}		Y_{1jb}
		Y_{2j1}	Y_{2j2}		Y_{2jk}		Y_{2jb}
A_j		$\vdots$	$\vdots$		$\vdots$		$\vdots$
		Y_{ij1}	Y_{ij2}		Y_{ijk}		Y_{ijb}
		$\vdots$	$\vdots$		$\vdots$		$\vdots$
$\vdots$							
		Y_{1a1}	Y_{1a2}		Y_{1ak}		Y_{1ab}
		Y_{2a1}	Y_{2a2}		Y_{2ak}		Y_{2ab}
		$\vdots$	$\vdots$		$\vdots$		$\vdots$
A_a		Y_{ia1}	Y_{ia2}		Y_{iak}		Y_{iab}
		$\vdots$	$\vdots$		$\vdots$		$\vdots$

cell. In this chapter, we consider the analysis of variance for this design and for similar designs that differ only in the number of independent variables.

5.2 A MODEL FOR THE COMPLETELY RANDOMIZED TWO-FACTOR DESIGN

The ab experimental groups of n subjects may be considered as ab random samples, one from each of ab treatment populations. These treatment populations are assumed to have been drawn from the same infinitely large parent population; thus, any differences among the ab distributions are attributable solely to differences among the treatment effects. The basic population parameters are the μ_{jk}, the mean of the treatment population defined by the jth level of A and the kth level of B; μ_j, the mean of all scores obtained under treatment A_j; μ_k, the mean of all scores obtained under B_k;

and μ, the expected value over all ab treatment populations or, equivalently, the mean of the original parent population.

The observed data are related to the population parameters by the following equation:

$$(5.1) \qquad Y_{ijk} = \mu + \alpha_j + \beta_k + (\alpha\beta)_{jk} + \varepsilon_{ijk}$$

where Y_{ijk} is the score of the ith subject in the jth treatment level of A and the kth treatment level of B,

$\alpha_j = \mu_j - \mu$, the *main effect* of treatment A_j

$\beta_k = \mu_k - \mu$, the *main effect* of treatment B_k

$(\alpha\beta)_{jk} = \mu_{jk} - \mu_j - \mu_k + \mu$, the *interaction effect* of treatments A_j and B_k

$\varepsilon_{ijk} = Y_{ijk} - [\mu + \alpha_j + \beta_k + (\alpha\beta)_{jk}]$

$\qquad = Y_{ijk} - \mu_{jk}$, the error component

The quantities involved in Equation (5.1) are similar to those encountered in Chapter 4, with the exception of the interaction, which therefore merits further definition. Note that

$$(5.2) \qquad (\mu_{jk} - \mu_j - \mu_k + \mu) = (\mu_{jk} - \mu) - (\mu_j - \mu) - (\mu_k - \mu)$$

In words, the interaction effect, $(\alpha\beta)_{jk}$, is the combined effect of the jth level of A and the kth level of B, which cannot be accounted for by α_j and β_k, the independent effects of A_j and B_k. The effects, α_j and β_k, are usually referred to as *main effects* in contrast with the *interaction effect*, $(\alpha\beta)_{jk}$. The interaction effect and its relation to main effects will be considered further in Section 5.3.2.

It follows from Equation (5.1) that the variability in our observed data matrix (Table 5–1) has several sources. These are:

(a) α_j, *the effect of treatment* A_j. This effect is a constant component of all scores obtained under A_j, but may vary over levels of j. A is assumed to be a fixed-effects variable; that is, the a levels are assumed to have been arbitrarily selected and the population of levels to have been exhausted. Therefore, $\sum_j \alpha_j = 0$. One null hypothesis that will be tested is

$$\alpha_1 = \alpha_2 = \cdots = \alpha_j = \cdots = \alpha_a = 0$$

(b) β_k, *the effect of treatment* B_k. This effect is a constant component of all scores obtained under B_k, but may vary over levels of k. We assume that B, like A, is a fixed-effect variable and, therefore, $\sum_k \beta_k = 0$. We will test the null hypothesis that

$$\beta_1 = \beta_2 = \cdots = \beta_k = \cdots = \beta_b = 0$$

(c) $(\alpha\beta)_{jk}$, *the interaction effect of* A_j *and* B_k. This effect is a constant component of all scores obtained under A_j and B_k, but may vary over the levels of j and k. Since A and B are both fixed-effect variables, $\sum_k (\alpha\beta)_{jk} = \sum_j (\alpha\beta)_{jk} = 0$. The relevant null hypothesis is that

$$(\alpha\beta)_{12} = (\alpha\beta)_{13} = \cdots = (\alpha\beta)_{jk} = \cdots = (\alpha\beta)_{ab} = 0$$

(d) ε_{ijk}, *the error component* that is not accounted for by the systematic manipulation of the variables A and B. This is the only source of variance within the cells that corresponds to the treatment combinations. We assume that the ε_{ijk} are independently and normally distributed with mean of zero and variance σ_e^2 within each treatment population defined by a combination of levels of A and B.

On the basis of the model presented, we may derive parameter estimates, expected mean squares, and F tests of the null hypotheses stated above. The development is similar to that of Section 4.2.1 for the one-factor design. The relevant parameter estimates are:

$$\hat{\mu} = \frac{\sum_i \sum_j \sum_k Y_{ijk}}{abn}$$

$$= \bar{Y}_{...}$$

$$\hat{\mu}_j = \frac{\sum_i \sum_k Y_{ijk}}{bn}$$

$$= \bar{Y}_{.j.}$$

(5.3)

$$\hat{\mu}_k = \frac{\sum_i \sum_j Y_{ijk}}{an}$$

$$= \bar{Y}_{..k}$$

$$\hat{\mu}_{jk} = \frac{\sum_i Y_{ijk}}{n}$$

$$= \bar{Y}_{.jk}$$

The expected mean squares and F tests will be considered in the next section as part of the general discussion of the analysis of variance.

5.3 THE ANALYSIS OF VARIANCE FOR THE COMPLETELY RANDOMIZED TWO-FACTOR DESIGN

The sources of variability in our data are specified by Equation (5.1). A general way of proceeding to obtain the appropriate sums of squares is to substitute parameter estimates for parameters in the basic equation. Thus, beginning with Equation (5.1),

$$Y_{ijk} = \mu + \alpha_j + \beta_k + (\alpha\beta)_{jk} + \varepsilon_{ijk}$$

and substituting from Equation (5.3), we arrive at the following identity:

(5.4) $Y_{ijk} = \overline{Y}_{...} + (\overline{Y}_{.j.} - \overline{Y}_{...}) + (\overline{Y}_{..k} - \overline{Y}_{...})$
$$+ (\overline{Y}_{.jk} - \overline{Y}_{.j.} - \overline{Y}_{..k} + \overline{Y}_{...}) + (Y_{ijk} - \overline{Y}_{.jk})$$

Subtracting $\overline{Y}_{...}$ from both sides of Equation (5.4) and squaring both sides yields

$$(Y_{ijk} - \overline{Y}_{...})^2 = (\overline{Y}_{.j.} - \overline{Y}_{...})^2 + (\overline{Y}_{..k} - \overline{Y}_{...})^2$$
$$+ (\overline{Y}_{.jk} - \overline{Y}_{.j.} - \overline{Y}_{..k} + \overline{Y}_{...})^2$$
$$+ (Y_{ijk} - \overline{Y}_{.jk})^2 + 2(\overline{Y}_{.j.} - \overline{Y}_{...})(\overline{Y}_{..k} - \overline{Y}_{...})$$

(5.5)
$$+ 2(\overline{Y}_{.j.} - \overline{Y}_{...})(\overline{Y}_{.jk} - \overline{Y}_{.j.} - \overline{Y}_{..k} + \overline{Y}_{...})$$
$$+ 2(\overline{Y}_{.j.} - \overline{Y}_{...})(Y_{ijk} - \overline{Y}_{.jk})$$
$$+ 2(\overline{Y}_{..k} - \overline{Y}_{...})(\overline{Y}_{.jk} - \overline{Y}_{.j.} - \overline{Y}_{..k} + \overline{Y}_{...})$$
$$+ 2(\overline{Y}_{..k} - \overline{Y}_{...})(Y_{ijk} - \overline{Y}_{.jk})$$
$$+ 2(\overline{Y}_{.jk} - \overline{Y}_{.j.} - \overline{Y}_{..k} + \overline{Y}_{...})(Y_{ijk} - \overline{Y}_{.jk})$$

After summing over the indices, i, j, and k, the cross-product terms vanish (the proof of this will be left to the student). They will therefore be ignored in the remainder of this derivation. Summing over i, we obtain

$$\sum_{i}^{n} (Y_{ijk} - \overline{Y}_{...})^2 = n(\overline{Y}_{.j.} - \overline{Y}_{...})^2 + n(\overline{Y}_{..k} - \overline{Y}_{...})^2$$

(5.6)
$$+ n(\overline{Y}_{.jk} - \overline{Y}_{.j.} - \overline{Y}_{..k} + \overline{Y}_{...})^2$$
$$+ \sum_{i} (Y_{ijk} - \overline{Y}_{.jk})^2$$

Next, summing over j, the result is

$$\sum_{i}^{n} \sum_{j}^{a} (Y_{ijk} - \overline{Y}_{...})^2 = n\sum_{j}^{a} (\overline{Y}_{.j.} - \overline{Y}_{...})^2 + an(\overline{Y}_{..k} - \overline{Y}_{...})^2$$

(5.7)
$$+ n\sum_{j}^{a} (\overline{Y}_{.jk} - \overline{Y}_{.j.} - \overline{Y}_{..k} + \overline{Y}_{...})^2$$
$$+ \sum_{i}^{n} \sum_{j}^{a} (Y_{ijk} - \overline{Y}_{.jk})^2$$

Finally, summing over k, we have

$$\sum_i^n \sum_j^a \sum_k^b (Y_{ijk} - \overline{Y}_{...})^2 = bn\sum_j^a (\overline{Y}_{.j.} - \overline{Y}_{...})^2$$

(sum of squares total, SS_{tot})   (sum of squares for A, SS_A)

$$+ an\sum_k^b (\overline{Y}_{..k} - \overline{Y}_{...})^2$$

(sum of squares for B, SS_B)

(5.8)

$$+ n\sum_j^a \sum_k^b (\overline{Y}_{.jk} - \overline{Y}_{.j.} - \overline{Y}_{..k} + \overline{Y}_{...})^2$$

(sum of squares for interaction, SS_{AB})

$$+ \sum_i^n \sum_j^a \sum_k^b (Y_{ijk} - \overline{Y}_{.jk})^2$$

(sum of squares within groups, $SS_{S/AB}$)

Equation (5.8) states that the total variability is to be partitioned into four parts: variability due to A, variability due to B, variability due to the AB interaction, and variability due to error. This analysis is not the only one possible, but rather the one appropriate for our particular model [Equation (5.1)]. For example, if we were to assume the absence of interaction, our model would assert that

(5.9) $$Y_{ijk} = \mu + \alpha_j + \beta_k + \varepsilon_{ijk}$$

and the identity that forms the basis for our analysis would be

(5.10) $$(Y_{ijk} - \overline{Y}_{...}) = (\overline{Y}_{.j.} - \overline{Y}_{...}) + (\overline{Y}_{..k} - \overline{Y}_{...})$$
$$+ (Y_{ijk} - \overline{Y}_{.j.} - \overline{Y}_{..k} + \overline{Y}_{...})$$

In this case, the total variability would be partitioned into three parts due to the A effect, the B effect, and error variability. The point of these comments is that the analysis of variance is not an arbitrary set of computations, but a logical consequence of the experimenter's assumptions about the relationships between his data and the parameters of the population from which he has sampled.

5.3.1 Summarizing the analysis of variance. Table 5–2 summarizes the analysis of variance for the completely randomized two-factor design. The source of variance (SV) column reflects the partitioning described above and, as was stated, is a direct consequence of Equation (5.1). For the A main effect there are $(a - 1)$ df because of the requirement that the sum of the deviations of the a treatment means about the grand mean must be zero. For a similar reason, the df for the B main effect are $b - 1$.

TABLE 5-2 Analysis of variance for a two-factor design

SV	df	SS	MS	EMS	F
Total	$abn - 1$	$\sum_i^n \sum_j^a \sum_k^b Y^2 - C*$			
A	$a - 1$	$\dfrac{\sum_j^a \left(\sum_i^n \sum_k^b Y\right)^2}{nb} - C$	$\dfrac{SS_A}{a - 1}$	$\sigma_e^2 + nb\theta_A^2$	$\dfrac{MS_A}{MS_{S/AB}}$
B	$b - 1$	$\dfrac{\sum_k^b \left(\sum_i^n \sum_j^a Y\right)^2}{na} - C$	$\dfrac{SS_B}{b - 1}$	$\sigma_e^2 + na\theta_B^2$	$\dfrac{MS_B}{MS_{S/AB}}$
AB	$(a - 1)(b - 1)$	$\dfrac{\sum_j^a \sum_k^b \left(\sum_i^n Y\right)^2}{n} - C - SS_A - SS_B$	$\dfrac{SS_{AB}}{(a - 1)(b - 1)}$	$\sigma_e^2 + n\theta_{AB}^2$	$\dfrac{MS_{AB}}{MS_{S/AB}}$
S/AB	$ab(n - 1)$	$SS_{\text{tot}} - SS_A - SS_B - SS_{AB}$	$\dfrac{SS_{S/AB}}{ab(n - 1)}$	σ_e^2	

$$* C = \frac{\left(\sum_i^n \sum_j^a \sum_k^b Y\right)^2}{nab}$$

In the case of the AB interaction, we begin with $(ab - 1)df$, because of the requirement that the sum of the deviations of the ab cell means about the grand mean must be zero. Removal of the variability among the cell means due to the A and B effects causes the loss of $(a - 1)\,df$ and $(b - 1)\,df$. Because of the removal of this variability, the row and column means of Table 5–1 are no longer free to vary, and the corresponding df are lost. Thus, restrictions placed on the variability of the cell means by the grand mean and by the removal of the A and B effects yield

$$(5.11) \qquad df_{AB} = (ab - 1) - (a - 1) - (b - 1)$$
$$= (a - 1)(b - 1)$$

The df associated with S/AB can be viewed as the result of taking the deviations of n scores about their cell mean, yielding $(n - 1)\,df$, then summing this result over the ab cells, yielding $ab(n - 1)\,df$. Alternatively, noting that the variability due to S/AB is a residual from the total after removal of the variability due to A, B, and AB, we have

$$(5.12) \qquad df_{S/AB} = (abn - 1) - (a - 1) - (b - 1) - (a - 1)(b - 1)$$
$$= ab(n - 1)$$

As a third alternative, we note that the total variability among scores must be due either to variability among cell means or variability among scores in the same cell. Therefore, the variability within cells equals the total minus the between-cells variability, or

$$(5.13) \qquad df_{S/AB} = (abn - 1) - (ab - 1)$$
$$= ab(n - 1)$$

The computational formulas in the SS column may be derived from the components of Equation (5.8), as was done in Chapter 4. Each term is expanded, the summations are carried out, raw score formulas are substituted for the means, and the expressions are simplified as far as possible. However, this lengthy procedure can be bypassed by again noting the isomorphism of df and squared quantities. The term $a - 1$ in the df column suggests the sum of a squared quantities minus the correction term, or

$$\sum_{j}^{a} (\quad)^2 - C$$

Remembering that any summations not represented outside the parentheses must appear inside, we have

$$\sum_{j}^{a} \left(\sum_{i}^{n} \sum_{k}^{b} Y_{ijk} \right)^2 - C$$

Finally, we divide by the number of scores inside the parentheses, yielding

$$SS_A = \frac{\sum_{j}^{a}(\sum_{i}^{n} \sum_{k}^{b} Y_{ijk})^2}{bn} - C$$

The SS_B follow in a similar manner. Two equivalent formulas are available for SS_{AB}. Expanding $(a - 1)(b - 1)$ yields $ab - a - b + 1$, which suggests

$$(5.15) \quad SS_{AB} = \frac{\sum_j^a \sum_k^b (\sum_i^n Y_{ijk})^2}{n} - \frac{\sum_j^a (\sum_i^n \sum_k^b Y_{ijk})^2}{bn} - \frac{\sum_k^b (\sum_i^n \sum_j^a Y_{ijk})^2}{an} + C$$

Since the df_{AB} also equals $(ab - 1) - (a - 1) - (b - 1)$, we might have

$$(5.16) \quad SS_{AB} = \frac{\sum_j^a \sum_k^b (\sum_i^n Y_{ijk})^2}{n} - C - SS_A - SS_B$$

The $ab(n - 1)$ df suggests calculating $SS_{S/AB}$ by computing a sum of squared deviations for each cell, and then pooling over cells. More simply, $abn - ab$ suggests

$$(5.17) \quad SS_{S/AB} = \sum_i^n \sum_j^a \sum_k^b Y_{ijk}^2 - \frac{\sum_j^a \sum_k^b (\sum_i^n Y_{ijk})^2}{n}$$

Alternatively, remembering that

$$df_{S/AB} = (abn - 1) - (a - 1) - (b - 1) - (a - 1)(b - 1)$$

we have

$$(5.18) \quad SS_{S/AB} = \sum_i^n \sum_j^a \sum_k^b Y_{ijk}^2 - C - SS_A - SS_B - SS_{AB}$$

The degrees of freedom not only suggest the proper computational formula, they suggest alternative formulas which, in turn, provide checks on our computation.

The MS column is again the result of dividing the entries in the SS column by the corresponding entries in the df column. The entries under EMS consist of σ_e^2, plus a term θ^2, representing the corresponding SV. The θ^2 term is multiplied by the number of levels of all variables that are not represented in the subscript. The θ^2 terms are defined as follows:

$$(5.19) \quad \theta_A^2 = \frac{\sum_j^a (\mu_j - \mu)^2}{a - 1}$$

$$(5.20) \quad \theta_B^2 = \frac{\sum_k^b (\mu_k - \mu)^2}{b - 1}$$

$$(5.21) \quad \theta_{AB}^2 = \frac{\sum_j^a \sum_k^b (\mu_{jk} - \mu_j - \mu_k + \mu)^2}{(a - 1)(b - 1)}$$

The notation θ^2 indicates that the subscripted variable is fixed. This follows from the definitions, Equations (5.19) to (5.21), where the squared deviations are pooled only over the levels of A appearing in the experiment. In contrast, σ_e^2 represents the variance of effects for a population of levels of which those in the experiment are only a random sample.

The appropriate F tests are also listed in Table 5–2. The logic is exactly that discussed in Chapter 4; we require an error term that consists of all the components of the numerator except the null hypothesis term. Then, if H_0 is true, the ratio of expected mean squares will equal 1. The null hypothesis that the μ_j are all equal (i.e., $\theta_A^2 = 0$) is tested by an F distributed on $(a - 1)$ df and $ab(n - 1)$ df. A significant value of F leads to a relatively restricted conclusion. It does not indicate that the levels of A differ in their effects at any given level of B or at any level of B other than those used in this particular experiment, but rather that, *averaged over the levels of B used in the present study*, there is an effect due to A. If the experimenter is interested in comparing the effects of A at a particular level of B, then a sum of squares of A should be computed only for the data at that level of B. This sum of squares should then be divided by $a - 1$ and tested against the usual error term $MS_{S/AB}$. Such a test is a test of a *simple effect* of A, as contrasted with the overall test of the main effect of A. The null hypothesis that θ_B^2 equals zero is tested by an F distributed on $b - 1$ and $ab(n - 1)$. Rejection of the null hypothesis indicates that the μ_k differ, where μ_k is the mean of the kth treatment population of B, obtained by averaging over all treatment populations of A. If we wish to compare the means of the B treatment populations at one particular level of A, we are again interested in testing simple effects, and the procedure previously described is applicable. The null hypothesis that θ_{AB}^2 equals zero is tested by an F distributed on $(a - 1)(b - 1)$ and $ab(n - 1)$ df. Its rejection indicates that variability remains among the μ_{jk} after removal of variability due to the effects of A and B.

5.3.2. Two-factor interaction. The two-factor interaction merits a more detailed interpretation than has thus far been provided. We begin with Equation (5.22), which follows from Equation (5.1); summing over i and dividing by n yields

$$(5.22) \qquad \overline{Y}_{.jk} = \mu + \alpha_j + \beta_k + (\alpha\beta)_{jk} + \frac{\sum_i \varepsilon_{ijk}}{n}$$

Substituting sample estimates of population parameters and transposing terms, we obtain

$$(5.23) \qquad \widehat{(\alpha\beta)}_{jk} = (\overline{Y}_{.jk} - \overline{Y}_{...}) - (\overline{Y}_{.j.} - \overline{Y}_{...}) - (\overline{Y}_{..k} - \overline{Y}_{...})$$

the estimate of the interaction effect of the jkth treatment combination (note that the estimate of $\sum_j \varepsilon_{ijk}/n$, $\sum_i(Y_{ijk} - \overline{Y}_{.jk})/n$, = 0). As an example, we next apply Equation (5.23) to the set of values of $\overline{Y}_{.jk} - \overline{Y}_{...}$ in Table 5–3. If, in accord with Equation (5.23), we remove the estimate of α_j from each cell mean, the result—Table 5–4—consists of the quantities $(\overline{Y}_{.jk} - \overline{Y}_{...}) - (\overline{Y}_{.j.} - \overline{Y}_{...})$. Note that the two df for A are literally lost; the row means are no longer free to vary but instead stand equal to each other and to the grand

TABLE 5–3 *A* 3 × 3 *table of values of* $\bar{Y}_{.jk} - \bar{Y}_{...}$

	B_1	B_2	B_3	$\bar{Y}_{.j.} - \bar{Y}_{...}$
A_1	−3.5	−3.0	−2.5	−3.0
A_2	−1.0	+1.0	+3.0	+1.0
A_3	−1.5	+2.0	+5.5	+2.0
$\bar{Y}_{..k} - \bar{Y}_{...}$	−2.0	0.0	+2.0	

TABLE 5–4 *Table 5–3 after removal of variability due to A*

	B_1	B_2	B_3	$\bar{Y}_{.j.} - \bar{Y}_{...}$
A_1	−0.5	0.0	+0.5	0.0
A_2	−2.0	0.0	+2.0	0.0
A_3	−3.5	0.0	+3.5	0.0
$\bar{Y}_{..k} - \bar{Y}_{...}$	−2.0	0.0	+2.0	

mean. Further, note that the column means are unchanged by this manipulation of the data, demonstrating the independence of row and column effects.

Next, remove ($\bar{Y}_{..k} - \bar{Y}_{...}$), the estimate of β_k, from each entry in Table 5–4. The result is Table 5–5, in which there are no *df* for rows or columns. The entries in this matrix are the estimates of $(\alpha\beta)_{jk}$. The important thing to note is that they are not all zero, that there is variability in the data not due to *A* and *B* effects. If this variability is significantly greater than that due to individual differences (*S/AB*), it is said that interaction effects are present in the data.

Tables 5–4 and 5–5 clarify one interpretation of interaction; namely, it is the variability among cell means that still remains when variability due to the main effects is removed. However, we ordinarily want to make some statement about interaction effects in terms of the tabulation or plot of the

TABLE 5–5 *Table 5–3 after removal of variability due to A and B*

	B_1	B_2	B_3	$\bar{Y}_{.j.} - \bar{Y}_{...}$
A_1	1.5	0	−1.5	0
A_2	0	0	0	0
A_3	−1.5	0	1.5	0
$\bar{Y}_{..k} - \bar{Y}_{...}$	0	0	0	

original cell means. Assuming that the data are error-free (or equivalently, that the cell means are means of populations rather than samples), what do the unadjusted means reveal about the presence or absence, as well as the direction, of interaction? To answer this question, consider a data set in which there are no interaction effects; ignoring error components, the appropriate structural equation is

(5.22′) $$\overline{Y}_{.jk} = \mu + \alpha_j + \beta_k$$

If, as in the previous example, the $\alpha_j = -3, 1, 2$ and the $\beta_k = -2, 0, 2$, and if μ is taken as, say, 6.0, the resulting matrix of cell means would be

$$\begin{bmatrix} 6 + (-3) + (-2) & 6 + (-3) + (0) & 6 + (-3) + (2) \\ 6 + (1) + (-2) & 6 + (1) + (0) & 6 + (1) + (2) \\ 6 + (2) + (-2) & 6 + (2) + (0) & 6 + (2) + (2) \end{bmatrix}$$

$$= \begin{bmatrix} 1 & 3 & 5 \\ 5 & 7 & 9 \\ 6 & 8 & 10 \end{bmatrix}$$

Plotting these values, as we have done in Figure 5–1, is instructive.

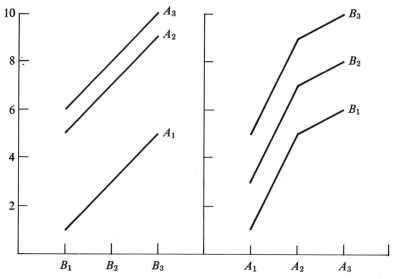

FIGURE 5–1 *A plot of cell means with no interaction present*

The notable aspect of the two plots is that the curves are parallel. That this must be true in the absence of interaction can be proven for any set of values. Consider four cell means: $\overline{Y}_{.jk}$, $\overline{Y}_{.j'k}$, $\overline{Y}_{.jk'}$, and $\overline{Y}_{.j'k'}$, where $j \neq j'$ and $k \neq k'$. From Equation (5.22′), it follows that $\overline{Y}_{.jk} - \overline{Y}_{.jk'} = \beta_k - \beta_{k'}$. However, $\overline{Y}_{.j'k} - \overline{Y}_{.j'k'}$, yields the same result. Thus, if there are *no* interaction effects, the simple effect of B is constant over levels of A. This is a verbal statement of the parallelism depicted in the right-hand panel of Figure 5–1. Similarly, if interaction is absent, the difference in the effects of A_j and $A_{j'}$ will be $\alpha_j - \alpha_{j'}$ at all levels of B; this is the situation depicted in the left-hand panel of Figure 5–1. Thus, *interaction is a significant departure from parallelism*. The spread among the means for the levels of one variable changes as a function of the level of the second variable. Alternatively, interaction may be viewed as significant variability among simple effects. The simple effects of one variable are not constant at all levels of the second and are therefore not all equal to the main effect.

Although the main and interaction effects are independent, an adequate interpretation of the data requires joint consideration of all three. Consider the two examples in Figure 5–2. For simplicity, again assume errorless data or, in other words, assume that the cell means are population means and any nonzero effects are therefore significant. In the left side of Figure 5–2 there are no A or AB effects, but the B effect is significant. Compare this with the right side, in which there is again no A effect, there is again a B effect, but this time there is also an AB interaction effect. The statement that there is no A effect has the same formal meaning for both data plots; averaging over both levels of B, the means for the three levels of A do not differ. However, in the data on the left there is no A effect at any level of B, while

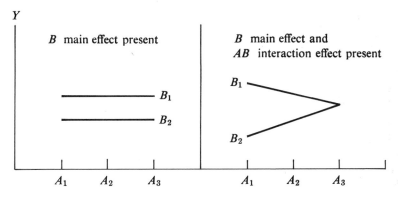

FIGURE 5–2 *Plots of two sets of means*

on the right there are simple A effects at both B_1 and B_2; these effects are in opposite directions and tend to cancel each other out. Clearly, if one is to obtain an accurate picture of the relationship among treatment population means, the total set of effects must be considered. Thus, knowing that B is a significant source of variance is not enough; it is only when the condition of the A and AB effects is noted that an intelligent discussion of the data can begin.

5.3.3 Behavioral and statistical hypotheses. An understanding of statistical effects is incomplete without some feeling for their relationship to behavioral hypotheses. Although there are no rules for the translation of hypotheses about behavior into hypotheses about statistical effects, we will attempt to demonstrate the process.

Consider the following experiment. On each of 100 trials, each subject has a choice between gambling and not gambling. There are two independent variables: A, the amount that can be won or lost on a gamble, and B, the consequences of not gambling. The levels of A are:

$A_1 = 5¢$ is won or lost on a gamble
$A_2 = 15¢$ is won or lost on a gamble
$A_3 = 25¢$ is won or lost on a gamble

On each gamble the subject has an equal chance of winning or losing. The levels of B are:

$B_1 =$ the subject always pays out $1¢$ on trials on which he does not gamble
$B_2 =$ the subject always receives $1¢$ on trials on which he does not gamble

One theory of choice behavior would predict the following:

(a) The percentage of gambles (P) will increase as the amount risked (the levels of A) increases when the consequence of not gambling is a sure gain (B_2).
(b) P will decrease as the amount risked increases when the consequences of not gambling is a sure loss (B_1).
(c) P will increase (when the alternative is B_2) over the levels of A at about the same rate that P will decrease (when the alternative is B_1) over the levels of A.
(d) P will be greater when the alternative to gambling is a sure loss (B_1) than when it is a sure gain (B_2), regardless of the level of A.

These hypotheses about the performances of subjects would be confirmed if the plot of the means for the AB combinations looked like the right-hand side of Figure 5–2. Thus, the translation into statistical hypotheses would be as follows:

(a') There will be a significant AB interaction (from (a) and (b) above).
(b') The A main effect will not be significant (from (c) above).
(c') The B main effect will be significant (from (d) above).

If hypotheses (a'), (b'), and (c') are verified and if, in addition, the effects are in the predicted direction, our behavioral hypotheses are supported.

Let us consider a second example of the relationship between behavioral and statistical hypotheses. Schizophrenics and normal subjects, equated on an intelligence test, are required to perform on a concept formation task. Two sets of stimuli are used; both involve the same concepts. One set contains pictures that express social approval (e.g., woman patting boy on head), the second set contains pictures expressing social disapproval (e.g., woman shaking finger at boy). Certain theories of personality and intellectual performance might lead to the following hypotheses:

(a) Normal subjects will take fewer trials to attain the relevant concept than will schizophrenics, whichever stimulus set is the basis for comparison.
(b) The performances of normal subjects will not be influenced by the type of stimulus set (approving or disapproving pictures).
(c) Schizophrenics will perform less well on the disapproving than on the approving set of pictures.

If all three hypotheses were correct, the data might look something like the plot in Figure 5-3. The following results are hypothesized for the statistical analysis:

(a′) There will be a significant main effect due to the personality variable (from (a) above).
(b′) There will be a significant main effect due to stimulus sets (from (b) and (c) above).
(c′) There will be a significant interaction of personality and stimulus (from (b) and (c) above).

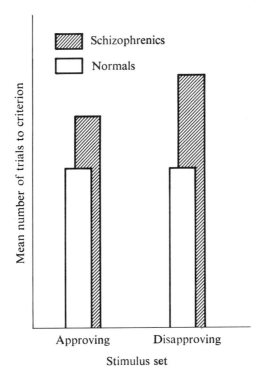

FIGURE 5-3 *Performance of schizophreni*ₛ *and normal subjects on a concept formation task*

It is a good idea to go through the process just discussed before the collection of data. The experimenter should try to visualize the sorts of results that might be obtained and should consider which sources of variance in the analysis will reflect these results. This type of activity will help ensure that the experimental design makes possible adequate tests of the behavioral hypotheses or provides answers to the questions posed. In addition, this type of thinking promotes a feeling for the relationship of data to psychological processes. Remember, the end goal of experimentation is not a statistical statement (e.g., the stimulus set variable is a significant source of variance), but rather a statement about behavior (e.g., stimuli that express social disapproval are less readily categorized than stimuli expressing approval by a population consisting of schizophrenic and normal subjects).

It is also important to plot the data once it is obtained and to study carefully the plot of data. Plot the same set of cell means several ways, as was done in Figure 5–1. Plot the means at the levels of B, averaging over the levels of A, in order to get a picture of the B main effect. Look at the A main effect by averaging over levels of B. Keep these plots at hand while reading the entries in the analysis of variance table.

5.3.4 A numerical example for the two-factor design, equal n. Table 5–6 presents data from an experiment involving two independent variables, A and B. As a first step in the analysis, obtain the sum of scores and the sum of squared scores for each cell (combination of j and k).

The sums of squares fomulas of Table 5–2 may now be applied. The correction term is again the sum of all scores squared, and divided by the total number of scores, thus,

$$C = \frac{(730)^2}{48}$$

$$= 11,102.080$$

The SS_{tot} is

$$SS_{tot} = 3,577 + \cdots + 2,725 - C$$

$$= 16,090 - C$$

$$= 4,987.920$$

Next,

$$SS_A = \frac{(343)^2 + (387)^2}{24} - C$$

$$= 11,142.417 - C$$

$$= 40.337$$

Similarly,

$$SS_B = \frac{(357)^2 + (224)^2 + (149)^2}{16} - C$$

$$= 12,489.125 - C$$

$$= 1,387.0450$$

The calculation of the total sum of squares is the same as it was for the one-factor design of Chapter 4. In fact, for any design, SS_{tot} is the sum of all squared scores minus the correction term. The sum of squares for the A treatment variable is also computed in the same manner as in Chapter 4. In

TABLE 5–6 *Data from a completely randomized two-factor experiment*

	A_1B_1	A_1B_2	A_1B_3	
	7	6	9	
	33	11	12	
	26	11	6	
	27	18	24	
	21	14	7	
	6	18	10	
	14	19	1	
	19	14	10	
$\sum_i Y_{i1k} = 153$		111	79	$\sum_i \sum_k Y_{i1k} = 343$
$\sum_i Y_{i1k}^2 = 3,577$		1,679	1,087	
	A_2B_1	A_2B_2	A_2B_3	
	42	28	13	
	25	6	18	
	8	1	23	
	28	15	1	
	30	9	3	
	22	15	4	
	17	2	6	
	32	37	2	
$\sum_i Y_{i2k} = 204$		113	70	$\sum_i \sum_k Y_{i2k} = 387$
$\sum_i Y_{i2k}^2 = 5,934$		2,725	1,088	
$\sum_j \sum_i Y_{ijk} = 357$		224	149	$\sum_i \sum_j \sum_k Y_{ijk} = 730$

general, the design is collapsed so that it consists of a groups of scores; then the SS_A is computed as would be done for a one-factor design. In our example, we ignore the B variable and view the design as consisting of two groups of 24 (in general, bn) scores each. To compute SS_B, we ignore the A variable and view the design as a one-factor design consisting of three groups of 16 scores (in general, b groups of an scores). This is exactly what has been done in the immediately preceding computations.

To calculate the SS_{AB}, one might first calculate the sum of squares for the cell means, which will be denoted by $SS_{\overline{AB}}$. Essentially, the design is again viewed as a one-factor design, this time consisting of six (ab) groups of eight (n) scores each. Consequently

$$SS_{\overline{AB}} = \frac{(153)^2 + (111)^2 + \cdots + (70)^2}{8} - C$$

$$= 12{,}657.000 - C$$

$$= 1{,}554.920$$

Next, remove that portion of the variability among cells due to A and B effects, resulting in

$$SS_{AB} = 1{,}554.920 - SS_A - SS_B$$

$$= 127.538$$

The above is the procedure described in the sum of squares column of Table 5–2. An alternative approach is to use Equation (5.15), which follows from the expanded df, $ab - a - b + 1$. Thus,

$$SS_{AB} = \frac{\sum_j^2 \sum_k^3 (\sum_i^8 Y_{ijk})^2}{8} - \frac{\sum_j^2 (\sum_i^8 \sum_k^3 Y_{ijk})^2}{24} - \frac{\sum_k^3 (\sum_i^8 \sum_j^2 Y_{ijk})^2}{16} + C$$

$$= 12{,}657.000 - 11{,}142.417 - 12{,}489.125 + 11{,}102.080$$

$$= 127.538$$

Now the sum of squares for the error term, S/AB, is required. This may be computed as a residual from the total variability; i.e.,

$$SS_{S/AB} = SS_{tot} - SS_A - SS_B - SS_{AB}$$

$$= 3{,}433.000$$

As an alternative, the error sum of squares might be computed as the difference between the total and between-cells variability, as in Equation (5.17):

$$SS_{S/AB} = SS_{tot} - SS_{\overline{AB}}$$

$$= 4{,}987.920 - 1{,}554.920$$

$$= 3{,}433.000$$

A third approach to the calculation of $SS_{S/AB}$ stems from the fact that the $MS_{S/AB}$ is equal to the average cell variance. Therefore,

$$(5.24) \qquad MS_{S/AB} = \sum_j \sum_k S^2_{jk/ab}$$

and by substituting and transposing terms,

$$SS_{S/AB} = (n-1)\sum_j \sum_k S^2_{jk}$$

$$(5.25) \qquad = 7(490.428)$$

$$= 3{,}432.996$$

which is correct within a slight rounding error. These alternative methods provide a *partial* check on the calculations. Thus, an error in calculating SS_A might affect the residual method of obtaining the error sum of squares and not affect the third method used. On the other hand, an error in squaring the individual scores would lead to the same result with all approaches.

Knowledge of these alternative calculations is also worthwhile because there are situations in which the within-cells variability must be obtained prior to the calculation of treatment effects. An instance of this will be noted in the discussion of calculations for the case of unequal n. An additional reason for demonstrating these alternative calculations is to emphasize the underlying identity of several conceptualizations of within-cells variability— it is residual when all treatment effects have been accounted for, it is the difference between total variability and variability among cell means, and it is the average within-cells variability.

The mean squares and the F ratios are now easily computed. Dividing sums of squares by the appropriate df, we obtain the entries in the MS column of Table 5–7. The A, B, and AB mean squares are each tested against the same error term, $MS_{S/AB}$, yielding the ratios reported in the F column. Only the B main effect is significant.

Turn now to Figure 5–4, which shows a plot of the group means. The source of the B main effect is clear; $\overline{Y}_{.k}$ consistently declines as k increases.

TABLE 5–7 *Analysis of variance for data from a two-factor experiment*

SV	df	SS	MS	F
Total	47	4,987.92		
A	1	40.34	40.34	.49
B	2	1,387.05	693.52	8.48*
AB	2	127.54	63.77	.78
S/AB	42	3,433.00	81.74	
				*p < .01

The figure also suggests that scores are higher under A_2 than under A_1 and that there is an AB interaction, the decline over levels of B appearing to be more rapid under A_2 than under A_1. This sort of result more than any other suggests the importance of the statistical analysis. While inspection of the figure suggests A and AB effects, the statistical analysis reveals that the A and AB variabilities are well within the range that might be expected on the basis of individual differences.

In connection with Figure 5–4, note that this method of graphing the data is appropriate only if B is a quantitative variable. If there is no rationale for ordering the levels of B, or if there is no rationale for spacing the levels along the abscissa, then a histogram rather than a line graph should be presented. The line graph suggests a function about which one can ask questions of shape and curvature, questions appropriate only when the independent variable is a quantitative scaled variable.

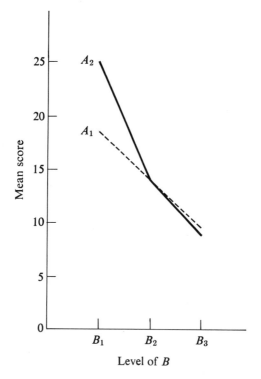

FIGURE 5–4 *Mean scores plotted as a function of B for both levels of A*

5.3.5 Unequal but proportional cell frequencies.

Thus far we have considered only the equal n case. If the cell frequencies are unequal but proportional, the analysis must be slightly modified. Proportionality is illustrated by the following set of cell frequencies:

$$
\begin{array}{lll|l}
n_{11} = 8 & n_{12} = 6 & n_{13} = 6 & n_{1.} = 20 \\
n_{21} = 4 & n_{22} = 3 & n_{23} = 3 & n_{2.} = 10 \\
\hline
n_{.1} = 12 & n_{.2} = 9 & n_{.3} = 9 &
\end{array}
$$

The critical feature of these n_{jk}s is that the ratio of row frequencies is constant over columns, $2:1$ in this example. Equivalently, the ratio of column frequencies is constant over rows, $4:3:3$ in the example.

The analysis of variance for the proportional but unequal n case is carried out as before, always dividing each squared quantity by the number of scores that have been summed to make up the quantity. The difference is that in the equal n case, for example,

$$
SS_A = \frac{\sum_j (\sum_i \sum_k Y_{ijk})^2}{bn} - C
$$

and in the proportional n case, we have

$$
SS_A = \frac{(\sum_i \sum_k Y_{i1k})^2}{\sum_k n_{1k}} + \cdots + \frac{(\sum_i \sum_k Y_{iak})^2}{\sum_k n_{ak}} - C
$$

If the n_{jk} are all equal,

$$
\sum_k n_{1k} = \cdots = \sum_k n_{ak} = bn
$$

and the equations for the equal and proportional cases are identical. The specific formulas for the analysis of variance, proportional n case, are:

(5.26)
$$
SS_{\text{tot}} = \sum_i \sum_j \sum_k Y_{ijk}^2 - C
$$

(5.27)
$$
SS_A = \sum_j \frac{(\sum_i \sum_k Y_{ijk})^2}{\sum_k n_{jk}} - C
$$

(5.28)
$$
SS_B = \sum_k \frac{(\sum_i \sum_j Y_{ijk})^2}{\sum_j n_{jk}} - C
$$

(5.29)
$$
SS_{AB} = \sum_j \sum_k \frac{(\sum_i Y_{ijk})^2}{n_{jk}} - C - SS_A - SS_B
$$

(5.30)
$$
SS_{S/AB} = SS_{\text{tot}} - SS_A - SS_B - SS_{AB}
$$

(5.31)
$$
C = \frac{(\sum_i \sum_j \sum_k Y_{ijk})^2}{\sum_j \sum_k n_{jk}}
$$

A word of caution about the preceding equations is in order. If proportionality occurs by chance and the cell frequencies are not truly representative of the frequencies in the treatment populations, the method described will improperly weight some treatment combinations. If, contrary to the evidence of the observed frequencies, it is believed that the treatment populations are of equal size, the experimenter is better advised to use the *method of unweighted means*, which will be described in the next section.

5.3.6 Disproportionate cell frequencies. This situation should be avoided whenever possible. As we noted in Chapter 4, unequal ns exaggerate the consequences of violations of the analysis of variance model. Furthermore, disproportionality requires somewhat different analyses from those considered thus far. The choice of analysis is not always simple, the interpretation is occasionally complicated, and the actual computational labor is always more than in the equal, or even proportional, case.

Something of the nature of the problem inherent in disproportionate cell frequencies is illustrated in Table 5–8. In the absence of interaction we

TABLE 5–8 *Example involving disproportionate cell frequencies*

		B_1	B_2	Totals	$\bar{Y}_{.j.}$
A_1	$\sum_i Y_{i1k}$	40	200	240	
	n_{1k}	2	8	10	
	$\bar{Y}_{.1k}$	20	25		24
A_2	$\sum_i Y_{i2k}$	40	20	60	
	n_{2k}	8	2	10	
	$\bar{Y}_{.2k}$	5	10		6
		80	220		
	Totals	10	10	$\bar{Y}_{...} = 15$	
	$\bar{Y}_{.k.}$	8	22		

would expect $\bar{Y}_{.2.} - \bar{Y}_{.1.}$ to equal $\bar{Y}_{.2k} - \bar{Y}_{.1k}$; however, $\bar{Y}_{.2.} - \bar{Y}_{.1.} = 18$ while the difference within each column is only 15. Similarly, $\bar{Y}_{..2} - \bar{Y}_{..1} = 14$, a difference considerably larger than the value of 5 obtained within each row. This clearly contradicts our earlier conclusion (p. 105) that, in the absence of interaction, the main and simple effects of a variable are the same. We obtain a result that appears even stranger if we carry out the analysis of variance. Using Equations (5.27), (5.28), and (5.29), we find that $SS_A = 1620$, $SS_B = 980$, and $SS_{AB} = (5080 - 3380) - 1620 - 980 = -900$. The minus sign is neither a computational nor a typographical error. Nevertheless, from the development of analysis of variance presented thus far, the result is utter nonsense. Judging by the cell means, the value of SS_{AB} should be zero; a negative value bears no interpretation within our framework.

In order to understand what is happening, let us reconsider the partitioning of $SS_{\overline{AB}}$, the variability among the cell means. We have

$$\sum_j \sum_k n_{jk}(\overline{Y}_{.jk} - \overline{Y}_{...})^2 = \sum_j \sum_k n_{jk}(\overline{Y}_{.j.} - \overline{Y}_{...})^2 + \sum_j \sum_k n_{jk}(\overline{Y}_{..k} - \overline{Y}_{...})^2$$

$$+ \sum_j \sum_k n_{jk}(\overline{Y}_{.jk} - \overline{Y}_{.j.} - \overline{Y}_{..k} + \overline{Y}_{...})^2$$

(5.32)
$$+ 2\sum_j \sum_k n_{jk}(\overline{Y}_{.j.} - \overline{Y}_{...})(\overline{Y}_{..k} - \overline{Y}_{...})$$

$$+ 2\sum_j \sum_k n_{jk}(\overline{Y}_{.j.} - \overline{Y}_{...})(\overline{Y}_{.jk} - \overline{Y}_{.j.} - \overline{Y}_{..k} + \overline{Y}_{...})$$

$$+ 2\sum_j \sum_k n_{jk}(\overline{Y}_{..k} - \overline{Y}_{...})(\overline{Y}_{.jk} - \overline{Y}_{.j.} - \overline{Y}_{..k} + \overline{Y}_{...})$$

If the n_{jk} are proportional or equal, the last three terms are zero. Otherwise, they can be either positive or negative. The reason $SS_{\overline{AB}} - SS_A - SS_B$ is negative in our example is that it includes the three cross-product terms, which have a relatively large negative total in this data set. The important point is that the cross-product terms are numerators of correlation coefficients, between the A and B effects, the A and AB effects, and the B and AB effects, respectively. If these correlations are not zero, then the three sum of squares terms are not independently distributed and the usual F test computations cannot provide independent assessments of the significance of main and interaction effects.

Despite our best efforts to avoid the problem, disproportionate ns do occur. In many instances, it is reasonable to assume that the treatment populations are of equal magnitude but ill luck caused disproportionate ns to appear. The air conditioning may fail and rats, ever contrary, die unequally over treatment groups. Or the apparatus is disbanded and, too late, we note that several subjects failed to appear in some groups, again not with equal frequencies. In such cases, we propose the *method of unweighted means*, which essentially treats the cell means as if they were based on equal ns. As we shall see, this method is reasonably straightforward to apply. However, the experimenter should question its application if the ns are very disparate. This would suggest that the populations are not of equal magnitude, that the ns are somehow related to the treatment level. In this case the unweighted mean analysis would be inappropriate.

There are frequent instances in which the most plausible, perhaps the only defensible, assumption is that unequal n typifies the treatment populations. This is very often the case when levels of the independent variable are not imposed by the experimenter, but rather are attributes of the subjects—for example, age, attitude, socio-economic level, personality classification. In a study of attitudes toward certain current problems, we might classify subjects in terms of political party and in terms of whether they consider themselves liberal or conservative. It is doubtful that the four treatment populations defined by these two variables are equal in magnitude; it is even doubtful that we have proportionality, equal proportions of Democrats and Republicans sharing a given political philosophy. If proportionality can reasonably be assumed, and if the obtained cell frequencies do not depart from proportionality so notably as to render the assumption suspect, the *method of expected cell frequencies* is appropriate. Otherwise, a *least-squares* analysis should be undertaken.

Method of unweighted means. This method, appropriate when treatment populations can be assumed to be equal in size, will be illustrated by application to the data of Table 5–9; these are a subset of the data of Table 5–6. We weight the treatment combinations equally by computing the main and interaction sums of squares as if each cell contained exactly one score, its mean. We then adjust the error variance appropriately.

STEP 1. Calculate SS_A, SS_B, and SS_{AB} for the cell means rather than for the original data.

$$SS_A = \frac{\sum_j(\sum_k \bar{Y}_{\cdot jk})^2}{b} - \frac{(\sum_j \sum_k \bar{Y}_{\cdot jk})^2}{ab}$$

$$= \frac{(43.125)^2 + (48.371)^2}{3} - \frac{(91.496)^2}{6}$$

$$= 1{,}399.840 - 1{,}395.263$$

$$= 4.577$$

$$SS_B = \frac{\sum_k(\sum_j \bar{Y}_{\cdot jk})^2}{a} - \frac{(\sum_j \sum_k \bar{Y}_{\cdot jk})^2}{ab}$$

$$= \frac{(44.696)^2 + (28.467)^2 + (18.333)^2}{2} - 1{,}395.263$$

$$= 176.838$$

$$SS_{AB} = \sum_j \sum_k \bar{Y}_{\cdot jk}^2 - \frac{(\sum_j \sum_k \bar{Y}_{\cdot jk})^2}{ab} - SS_A - SS_B$$

$$= (19.125)^2 + \cdots + (8.000)^2 - 1{,}395.263 - 4.577 - 176.838$$

$$= 19.561$$

Compute mean squares as in the proportional case, dividing each sum of squares by its *df*. The error mean square is ordinarily the average within-cell variance. In the present case, each cell contains only one "score," the mean of the original entries. Therefore, we require the average variance of the cell means. Since the variance of a mean is known to be the variance of the scores

TABLE 5–9 *A two-dimensional data matrix with disproportionate cell frequencies*

	A_1B_1	A_1B_2	A_1B_3	
	7	6	9	
	33	11	12	
	26	18	6	
	27	14	24	
	21	19	1	
	6	14	10	
	14			
	19			
$\sum_i Y_{i1k} = 153$	82	62	$\sum_k \sum_i Y_{ijk} = 297$
$\bar{Y}_{1k} = 19.125$	13.667	10.333	$\sum_k \bar{Y}_{1k} = 43.125$
$\sum_i Y_{i1k}^2 = 3{,}577$	1,234	938	

	A_2B_1	A_2B_2	A_2B_3	
	42	28	13	
	8	6	10	
	28	1	1	
	30	2	6	
	22	37	10	
	17			
	32			
$\sum_i Y_{i2k} = 179$	74	40	$\sum_j \sum_i Y_{ijk} = 293$
$\bar{Y}_{2k} = 25.571$	14.800	8.000	$\sum_k \bar{Y}_{2k} = 48.371$
$\sum_i Y_{i2k}^2 = 5{,}309$	2,194	406	
$\sum_j \bar{Y}_{.jk} = 44.696$	28.467	18.333	$\sum_j \sum_k \bar{Y}_{.jk} = 91.496$
$\sum_j \sum_i Y_{ijk} = 332$	156	109	

divided by the cell frequency; that is, $\sigma_{\bar{Y}}^2 = \sigma^2/n$—we compute $MS'_{S/AB}$, the average variance of the cell means, as

(5.33) $$MS'_{S/AB} = \frac{1}{ab} \sum_j \sum_k \frac{\hat{\sigma}_{jk}^2}{n_{jk}}$$

Assuming homogeneity of variance, the best estimate of σ_{jk}^2, for all j and k, is $MS_{S/AB}$, the average within-cell variance for the original data.

STEP 2. Compute $SS_{S/AB}$ from the data of Table 5-9.

$$SS_{S/AB} = SS_{tot} - SS_{\overline{AB}}$$

$$= \sum_k^b \sum_j^a \sum_i^{n_{jk}} Y_{ijk}^2 - \sum_k^b \sum_j^a \frac{(\sum_i Y_{ijk})^2}{n_{jk}}$$

$$= (3,577 + 1,234 + \cdots + 406) - \left[\frac{(153)^2}{8} + \frac{(82)^2}{6} + \cdots + \frac{(40)^2}{5} \right]$$

$$= 13,658 - 10,679.915$$

$$= 2,978.085$$

STEP 3. Compute the reciprocal of the harmonic mean of the cell entries.

$$\frac{1}{\tilde{n}_h} = \left(\frac{1}{ab}\right)\left(\sum_j \sum_k \frac{1}{n_{jk}}\right)$$

$$= \frac{1}{6}\left(\frac{1}{8} + \frac{1}{6} + \cdots + \frac{1}{5}\right)$$

$$= .163$$

STEP 4. It follows from Equation (5.33) and the use of $MS_{S/AB}$ as the estimate of σ_{jk}^2 that

$$MS'_{S/AB} = \frac{1}{\tilde{n}_h} MS_{S/AB}$$

$$= (.163)(2,978.085/31.0)$$

$$= 15.65$$

Note that the error *df* are

$$df_{S/AB} = df_{total} - df_{between\ cells}$$

$$= (37 - 1) - (6 - 1)$$

$$= 31$$

The *F* ratios are formed in the usual manner. The results of the analysis are summarized in Table 5–10 and lead to the same conclusions as does the analysis reported in Table 5–7.

Method of expected cell frequencies. This method, appropriate when the observed cell frequencies are almost proportional and there is good reason to assume proportionality in the populations, will also be applied to the data of Table 5–9.

STEP 1. Compute the $SS_{S/AB}$ as in Step 1 of the unweighted analysis.

$$SS_{S/AB} = SS_{tot} - SS_{\overline{AB}}$$

$$= 2,978.085$$

TABLE 5–10 *Analysis of variance for the data of Table 5–8 using the method of unweighted means*

SV	df	SS	MS	F
A	1	4.58	4.58	.29
B	2	176.84	88.42	5.64*
AB	2	19.56	9.78	.62
S/AB(adj)	31	485.43	15.65	
				*p < .01

STEP 2. Compute the expected cell frequencies. If $n_{j.}/n_{..}$ is the probability of sampling an individual from population A_j, and if $n_{.k}/n_{..}$ is the probability of sampling an individual from population B_k, then

$$
E(n_{jk}) = \left(\frac{n_{j.}}{n_{..}}\right)\left(\frac{n_{.k}}{n_{..}}\right)(n_{..})
$$

(5.34)

$$
= \frac{n_{j.}n_{.k}}{n_{..}}
$$

where $n_{j.}$ and $n_{.k}$ have been previously defined, and $n_{..}$ is the total number of observations in the experiment. These frequencies are tabulated in Table 5–11.

STEP 3. Compute the expected cell totals. Multiply the obtained cell mean (Table 5–9) by its expected denominator, obtained in the previous step. The result is an estimate of what the sum of scores should be for each cell, given the denominator calculated by Equation (5.34). These expected totals are also tabulated in Table 5.11.

STEP 4. Apply Equations (5.27), (5.28), and (5.29), using the frequencies and sums of Table 5–11. Thus,

$$
SS_A = \frac{(297.88)^2}{20} + \frac{(291.32)^2}{17} - \frac{(589.20)^2}{37}
$$

$$
= 46.216
$$

$$
SS_B = \frac{(331.32)^2}{15} + \frac{(156.08)^2}{11} + \frac{(101.80)^2}{11} - \frac{(589.20)^2}{37}
$$

$$
= 1,092.330
$$

$$
SS_{AB} = \frac{(155.14)^2}{8.11} + \cdots + \frac{(40.40)^2}{5.05} - \frac{(589.20)^2}{37} - 46.216 - 1,092.330
$$

$$
= 126.504
$$

The mean squares and *F* ratios are presented in Table 5–12, and are consistent with those of preceding analysis for these data.

TABLE 5-11 *Expected cell frequencies and sums for the data of Table 5-8*

	B_1	B_2	B_3	
A_1				
$E(n_{1k})$	$\dfrac{(20)(15)}{37} = 8.11$	$\dfrac{(20)(11)}{37} = 5.95$	$\dfrac{(20)(11)}{37} = 5.95$	$n_{1.} = 20$
$E\left(\sum_i Y_{i1k}\right)$	$(19.13)(8.11) = 155.14$	$(13.67)(5.95) = 81.34$	$(10.33)(5.95) = 61.40$	$E\left(\sum_i \sum_k Y_{i1k}\right) = 297.88$
A_2				
$E(n_{2k})$	$\dfrac{(17)(15)}{37} = 6.89$	$\dfrac{(17)(11)}{37} = 5.05$	$\dfrac{(17)(11)}{37} = 5.05$	$n_{2.} = 17$
$E\left(\sum_j Y_{i2k}\right)$	$(25.57)(6.89) = 176.18$	$(14.80)(5.05) = 74.74$	$(8.00)(5.05) = 40.40$	$E\left(\sum_i \sum_k Y_{i2k}\right) = 291.32$
$n_{.k}$	15	11	11	$n_{..} = 37$
$E\left(\sum_i \sum_j Y_{ijk}\right)$	331.32	156.08	101.80	$E\left(\sum_i \sum_j \sum_k Y_{ijk}\right) = 589.20$

TABLE 5–12 *Analysis of variance for the data of Table 5–8 using the method of expected cell frequencies*

SV	df	SS	MS	F
A	1	46.216	46.216	.48
B	2	1,092.330	546.165	5.69*
AB	2	126.504	63.252	.58
S/AB	31	2,978.085	96.067	
				*p < .01

Method of least-squares. This method is applicable when neither equality nor proportionality of the *ab* treatment population sizes can reasonably be assumed. It involves obtaining estimates of the parameters μ, α_j, β_k, and $(\alpha\beta)_{jk}$ such that the quantity

$$\sum_j \sum_k [\bar{Y}_{.jk} - \hat{\mu} - \hat{\alpha}_j - \hat{\beta}_k - \widehat{(\alpha\beta)}_{jk}]^2$$

will be minimized. If we differentiate this quantity with respect to the parameters of interest, we obtain a set of equations whose solution gives the required least-squares estimators. These can then be used to obtain main and interaction sums of squares, which are adjusted for their correlation with other terms in the analysis. Ordinarily, the sum of squares for each main effect is adjusted for the influence of the remaining main effects while interaction sums of squares are adjusted for the influence of all main effects. However, there are other alternatives, all requiring different interpretations. The point is discussed by Overall and Spiegel (1969), who also consider the rationale of the least-squares analysis in some detail. They provide numerical examples as do also Snedecor and Cochran (1967). More mathematical treatments of the last-squares approach to analysis of variance may be found in several texts; for example, Kempthorne (1952, pp. 79–87) and Peng (1967, pp. 173–178). The actual computations involved in this approach, particularly in the solving of the set of normal equations, is extremely tedious. Since programs are generally available at computing centers, we will move on to other topics.

5.4 A MODEL FOR THE COMPLETELY RANDOMIZED THREE-FACTOR DESIGN

The approach developed in the preceding part of this chapter can be extended to designs involving more than two factors. The layout of a three-factor design is presented in Table 5–13. The independent variables are *A*, *B*, and *C*, and the relevant indices are

$$i = 1, 2, \ldots, n$$
$$j = 1, 2, \ldots, a$$
$$k = 1, 2, \ldots, b$$
$$m = 1, 2, \ldots, c$$

The abc experimental groups of n subjects may be viewed as abc random samples, one from each of abc populations, which systematically differ among themselves only with respect to the treatment combination applied. The mean of the treatment population defined by the jth level of A, the kth level of B, and the mth level of C is μ_{jkm}. The mean of the population of all scores obtained under the jth level of A is μ_j, the expected value over the

TABLE 5–13 *Data matrix for a three-dimensional design*

		B_1	$\cdots$	B_k	$\cdots$	B_b
		Y_{1111}		Y_{11k1}		Y_{11b1}
		$\vdots$		$\vdots$		$\vdots$
	A_1	Y_{i111}		Y_{i1k1}		Y_{i1b1}
		$\vdots$		$\vdots$		$\vdots$
		Y_{n111}		Y_{n1k1}		Y_{n1b1}
	$\vdots$					
C_1	$\vdots$	$\vdots$		$\vdots$		$\vdots$
	A_j	Y_{1j11}		Y_{ijk1}		Y_{ijb1}
	$\vdots$					
		$\vdots$		$\vdots$		$\vdots$
	A_a	Y_{ia11}		Y_{iak1}		Y_{iab1}
		$\vdots$		$\vdots$		$\vdots$
	$\vdots$					
C_m	A_j	Y_{ij1m}		Y_{ijkm}		Y_{ijbm}
		$\vdots$		$\vdots$		$\vdots$
$\vdots$						
C_c	A_j	Y_{ij1c}		Y_{ijkc}		Y_{ijbc}
		$\vdots$		$\vdots$		$\vdots$

bc populations to which treatment A_j has been applied. Similarly, μ_k is the expected value computed over the *ac* treatment populations to which B_k has been applied and μ_m is the expected value over the *ab* treatment populations to which C_m has been applied. In addition, we can conceptualize the expected value of all scores under the treatment combinations $A_j B_k$, $A_j C_m$, and $B_k C_m$; such means will be denoted by μ_{jk}, μ_{jm}, and μ_{km}, respectively. Finally, the expected value over all *abc* treatment populations is denoted by μ.

The observed data are related to the population parameters by the equation

$$(5.35) \qquad Y_{ijkm} = \mu + \alpha_j + \beta_k + \gamma_m + (\alpha\beta)_{jk} + (\alpha\gamma)_{jm} + (\beta\gamma)_{km}$$
$$+ (\alpha\beta\gamma)_{jkm} + \varepsilon_{ijkm}$$

where

$\alpha_j = \mu_j - \mu$, the main effect of treatment A_j

$\beta_k = \mu_k - \mu$, the main effect of treatment B_k

$\gamma_m = \mu_m - \mu$, the main effect of treatment C_m

$(\alpha\beta)_{jk} = \mu_{jk} - \mu_j - \mu_k + \mu$, the interaction effect of A_j and B_k

$(\alpha\gamma)_{jm} = \mu_{jm} - \mu_j - \mu_m + \mu$, the interaction effect of A_j and C_m

$(\beta\gamma)_{km} = \mu_{km} - \mu_k - \mu_m + \mu$, the interaction effect of B_k and C_m

$(\alpha\beta\gamma)_{jkm} = \mu_{jkm} - \mu_{jk} - \mu_{jm} - \mu_{km} + \mu_j + \mu_k + \mu_m - \mu$, the interaction effect of A_j, B_k, and C_m

$\varepsilon_{ijkm} = Y_{ijkm} - [\mu + \alpha_j + \beta_k + \gamma_m + (\alpha\beta)_{jk} + (\alpha\gamma)_{jm} + (\beta\gamma)_{km} + (\alpha\beta\gamma)_{jkm}]$

$\qquad = Y_{ijkm} - \mu_{jkm}$, the error component

If the levels of all three independent variables are arbitrarily chosen, then the sum of *any* effect over *any* of its indices is zero; e.g.,

$$\sum_j (\alpha\beta\gamma)_{jkm} = \sum_k (\alpha\beta\gamma)_{jkm} = \sum_m (\alpha\beta\gamma)_{jkm} = 0$$

and similarly for all main and first-order (two-factor) interaction effects. The error component, ε_{ijkm}, is assumed to be independently and normally distributed, with zero mean and variance σ_e^2 within each of the *abc* treatment populations defined by the selected treatment combinations. As in the simpler designs already considered, the error component is that component of each score not accounted for by the contribution of the grand mean, μ, or by the systematic manipulation of the independent variables.

With one exception, the quantities defined above have all been encountered in the preceding sections of this chapter. The exception is $(\alpha\beta\gamma)_{jkm}$, which is designated the *second-order interaction effect* to distinguish it from the first-order interactions, which involve only two independent variables. Note that

(5.36) $(\alpha\beta\gamma)_{jkm} = (\mu_{jkm} - \mu) - [\alpha_j + \beta_k + \gamma_m + (\alpha\beta)_{jk}$

$$+ (\alpha\gamma)_{jm} + (\beta\gamma)_{km}]$$

The interaction effect, $(\alpha\beta\gamma)_{jkm}$, is the combined effect of the jth level of A, the kth level of B, and the mth level of C adjusted for the independent contributions of the main and interaction effects that have been removed in Equation (5.36).

5.5 THE ANALYSIS OF VARIANCE FOR THE COMPLETELY RANDOMIZED THREE-FACTOR DESIGN

The relationship between any single score, Y_{ijkm}, and the single and joint effects of the variables A, B, and C is described in the identity of Equation (5.37).

$$
\begin{aligned}
Y_{ijkm} - \overline{Y}_{....} = {}& (Y_{ijkm} - \overline{Y}_{.jkm}) + (\overline{Y}_{.j..} - \overline{Y}_{....}) \\
& + (\overline{Y}_{..k.} - \overline{Y}_{....}) + (\overline{Y}_{...m} - \overline{Y}_{....}) \\
& + (\overline{Y}_{.jk.} - \overline{Y}_{.j..} - \overline{Y}_{..k.} + \overline{Y}_{....}) \\
& + (\overline{Y}_{.j.m} - \overline{Y}_{.j..} - \overline{Y}_{...m} + \overline{Y}_{....}) \\
& + (\overline{Y}_{..km} - \overline{Y}_{..k.} - \overline{Y}_{...m} + \overline{Y}_{....}) \\
& + (\overline{Y}_{.jkm} + \overline{Y}_{.j..} + \overline{Y}_{..k.} + \overline{Y}_{...m} \\
& \quad - \overline{Y}_{.jk.} - \overline{Y}_{.j.m} - \overline{Y}_{..km} - \overline{Y}_{....})
\end{aligned}
$$

(5.37)

Only one of the terms in the right-hand side of Equation (5.37) is new. The expression $(\overline{Y}_{.jkm} + \overline{Y}_{.j..} + \overline{Y}_{..k.} + \overline{Y}_{...m} - \overline{Y}_{.jk.} - \overline{Y}_{.j.m} - \overline{Y}_{..km} - \overline{Y}_{....})$ is the least-squares estimate of $(\alpha\beta\gamma)_{jkm}$, the joint effect of the jth, kth, and mth levels of the variables A, B, and C adjusted for the main effects of these variables and for all first-order interaction effects of these variables. Equation (5.38) expresses this interpretation of the second-order, or three-variable, interaction effect more explicitly.

$$(\overline{Y}_{.jkm} + \overline{Y}_{.j..} + \overline{Y}_{..k.} + \overline{Y}_{...m} - \overline{Y}_{.jk.} - \overline{Y}_{.j.m} - \overline{Y}_{..km} - \overline{Y}_{....})$$

$$= (\overline{Y}_{.jkm} - \overline{Y}_{....}) - (\overline{Y}_{.j..} - \overline{Y}_{....}) - (\overline{Y}_{..k.} - \overline{Y}_{....}) - (\overline{Y}_{....m} - \overline{Y}_{....})$$

(5.38) $- (\overline{Y}_{.jk.} - \overline{Y}_{.j..} - \overline{Y}_{..k.} + \overline{Y}_{....}) - (\overline{Y}_{.j.m} - \overline{Y}_{.j..} - \overline{Y}_{...m} + \overline{Y}_{....})$

$$- (\overline{Y}_{..km} - \overline{Y}_{..k.} - \overline{Y}_{...m} + \overline{Y}_{....})$$

Squaring both sides of Equation (5.37) and summing over all indices, as was done in the early part of this chapter and in Chapter 4, results in Equation (5.39) (note that the cross-product terms have again vanished).

$$\sum_i^n \sum_j^a \sum_k^b \sum_m^c (Y_{ijkm} - \overline{Y}_{....})^2 = nbc \sum_j^a (\overline{Y}_{.j..} - \overline{Y}_{....})^2 + nac \sum_k^b (\overline{Y}_{..k.} - \overline{Y}_{....})^2$$

$$(SS_{\text{tot}}) \qquad\qquad\qquad (SS_A) \qquad\qquad\qquad (SS_B)$$

$$+ nab \sum_m^c (\overline{Y}_{...m} - \overline{Y}_{....})^2$$

$$(SS_C)$$

$$+ nc \sum_j^a \sum_k^b (\overline{Y}_{.jk.} - \overline{Y}_{.j..} - \overline{Y}_{..k.} + \overline{Y}_{....})^2$$

$$(SS_{AB})$$

(5.39)
$$+ nb \sum_j^a \sum_m^c (\overline{Y}_{.j.m} - \overline{Y}_{.j..} - \overline{Y}_{...m} + \overline{Y}_{....})^2$$

$$(SS_{AC})$$

$$+ na \sum_k^b \sum_m^c (\overline{Y}_{..km} - \overline{Y}_{..k.} - \overline{Y}_{...m} + \overline{Y}_{....})^2$$

$$(SS_{BC})$$

$$+ n \sum_j^a \sum_k^b \sum_m^c (\overline{Y}_{.jkm} + \overline{Y}_{.j..} + \overline{Y}_{..k.} + \overline{Y}_{...m}$$

$$- \overline{Y}_{.jk.} - \overline{Y}_{.j.m} - \overline{Y}_{..km} - \overline{Y}_{....})^2$$

$$(SS_{ABC})$$

$$+ \sum_i^n \sum_j^a \sum_k^b \sum_m^c (Y_{ijkm} - \overline{Y}_{.jkm})^2$$

$$(SS_{S/ABC})$$

We turn next to Table 5–14, which summarizes the analysis of variance for the three-dimensional case. The sources of variance follow immediately from Equation (5.39). The *df* are derived as before; for example, to obtain df_{ABC} we refer to Equation (5.38), which states that the second-order interaction effect is the cell effect (deviation of the cell mean from the grand mean) adjusted for first-order interaction effects and for main effects. This suggests

$$df_{ABC} = (abc - 1) - (a - 1)(b - 1) - (a - 1)(c - 1)$$

(5.40)
$$- (b - 1)(c - 1) - (a - 1) - (b - 1) - (c - 1)$$

$$= (a - 1)(b - 1)(c - 1)$$

Table 5-14 *Analysis of variance for a three-factor design*

SV	df	SS	EMS	F
A	$a - 1$	$\dfrac{\sum_j^a \left(\sum_i^n \sum_k^b \sum_m^c Y\right)^2}{nbc} - C$	$\sigma_e^2 + nbc\theta_A^2$	$\dfrac{MS_A}{MS_{S/ABC}}$
B	$b - 1$	$\dfrac{\sum_k^b \left(\sum_i^n \sum_j^a \sum_m^c Y\right)^2}{nac} - C$	$\sigma_e^2 + nac\theta_B^2$	$\dfrac{MS_B}{MS_{S/ABC}}$
C	$c - 1$	$\dfrac{\sum_m^c \left(\sum_i^n \sum_j^a \sum_k^b Y\right)^2}{nab} - C$	$\sigma_e^2 + nab\theta_C^2$	$\dfrac{MS_C}{MS_{S/ABC}}$
AB	$(a - 1)(b - 1)$	$\dfrac{\sum_j^a \sum_k^b \left(\sum_i^n \sum_m^c Y\right)^2}{nc} - C - SS_A - SS_B$	$\sigma_e^2 + nc\theta_{AB}^2$	$\dfrac{MS_{AB}}{MS_{S/ABC}}$
AC	$(a - 1)(c - 1)$	$\dfrac{\sum_j^a \sum_m^c \left(\sum_i^n \sum_k^b Y\right)^2}{nb} - C - SS_A - SS_C$	$\sigma_e^2 + nb\theta_{AC}^2$	$\dfrac{MS_{AC}}{MS_{S/ABC}}$
BC	$(b - 1)(c - 1)$	$\dfrac{\sum_k^b \sum_m^c \left(\sum_i^n \sum_j^a Y\right)^2}{na} - C - SS_B - SS_C$	$\sigma_e^2 + na\theta_{BC}^2$	$\dfrac{MS_{BC}}{MS_{S/ABC}}$
ABC	$(a - 1)(b - 1)(c - 1)$	$\dfrac{\sum_j^a \sum_k^b \sum_m^c \left(\sum_i^n Y\right)^2}{n} - C - SS_A - SS_B - SS_C - SS_{BC} - SS_{AB} - SS_{AC}$	$\sigma_e^2 + n\theta_{ABC}^2$	$\dfrac{MS_{ABC}}{MS_{S/ABC}}$
S/ABC	$abc(n - 1)$	$\sum_i^n \sum_j^a \sum_k^b \sum_m^c Y^2 - \dfrac{\sum_j^a \sum_k^b \sum_m^c \left(\sum_i^n Y\right)^2}{n}$	σ_e^2	

The various sums of squares can be obtained by appropriate substitution in, and expansion of, Equation (5.39). Since this is extremely laborious, it is preferable to utilize the isomorphism that is known to exist between *df* and squared quantities. For example, the $(a - 1)$ *df* for the A main effect suggests that the sum of squares expression consists of $a - 1$ squared quantities. We therefore have

$$\sum_{j}^{a} (\quad)^2 - C$$

where C is, as always, the squared sum of all scores, divided by the total number of scores; i.e.,

$$C = \frac{(\sum_{i}^{n} \sum_{j}^{a} \sum_{k}^{b} \sum_{m}^{c} Y_{ijkm})^2}{nabc}$$

As before, any summations not represented outside the parentheses must be represented within the parentheses. We now have

$$\sum_{j}^{a} \left(\sum_{i}^{n} \sum_{k}^{b} \sum_{m}^{c} Y_{ijkm} \right)^2 - C$$

Finally, we divide by the number of scores within the parentheses, yielding

$$SS_A = \frac{\sum_{j}^{a} (\sum_{i}^{n} \sum_{k}^{b} \sum_{m}^{c} Y_{ijkm})^2}{nbc} - C$$

The above operations may be stated verbally: all scores at a given level of A are summed, the sum is squared and divided by the number of scores that have been summed, the process is repeated at the next level of A, and the next, until all levels of A have been exhausted; the a quantities are then added, and the correction term is subtracted from this total. The student should become so familiar with the notational language developed in Chapter 2 and used repeatedly in Chapters 4 and 5 that the much more succinct statement embodied in the computational formula becomes as meaningful to him as the lengthy verbal statement just completed. The SS_B and SS_C are obtained in a similar manner.

We next consider the sum of squares for the first-order interactions, taking for an example SS_{AB}. Expanding the *df* for AB, we obtain $ab - a - b + 1$. This suggests

$$\sum_{j}^{a} \sum_{k}^{b} (\quad)^2 - \sum_{j}^{a} (\quad)^2 - \sum_{k}^{b} (\quad)^2 + C$$

Remembering that all summation signs must be represented in each term, we have

$$\sum_{j}^{a} \sum_{k}^{b} \left(\sum_{i}^{n} \sum_{m}^{c} Y \right)^2 - \sum_{j}^{a} \left(\sum_{i}^{n} \sum_{k}^{b} \sum_{m}^{c} Y \right)^2 - \sum_{k}^{b} \left(\sum_{i}^{n} \sum_{j}^{a} \sum_{m}^{c} Y \right)^2 + C$$

The third step is division by the number of scores within each set of parentheses. We then have

(5.41)

$$SS_{AB} = \frac{\sum_j^a \sum_k^b (\sum_i^n \sum_m^c Y)^2}{nc} - \frac{\sum_j^a (\sum_i^n \sum_k^b \sum_m^c Y)^2}{nbc} - \frac{\sum_k^b (\sum_i^n \sum_j^a \sum_m^c Y)^2}{nac} + C$$

By subtracting and adding the correction term, C, the equation is fundamentally unchanged, but we may now write

(5.42)
$$SS_{AB} = \left[\frac{\sum_j \sum_k (\sum_i \sum_m Y)^2}{nc} - C \right]$$
$$- \left[\frac{\sum_j (\sum_i \sum_k \sum_m Y)^2}{nbc} - C \right] - \left[\frac{\sum_k (\sum_i \sum_j \sum_m Y)^2}{nac} - C \right]$$

which is basically the form of the entry in Table 5–14. This form of the expression clearly demonstrates that the SS_{AB} is the variability among the means for the treatment combinations defined by A and B, adjusted for variability due to A and B effects. It is helpful to be able to think in terms of both Equation (5.41) and Equation (5.42). The approach underlying Equation (5.42) requires an understanding of the interpretation of interaction as a residual variability among cell means; that underlying Equation (5.41) is very general and readily provides computational formulas for even the most complex of designs.

The student who is still not comfortable with the notational language developed in Chapter 2 would do well to ponder the verbal translation of some of the expressions presented in this section. For example,

$$\frac{\sum_j^a \sum_k^b (\sum_i^n \sum_m^c Y)^2}{nc}$$

implies that we consider a single combination of levels of A and B (i.e., hold j and k constant), sum all scores at this combination of levels, square this sum, divide by the number of scores that have been summed, repeat this same process for all ab combinations of levels of A and B, and then sum the ab quantities that have been computed.

An expression for SS_{ABC} is no more difficult to arrive at than those previously considered; it is merely longer. Expanding the df results in $abc + a + b + c - ab - ac - bc - 1$, which suggests

$$\sum_j^a \sum_k^b \sum_m^c (\quad)^2 + \sum_j^a (\quad)^2 + \sum_k^b (\quad)^2 + \sum_m^c (\quad)^2$$
$$- \sum_j^a \sum_k^b (\quad)^2 - \sum_j^a \sum_m^c (\quad)^2 - \sum_k^b \sum_m^c (\quad)^2 - C$$

Introducing the remaining summations within the parentheses, we obtain

$$\sum_{j}^{a}\sum_{k}^{b}\sum_{m}^{c}\left(\sum_{i}^{n}Y\right)^{2} + \sum_{j}^{a}\left(\sum_{i}^{n}\sum_{k}^{b}\sum_{m}^{c}Y\right)^{2} + \sum_{k}^{b}\left(\sum_{i}^{n}\sum_{j}^{a}\sum_{m}^{c}Y\right)^{2} + \sum_{m}^{c}\left(\sum_{i}^{n}\sum_{j}^{a}\sum_{k}^{b}Y\right)^{2}$$

$$- \sum_{j}^{a}\sum_{k}^{b}\left(\sum_{i}^{n}\sum_{m}^{c}Y\right)^{2} - \sum_{j}^{a}\sum_{m}^{c}\left(\sum_{i}^{n}\sum_{k}^{b}Y\right)^{2} - \sum_{k}^{b}\sum_{m}^{c}\left(\sum_{i}^{n}\sum_{j}^{a}Y\right)^{2} - C$$

Finally, we divide by the number of scores within each set of parentheses and obtain

$$SS_{ABC} = \frac{\sum_{j}\sum_{k}\sum_{m}(\sum_{i}Y)^{2}}{n} + \frac{\sum_{j}(\sum_{i}\sum_{k}\sum_{m}Y)^{2}}{nbc} + \frac{\sum_{k}(\sum_{i}\sum_{j}\sum_{m}Y)^{2}}{nac}$$

(5.43)
$$+ \frac{\sum_{m}(\sum_{i}\sum_{j}\sum_{k}Y)^{2}}{nab} - \frac{\sum_{j}\sum_{k}(\sum_{i}\sum_{m}Y)^{2}}{nc}$$

$$- \frac{\sum_{j}\sum_{m}(\sum_{i}\sum_{k}Y)^{2}}{nb} - \frac{\sum_{k}\sum_{m}(\sum_{i}\sum_{j}Y)^{2}}{na} - C$$

The right-hand side of Equation (5.43) is algebraically identical to the expression for SS_{ABC} in Table 5–12. That expression is of the form of Equation (5.40) for degrees of freedom, and follows from the definition of interaction effect provided by Equation (5.38); i.e., the second-order interaction effect is the cell effect adjusted for the first-order interaction and main effects contributing to the cell effect. If all main and first-order effects have been previously computed, the only new term to calculate is

$$\frac{\sum_{j}\sum_{k}\sum_{m}(\sum_{i}Y)^{2}}{n}$$

regardless of which expression for SS_{ABC} is used.

The $SS_{S/ABC}$ may again be computed as a residual term obtained by subtracting all main and interaction terms from the total. Alternatively, we note that $SS_{S/ABC}$ is the sum of the variability within the cells defined by the three independent variables. Consequently, it is the difference between the total variability and the variability among cell means. Therefore,

$$SS_{S/ABC} = SS_{tot} - SS_{\overline{ABC}}$$

(5.44)
$$= \left(\sum_{i}\sum_{j}\sum_{k}\sum_{m}Y^{2} - C\right) - \left[\frac{\sum_{j}\sum_{k}\sum_{m}(\sum_{i}Y)^{2}}{n} - C\right]$$

$$= \sum_{i}\sum_{j}\sum_{k}\sum_{m}Y^{2} - \frac{\sum_{j}\sum_{k}\sum_{m}(\sum_{i}Y)^{2}}{n}$$

which is the expression in Table 5–14 and which is clearly on $nabc - abc = abc(n - 1)$ degrees of freedom.

The mean square column has been omitted since it has been obtained, as previously, by dividing the entries in the *SS* column by the corresponding *df*. The entries in the *EMS* column are similar in form to those previously presented. With the exception of the entry for S/ABC, they consist of two components, σ_e^2 and some term involving a θ^2 quantity. The subscript of the θ^2 term is always the entry in the *SV* column for that line of the analysis of variance table. The multiplier of the θ^2 term consists of the product of all dimensions of the design not included in the subscript. Thus, θ_{BC}^2 does not include S or A in the subscripts and is therefore multiplied by the levels of these, n and a. The definitions of the θ^2 term are similar to those encountered previously:

$$\theta_A^2 = \frac{\sum_j (\mu_j - \mu)^2}{a - 1}$$

$$\theta_B^2 = \frac{\sum_k (\mu_k - \mu)^2}{b - 1}$$

$$\theta_C^2 = \frac{\sum_m (\mu_m - \mu)^2}{c - 1}$$

$$\theta_{AB}^2 = \frac{\sum_j \sum_k (\mu_{jk} - \mu_j - \mu_k + \mu)^2}{(a - 1)(b - 1)}$$

(5.45)

$$\theta_{AC}^2 = \frac{\sum_j \sum_m (\mu_{jm} - \mu_j - \mu_m + \mu)^2}{(a - 1)(c - 1)}$$

$$\theta_{BC}^2 = \frac{\sum_k \sum_m (\mu_{km} - \mu_k - \mu_m + \mu)^2}{(b - 1)(c - 1)}$$

$$\theta_{ABC}^2 = \frac{\sum_j \sum_k \sum_m (\mu_{jkm} + \mu_j + \mu_k + \mu_m - \mu_{jk} - \mu_{jm} - \mu_{km} - \mu)^2}{(a - 1)(b - 1)(c - 1)}$$

Tests of null hypotheses about the various θ^2s are readily made by dividing the corresponding mean squares by $MS_{S/ABC}$. Since the *EMS* for this term differs from the expectations for all other terms only by the null hypothesis component, $MS_{S/ABC}$ is clearly the appropriate error term. We again note that this situation holds only when the levels of the independent variables have been arbitrarily selected; when one or more of the independent variables are random (in the sense defined in Chapter 1), different expectations will be encountered and, consequently, error terms other than the within-cells mean square may be required.

5.6 INTERACTION EFFECTS IN THE THREE-FACTOR DESIGN

The interpretation of first-order interactions follows that of the previous sections in which we dealt with two-factor designs. A significant *AB* interaction again indicates that the differences among the means of the populations defined by the levels of A change as a function of the level of B (or equivalently, that the differences among the means for the levels of B change as a function of the level of A). For example, assume that the entries in the $2 \times 2 \times 2$ (2^3) design of Table 5–15 are treatment population means. To investigate the AB interaction, we average over the levels of C, obtaining

$$\begin{array}{cc} & B_1 \quad B_2 \\ \begin{array}{c} A_1 \\ A_2 \end{array} & \begin{bmatrix} 12 & 6 \\ 8 & 18 \end{bmatrix} \end{array}$$

An AB interaction is clearly present, since

$$(\bar{Y}_{.11.} - \bar{Y}_{.21.}) - (\bar{Y}_{.12.} - \bar{Y}_{.22.}) = (12 - 8) - (6 - 18) = 16$$

rather than zero, as would be the case if all AB interaction effects were absent. The AC and BC interactions also contribute to the variability among cell means; the relevant matrices are

$$\begin{array}{cc} & C_1 \quad C_2 \\ \begin{array}{c} A_1 \\ A_2 \end{array} & \begin{bmatrix} 11 & 7 \\ 9 & 17 \end{bmatrix} \end{array} \quad \text{and} \quad \begin{array}{cc} & C_1 \quad C_2 \\ \begin{array}{c} B_1 \\ B_2 \end{array} & \begin{bmatrix} 10 & 10 \\ 10 & 14 \end{bmatrix} \end{array}$$

A first-order interaction is significant when the differences among simple effects of one variable change significantly over the levels of the second variable. Similarly, a second-order interaction effect is significant when the simple interaction effects of two variables change as a function of the level of the third variable. This is easily investigated for the 2^3 design. For the data of Table 5–15, the measure of the simple AB interaction at C_1 is $(15 - 5) - (7 - 13) = 16$ and at C_2, $(9 - 11) - (5 - 23) = 16$. According to the definition just given, the ABC interaction does not contribute to the variability among cell means, since the measure of the AB interaction is the same at both levels of the third variable, C. It is important to realize that the same conclusion follows if the AC interaction is considered at each level of B, or the BC interaction at each level of A. These are all equivalent approaches, yielding identical information.

The approach just exemplified may be extended to any three-factor

TABLE 5–15 *A set of treatment population means*

	C_1		C_2	
	B_1	B_2	B_1	B_2
A_1	15	7	9	5
A_2	5	13	11	23

design, regardless of the number of treatment levels involved. Assuming that no error variance exists, if there is no second-order interaction

(5.46) $[(\overline{Y}_{.jkm} - \overline{Y}_{.j'km}) - (\overline{Y}_{.jk'm} - \overline{Y}_{.j'k'm})]$

$$- [(\overline{Y}_{.jkm'} - \overline{Y}_{.j'km'}) - (\overline{Y}_{.jk'm'} - \overline{Y}_{.j'k'm'})] = 0$$

for all values of j, j', k, k', m, and m'. Suppose a third level of A is added to the data of Table 5–15, yielding

	C_1		C_2	
	B_1	B_2	B_1	B_2
A_1	15	7	9	5
A_2	5	13	11	23
A_3	8	3	10	9

Applying Equation (5.46), when $j = 1$, $j' = 2$, $k = 1$, $k' = 2$, $m = 1$, and $m' = 2$,

$$[(15 - 5) - (7 - 13)] - [(9 - 11) - (5 - 23)] = 16 - 16 = 0$$

and when $j = 1$, $j' = 3$, $k = 1$, $k' = 2$, $m = 1$, and $m' = 2$,

$$[(15 - 8) - (7 - 3)] - [(9 - 10) - (5 - 9)] = 3 - 3 = 0$$

and when $j = 2$, $j' = 3$, $k = 1$, $k' = 2$, $m = 1$, and $m' = 2$,

$$[(5 - 8) - (13 - 3)] - [(11 - 10) - (23 - 9)] = -13 - (-13) = 0$$

One can again conclude that there is no ABC interaction.

An alternative method of investigating the ABC interaction is suggested by Tables 5–3, 5–4, and 5–5, which are the result of a direct application of the analysis of variance model to first-order interaction effects. For second-order effects, Equation (5.36) could be applied, removing all main and first-order interaction effects from the three-dimensional matrix of means. If the ABC interaction is not significant, the variance among the adjusted cell means would be insignificant.

To further promote understanding of interaction effects in three-factor designs, we next consider graphs of several sets of means for various combinations of effects. Assuming that no error variance is present, the data of Figure 5–5 clearly reflect the presence of A, B, and C main effects. Averaging over the combinations of B and C, we have

A_1	A_2	A_3
25	20	18

averaging over the combinations of A and C, we have

B_1	B_2	B_3
27	17	19

and averaging over the combinations of A and B, we have

$$
\begin{array}{ccc}
C_1 & C_2 & C_3 \\
[16 & 24 & 23]
\end{array}
$$

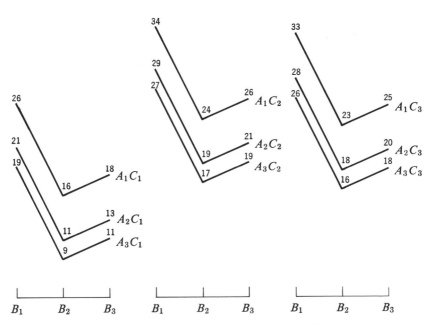

FIGURE 5–5 *Data from a 3^3 design with only main effects present*

We next note that at each level of C the three curves are parallel; there are no simple AB interaction effects at any of the levels of C. This absence of simple interaction effects is a *sufficient* condition for the absence of an overall AB interaction, since the overall effect is merely an average of the simple effects at the different C levels and the average of a set of zeros is just zero. The absence of simple effects is also a sufficient condition for the conclusion that there is no ABC interaction. The existence of the ABC interaction requires that the simple interaction effects of two variables change as a function of the level of the third variable. This condition is not met, since the AB interaction effects are zero at all levels of C. Using other sets of data, we will soon illustrate the fact that the absence of simple interaction effects is not a *necessary* condition for overall zero AB or ABC interactions.

The status of the AC and BC interactions is not so clear as that of the AB. Upon shifting the curves in Figure 5–5 so that the three curves in each set are similar with respect to the level of A, it becomes clearer that no BC interaction exists. Alternatively, we can average over the levels of A to obtain the matrix

$$
\begin{array}{c c c c}
 & C_1 & C_2 & C_3 \\
B_1 & \begin{bmatrix} 22 & 30 & 29 \\ 12 & 20 & 19 \\ 14 & 22 & 21 \end{bmatrix}
\end{array}
$$

A plot of this set of means will result in three parallel lines. This is also true of

$$
\begin{array}{c c c c}
 & A_1 & A_2 & A_3 \\
C_1 & \begin{bmatrix} 20 & 15 & 13 \\ 28 & 23 & 21 \\ 27 & 22 & 20 \end{bmatrix}
\end{array}
$$

the matrix of means for the AC combinations.

Two plots of another data set are presented in the upper and lower halves of Figure 5–6. It is again assumed that the means are populations means; that there is no error variance in the data set. Considering first the upper half of the figure, we find that an A main effect exists (the means are lower under the A_3 conditions than under the A_1 and A_2 conditions at all combinations of levels of B and C); that a B main effect exists (the B_1 means are highest, B_2 next, and B_3 lowest at combinations of A and C levels); and that a C main effect exists (the means generally increase as the level of C does). The status of interaction effects is more difficult to determine than it was for the data of Figure 5–5. The added complication is the presence of simple interaction effects; there is definitely an AB interaction at each level of C. The replot of the data in the bottom half of Figure 5–6 provides some illumination in this case. Shifting curves so that each set of three is at a single level of A, we find that there are no simple BC interaction effects. At each level of A, the spread among the curves does not change as a function of the level of B. As in our discussion of the data of Figure 5–5, we may draw two conclusions from the absence of BC effects at the levels of A: (a) there is no overall BC interaction, and (b) there is no overall ABC interaction. A plot of means for the BC combinations, averaging over levels of A, will further verify (a), and the truth of (b) may be tested by investigating the validity of Equation (5.46) for the data of Figure 5–6. The appropriate data plots will also verify that AC and AB interactions do exist.

It has been seen that the absence of simple first-order interaction effects means that the overall first-order interaction, as well as the second-order interaction, cannot be a source of variance. The next question is whether it is possible to have simple interaction effects, but no overall interaction, and what the implications of such results are for the second-order interaction. The relevant data are plotted in Figure 5–7. At each level of C, AB interaction effects are present. Alternative plots of the data will verify that simple AC and BC interaction effects also exist. However, when we table the means for

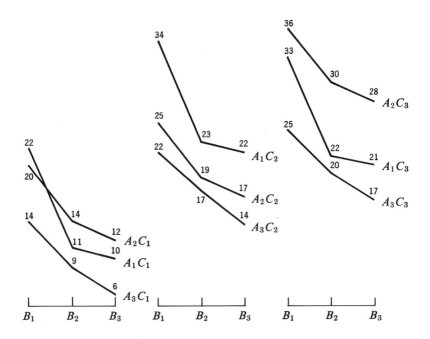

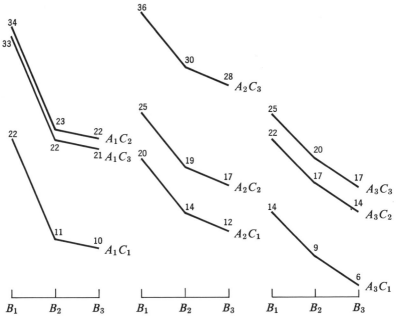

FIGURE 5–6 *Two plots of a set of data with A, B, C, AB, and AC effects present*

the AB combinations, first averaging over levels of C, we obtain

$$
\begin{array}{c c c c}
 & B_1 & B_2 & B_3 \\
A_1 & \begin{bmatrix} 18 \\ 18 \\ 24 \end{bmatrix} & \begin{matrix} 15 \\ 15 \\ 21 \end{matrix} & \begin{matrix} 12 \\ 12 \\ 18 \end{matrix} \\
A_2 & & & \\
A_3 & & &
\end{array}
$$

which, when plotted, results in a set of parallel curves. Similarly, we obtain

$$
\begin{array}{c c c c}
 & C_1 & C_2 & C_3 \\
A_1 & \begin{bmatrix} 15 \\ 15 \\ 21 \end{bmatrix} & \begin{matrix} 15 \\ 15 \\ 21 \end{matrix} & \begin{matrix} 15 \\ 15 \\ 21 \end{matrix} \\
A_2 & & & \\
A_3 & & &
\end{array}
$$

demonstrating the absence of an AC interaction. A third tabulation demonstrates that no BC interaction exists:

$$
\begin{array}{c c c c}
 & B_1 & B_2 & B_3 \\
C_1 & \begin{bmatrix} 20 \\ 20 \\ 20 \end{bmatrix} & \begin{matrix} 17 \\ 17 \\ 17 \end{matrix} & \begin{matrix} 14 \\ 14 \\ 14 \end{matrix} \\
C_2 & & & \\
C_3 & & &
\end{array}
$$

Thus it is possible that simple two-variable interaction effects may exist at each level of the third variable, but vary in such a way over levels of the third variable as to cancel each other out. The result is that there is no overall first-order interaction. An additional consequence of this situation is that a second-order, or three-variable, interaction must exist since, by definition, if

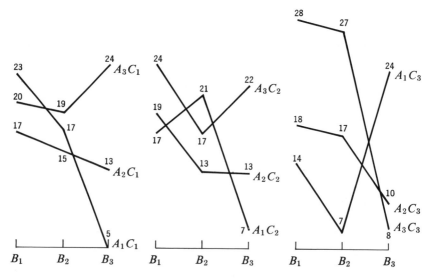

FIGURE 5–7 *Data from a 3^3 design with A, B, and ABC effects present*

the interaction effects of two variables change over the levels of the third variable, there is a second-order interaction. Independent verification of the hypothesis of an *ABC* interaction may be obtained by applying Equation (5.46) to the data of Figure 5–7. For example, let

$$j = 1 \qquad j' = 3$$
$$k = 1 \qquad k' = 2$$
$$m = 1 \qquad m' = 3$$

Then,

$$[(\overline{Y}_{jkm} - \overline{Y}_{j'km}) - (\overline{Y}_{jk'm} - \overline{Y}_{j'k'm})] - [(\overline{Y}_{jkm'} - \overline{Y}_{j'km'}) - (\overline{Y}_{jk'm'} - \overline{Y}_{j'k'm'})]$$
$$= [(23 - 20) - (17 - 19)] - [(14 - 28) - (7 - 27)]$$
$$= 5 - 6 = -1$$

Since the above result is not equal to zero, the *ABC* interaction contributes to the variability among cell means.

Still assuming errorless data (or equivalently, that we are dealing with population means), the preceding results may be summarized in a very general way. Let *X* be any source of variance (e.g., *A*, *AB*, *ABC*) and *Y* be some variable not involved in the effect. Then it is possible to speak of the simple *X* effects at the levels of *Y*; the overall *X* effects obtained by averaging over the levels of *Y*; and the interaction of *X* and *Y*:

(a) If the *X* effects are zero at all levels of *Y*, the overall *X* effects will be zero. Thus, if there is no effect due to *A* at any level of *B*, there will be no *A* main effect; if there are no *AB* interactions at any level of *C*, there will be no overall *AB* effects.

(b) If the *X* effects at all levels of *Y* are zero, the overall interaction of *X* and *Y* will be zero. Thus, if the simple effects due to *A* are zero at all levels of *B*, the overall *AB* effects will be zero; if the effects due to *AB* are zero at all levels of *C*, the overall effects due to *ABC* will be zero.

(c) If the overall effects due to *X* are zero but some of the simple effects of *X* at the levels of *Y* are not zero, effects due to the interaction *XY* will not be zero. Thus, if the main effects due to *A* are zero but simple effects at various levels of *B* are not, the *AB* interaction effects will not be zero; if the overall *AB* interaction effects are zero and if the simple *AB* effects at levels of *C* are not, the *ABC* interaction effects will be non-zero.

(d) If the interaction effects of *X* and *Y* are zero and if the overall effects of *X* are zero, then the simple effects of *X* at each level of *Y* will be zero. Thus, if the overall *A* main effects and the overall *AB* interaction effects are zero, the simple effects of *A* at each level of *B* will be zero; If the overall *AB* and *ABC* effects are zero, the *AB* simple interaction effects at each level of *C* will all be zero.

The preceding discussion has dealt solely with relationships among population means. With actual data, apparent violations of our conclusions will occasionally occur. For example, overall tests of A and AB effects may yield nonsignificant results, but tests of simple effects of A may prove significant at one or more levels of B. This set of statistical outcomes is inconsistent with statement (d), which is a true statement about population means. Such inconsistencies between the results of the analysis and our conclusions about relationships in the population serve as a warning to view our inferences with more than the usual skepticism. In the present example, either a Type II error has occurred in testing the overall effects or a Type I error has occurred in testing the simple effects.

5.7 A NUMERICAL EXAMPLE FOR THE THREE-FACTOR DESIGN

Assume that 120 subjects are each given 45 trials on a discrimination problem. The subjects are equally divided with respect to age (A: 6, 8, 10 years), amount of reward (R: 1¢, 5¢), and delay of reward (D: 0 seconds, 10 seconds). The data have been summarized in Table 5–16 in a manner most convenient for the analysis of variance. The indices of notation are to be interpreted as follows:

i indexes the scores in each cell: $i = 1, 2, \ldots, 10$
j indexes the age levels: $j = 1$ (6 years), 2 (8 years), 3 (10 years)
k indexes the reward levels: $k = 1$ (1¢), 2 (5¢)
m indexes the delay levels: $m = 1$ (0 seconds), 2 (10 seconds).

We have the sum of scores and squared scores for each combination of A, R, and D ($\sum_i Y$, $\sum_i Y^2$); the sum of scores for each combination of R and D ($\sum_i \sum_j Y$), for each combination of A and D ($\sum_i \sum_k Y$), for each combination of A and R ($\sum_i \sum_m Y$), for each level of A ($\sum_i \sum_k \sum_m Y$), for each level of R ($\sum_i \sum_j \sum_m Y$), for each level of D ($\sum_i \sum_j \sum_k Y$); and the sum of all scores and of all squared scores ($\sum_i \sum_j \sum_k \sum_m Y$; $\sum_i \sum_j \sum_k \sum_m Y^2$). With these quantities calculated, it is necessary only to substitute in the formulas of Table 5–14 to complete the analysis of variance. These totals will also be found useful in subsequent tests, which will be described in Chapters 13 and 14, and in tests of simple effects.

The correction term, C, is again the sum of all scores, squared and divided by the total number of scores.

$$C = \frac{(\sum_i \sum_j \sum_k \sum_m Y)^2}{120}$$

$$= \frac{(3,578)^2}{120}$$

$$= 106,684.03$$

The SS_{tot} is again

$$SS_{tot} = \sum_i \sum_j \sum_k \sum_m Y^2 - C$$

$$= 114{,}030 - 106{,}684.03$$

$$= 7{,}345.97$$

TABLE 5–16 *Data from a three-factor experiment*

			6 Years	8 Years	10 Years	
0 sec.	1¢	$\sum_i Y =$	258	329	368	$\sum_j \sum_i Y = 955$
		$\sum_i Y^2 =$	6,864	11,209	13,704	
	5¢	$\sum_i Y =$	262	351	383	$\sum_j \sum_i Y = 996$
		$\sum_i Y^2 =$	7,306	12,585	14,789	
		$\sum_k \sum_i Y =$	520	680	751	$\sum_k \sum_j \sum_i Y = 1{,}951$
		$\sum_k \sum_i Y^2 =$	14,166	23,794	28,493	
10 sec.	1¢	$\sum_i Y =$	217	238	308	$\sum_j \sum_i Y = 763$
		$\sum_i Y^2 =$	4,887	5,928	9,696	
	5¢	$\sum_i Y =$	220	263	381	$\sum_j \sum_i Y = 864$
		$\sum_i Y^2 =$	5,220	7,221	14,625	
		$\sum_k \sum_i Y =$	437	501	689	$\sum_k \sum_j \sum_i Y = 1{,}627$
		$\sum_k \sum_i Y^2 =$	10,107	13,149	24,321	
		$\sum_m \sum_k \sum_i Y =$	957	1,181	1,440	$\sum_m \sum_k \sum_j \sum_i Y = 3{,}578$
	1¢	$\sum_m \sum_i Y =$	675	567	676	$\sum_m \sum_j \sum_i Y = 1{,}718$
	5¢	$\sum_m \sum_i Y =$	482	614	764	$\sum_m \sum_j \sum_i Y = 1{,}860$
		$\sum_m \sum_k \sum_i Y =$	957	1,181	1,440	$\sum_m \sum_k \sum_j \sum_i Y = 3{,}578$
						$\sum_m \sum_k \sum_j \sum_i Y^2 = 114{,}030$

To obtain the sum of squares for *any* main effect:

    (a) square the sum for each level of the variable,
    (b) sum the squared quantities,
    (c) divide by the number of scores at each level,
    (d) subtract C.

Thus,

$$SS_A = \frac{\sum_j (\sum_i \sum_k \sum_m Y)^2}{40} - C$$

$$= \frac{(957)^2 + (1,181)^2 + (1,440)^2}{40} - \frac{(3,578)^2}{120}$$

$$= 109,605.25 - 106,684.03$$

$$= 2,921.22$$

$$SS_R = \frac{\sum_k (\sum_i \sum_j \sum_m Y)^2}{60} - C$$

$$= \frac{(1,718)^2 + (1,860)^2}{60} - \frac{(3,578)^2}{120}$$

$$= 106,852.06 - 106,684.03$$

$$= 168.03$$

$$SS_D = \frac{\sum_m (\sum_i \sum_j \sum_k Y)^2}{40} - C$$

$$= \frac{(1,951)^2 + (1,627)^2}{60} - \frac{(3,578)^2}{120}$$

$$= 107,558.83 - 106,684.03$$

$$= 874.80$$

To obtain the sums of squares for *any* interaction:

    (a) square the sum for each combination of levels of the variables involved in the interaction,
    (b) sum the squared quantities,
    (c) divide by the number of scores in each combination of levels,
    (d) subtract C,
    (e) subtract the sums of squares for all effects contributing to the variability among the means for the treatment combinations involved in the interaction; e.g., to obtain the ARD interaction sum of squares, we remove the sums of squares for A, R, D, AR, AD, and DR.

Thus,

$$SS_{AR} = \frac{\sum_j \sum_k (\sum_i \sum_m Y)^2}{20} - C - SS_A - SS_R$$

$$= \frac{(675)^2 + \cdots + (764)^2}{20} - \frac{(3,578)^2}{120} - 2,921.22 - 168.03$$

$$= 109,855.30 - 106,684.03 - 2,921.22 - 168.03$$

$$= 82.02$$

$$SS_{AD} = \frac{\sum_j \sum_m (\sum_i \sum_k Y)^2}{20} - C - SS_A - SS_D$$

$$= \frac{(520)^2 + \cdots + (689)^2}{20} - \frac{(3,578)^2}{120} - 2,921.22 - 874.80$$

$$= 110,674.60 - 106,684.03 - 2,921.22 - 874.80$$

$$= 194.55$$

$$SS_{RD} = \frac{\sum_k \sum_m (\sum_i \sum_j Y)^2}{30} - C - SS_R - SS_D$$

$$= \frac{(955)^2 + \cdots + (864)^2}{30} - \frac{(3,578)^2}{120} - 168.03 - 874.80$$

$$= 107,756.86 - 106,684.03 - 168.03 - 874.80$$

$$= 30.00$$

$$SS_{ARD} = \frac{\sum_j \sum_k \sum_m (\sum_i Y)^2}{10} - C - SS_A - SS_R - SS_D - SS_{AR} - SS_{AD}$$

$$- SS_{RD}$$

$$= \frac{(258)^2 + \cdots + (381)^2}{10} - \frac{(3,578)^2}{120} - 2,921.22 - 168.03$$

$$- 874.80 - 82.02 - 194.55 - 30.00$$

$$= 111,009.00 - 106,684.03 - 2,921.22 - 168.03 - 874.80 - 82.02$$

$$- 194.55 - 30.00$$

$$= 54.35$$

The sum of squares for *subjects-within-ARD treatment combinations* may be most simply computed as

$$SS_{S/ARD} = SS_{tot} - SS_{\overline{ARD}}$$

$$= 7,345.97 - (111,009.00 - 106,684.03)$$

$$= 3,021.00$$

The results of the complete analysis of variance are summarized in Table 5–17. It is clear that performance improves with age and deteriorates

TABLE 5–17 *Analysis of variance for data from a three-factor experiment*

SV	df	SS	MS	F
A	2	2,921.22	1,460.61	52.33*
R	1	168.03	168.03	6.02
D	1	874.80	874.80	31.34*
AR	2	82.02	41.01	1.47
AD	2	194.55	97.28	3.49
RD	1	30.00	30.00	1.07
ARD	2	54.35	27.18	.97
S/ARD	108	3,021.00	27.91	
				*p < .01

when reinforcement is delayed. The main effect of reward is less clearly established, but the *F* ratio is almost significant at the 0.1 level, and the effects are in the direction that theory and experimental data would suggest (i.e., performance is better under the higher reward). It appears reasonable to conclude that the levels of reward used in this study do differ in their effects. The failure to obtain any significant interactions suggests that the effects of any one variable do not change markedly over levels of the other two variables.

5.8 MORE THAN THREE INDEPENDENT VARIABLES

The analysis and interpretation of data for completely randomized designs involving four or more independent variables are in all respects straightforward generalizations of the material previously presented in this chapter. Each variable and each possible combination of variables is a potential contributor to the total variability, as is the variability among subjects within each combination of variables.

The *df* for main effects are again the number of levels of the variable minus 1; those for an interaction are the product of the *df* for the variables entering into the interaction. For example, assume a design involving six independent variables labeled $A, B, \ldots, F$. Then the *df* for the *ABDE* interaction are $(a - 1)(b - 1)(d - 1)(e - 1)$, where $a, b, d,$ and e are the numbers of levels of $A, B, D,$ and E, respectively, following our previous notational usage.

The rules for computing *SS* for main (p. 140) and for interaction (p. 141) effects apply in all designs. Thus, to obtain the *SS* for the *ABDE* interaction:

(a) square the sum of scores for each of the *abde* combinations of levels of the four variables,
(b) sum the *abde* squared quantities,
(c) divide by the number of scores obtained in each of the *abde* combinations,
(d) subtract C,
(e) subtract the sums of squares for all main and interaction effects imbedded in the *ABDA* combination.

The above steps may be summarized by

$$(5.47) \qquad SS_{ABDE} = \frac{\overset{a}{\sum}\overset{b}{\sum}\overset{d}{\sum}\overset{e}{\sum}(\overset{n}{\sum}\overset{c}{\sum}\overset{f}{\sum} Y)^2}{ncf} - C - SS_A - SS_B$$

$$- SS_D - SS_E - SS_{AB} - SS_{AD} - SS_{AE} - SS_{BD}$$

$$- SS_{BE} - SS_{DE} - SS_{ABD} - SS_{ADE} - SS_{BDE}$$

Note that if each term in Equation (5.47) is replaced by the corresponding *df*, the result is

$$(abde - 1) - (a - 1) - (b - 1) - \cdots - (b - 1)(d - 1)(e - 1)$$

$$= (a - 1)(b - 1)(d - 1)(e - 1)$$

The computations illustrated above are applicable when all cells contain equal numbers of subjects. Alternative analyses have been previously presented in the two-factor case for unequal but proportionate *n*s and for disproportionate *n*s. These analyses are easily extended to data for designs involving more than two factors, and the details are therefore omitted.

In all designs, regardless of the number of variables, mean squares are ratios of sums of squares to degrees of freedom. Still assuming completely randomized designs in which the levels of all variables have been arbitrarily chosen, the *EMS* will consist of σ_e^2 plus some quantity, θ^2 with a subscript that denotes the source of variance being referred to and is multiplied by n (where n is the number of subjects in each cell) and by the numbers of levels of those variables in the design which are not in the subscript. For example, the *EMS* for the *ABDE* interaction in the six-variable experiment referred to before is $\sigma_e^2 + cfn\theta_{ABDE}^2$, where n is the number of subjects in each of the *abcdef* cells. It follows from this discussion that the mean square for subjects within cells will be the error term (i.e., the denominator) for all F tests to be made in completely randomized designs. The assumptions underlying the F test for multi-factor designs parallel those discussed previously for the two- and three-factor designs. Specifically, we assume that the error components of the scores are independently and normally distributed, with mean of zero and variance σ_e^2 within each population defined by the total number of treatment combinations.

5.9 POOLING IN MULTIFACTOR DESIGNS

Experiments frequently involve sources of variance which, while necessary to the design, are really of no interest to the investigator. For example, an experiment might be designed to study the effects of several drugs (D) and several drug concentrations (C) upon maze learning performance in rats. Position (P) of the food reinforcement is carefully balanced, the reward appearing in the right arm for half of the subjects in each CD combination and in the left arm for the other half of the subjects. Because running the experiment is time-consuming, two experimenters (E) share the work; each one runs exactly half of the subjects undergoing each CDP combination. We now have a four-factor design; two of the factors, C and D, are of interest while P has been introduced as a control for position preferences and E has been introduced for practical reasons. We have three major options in analyzing this design. One possibility is to perform a complete analysis, testing separately each of the four main effects and all possible interactions. Excluding the within-cell error term, there are 15 such terms. There are three objections to this procedure: (1) it involves computational labor, which our interests do not require; (2) we are performing 12 F tests that are of no interest (assuming that only the C, D, and CD effects are of interest) and are greatly increasing the probability of obtaining significance by chance, thereby making it more difficult to interpret the three F tests that are of interest; and (3) we are cluttering the presentation of the analysis of variance results with 12 added terms, making it more difficult to pick out those terms that are of interest. A second possible analysis, which avoids the difficulties cited, is to compute SS_C, SS_D, SS_{CD}, and a residual error sum of squares; the residual is equal to the difference between SS_{total} and the first three terms. This is frequently done but it is a very dangerous procedure. If the null hypothesis is false for any of the 12 terms whose sums of squares are added to $SS_{S/CDPE}$, then the error mean square estimates σ_e^2 *and variability due to treatment effects as well*. In testing C, D, or CD, the ratio of expected mean squares will be less than 1 if H_0 is true; that is, the F test will be negatively biased. This point will be discussed at greater length in Chapter 11.

There is a third procedure that avoids the pitfalls of the first two. In the present example, it entails calculating sums of squares for C, D, CD, $S/CDPE$, and one for the pool of all the 12 terms involving P and E. This combined, or pooled, sum of squares may be readily obtained by subtraction as may its associated df. Thus

$$df_{P,E \text{ pool}} = (cdpen - 1) - [(c - 1) + (d - 1) + (c - 1)(d - 1)$$
$$+ cdpe(n - 1)]$$
$$= cd(pe - 1)$$

This result suggests a computational check. As we have pointed out before, calculations of sums of squares follow from knowledge of the df. Thus $SS_{P,E \text{ pool}}$ may be calculated directly as

$$\sum^{c}\sum^{d}\sum^{p}\sum^{e}\left(\sum^{n} Y\right)^2/n - \sum^{c}\sum^{d}\left(\sum^{p}\sum^{e}\sum^{n} Y\right)^2/pen$$

The approach taken here is quite flexible. We might be interested in the P and E main effects but not in their interactions. These would then be calculated separately and the pool based on only 10 terms.

5.10 COMPUTATIONS FOR SINGLE df EFFECTS

Consider an experimental design in which there are two levels of the variables, A and B. We then have four treatment combinations: A_1B_1, A_1B_2, A_2B_1, and A_2B_2. Designate the sum of n scores in each cell as T_{11}, T_{12}, T_{21}, and T_{22}, respectively. The usual computational formula for the SS_A is then

(5.48)

$$SS_A = \frac{(T_{11} + T_{12})^2}{2n} + \frac{(T_{21} + T_{22})^2}{2n} - \frac{(T_{11} + T_{12} + T_{21} + T_{22})^2}{4n}$$

Let

$$T_{1.} = T_{11} + T_{12} = \sum_{i=1}^{n}\sum_{k=1}^{2} Y_{i1k}$$

(5.49)

$$T_{2.} = T_{21} + T_{22} = \sum_{i=1}^{n}\sum_{k=1}^{2} Y_{i2k}$$

Substituting Equation (5.49) into Equation (5.48) and expanding,

$$SS_A = \frac{T_{1.}^2 + T_{2.}^2}{2n} - \frac{T_{1.}^2 + T_{2.}^2 + 2T_{1.}T_{2.}}{4n}$$

(5.50)

$$= \frac{T_{1.}^2 + T_{2.}^2 - 2T_{1.}T_{2.}}{4n}$$

$$= \frac{(T_{1.} - T_{2.})^2}{4n}$$

Equation (5.50) provides a much faster way of calculating SS_A than does Equation (5.48). The reduction in computational effort is even more marked when we consider the shortcut formula for SS_{AB}:

$$(5.51) \qquad SS_{AB} = \frac{(T_{11} + T_{22} - T_{12} - T_{21})^2}{4n}$$

Note that there is no need to subtract sums of squares for main effects or to remove a correction term. Equation (5.51) may be derived from the general interaction formula presented earlier, as was done for Equation (5.49). A quick and very general method for arriving at the sum of squares formulas for any single *df* term is available if we redesignate the treatment combinations:

$$A_1 B_1 = ab$$
$$A_1 B_2 = a$$
$$A_2 B_1 = b$$
$$A_2 B_2 = (1)$$

The subscript 1 indicates the presence of the lower-case letter, and the subscript 2 indicates the absence of the letter. If all subscripts are 2s, the new designation is (1). This notation can be extended to any number of variables, all of which are tested at two levels. Thus, the cell $A_1 B_2 C_2 D_1$ in a 2^4 design is relabeled *ad*. Given these new labels, we may readily state the appropriate shortcut formula. The following steps are involved:

(a) List the new designations for all cells. For a 2^3 design, these would be

$abc \,(= A_1 B_1 C_1)$	$a \,(= A_1 B_2 C_2)$
$ab \,(= A_1 B_1 C_2)$	$b \,(= A_2 B_1 C_2)$
$ac \,(= A_1 B_2 C_1)$	$c \,(= A_2 B_2 C_1)$
$bc \,(= A_2 B_1 C_1)$	$(1) \,(= A_2 B_2 C_2)$

(b) Divide the cells into two classes, those that have an even number of letters in common with the effect and those that have an odd number of letters in common with the effect. For the AC interaction, the two classes are

1	2
abc	*a*
ac	*c*
b	*ab*
(1)	*bc*

Each of the designations in class 1 contains either two or none of the letters *a* and *c*. All designations in class 2 contain one of the letters *a* and *c*.

(c) Obtain the sum of scores for classes 1 and 2 above and then square the difference in the sums. Divide this quantity by the total number of scores.

In the preceding 2^3 example, we might represent the operation of subtracting class 2 from class 1 totals by

$$[abc + ac + b + (1)] - [a + c + ab + bc]$$

which equals

$$[(abc + ac) - (ab + a)] - \{(bc + c) - [b + (1)]\}$$

This last quantity is the difference between C_1 and C_2 totals at A_1 minus the difference between C_1 and C_2 totals at A_2, which we had previously defined (Section 5.3.2) as a measure of first-order interaction. To further show the development of single *df* formulas and their relationship to the meaning of interaction, we next consider the *ABC* interaction for a 2^3 design. According to the development of Section 5.6 [in particular, see Equation (5.46)], the *ABC* interaction is a measure of the variation in the interaction of two variables that occurs over the levels of the third variable. In our new notation, the appropriate contrast is

$$[(abc - ac) - (bc - c)] - \{(ab - a) - [b - (1)]\}$$

which equals

$$[abc + a + b + c] - [ab + ac + bc + (1)]$$

exactly the contrast that the odd-even rule, rule (b) above, demands. All designations to the left of the minus sign have one or three letters in common with *ABC*. To obtain SS_{ABC} we therefore insert the appropriate cell totals in the preceding formula for the contrast, square the result, and divide by $8n$, the total number of measurements.

As an alternative to the odd-even rule, consider this algebraic technique for generating contrasts. If the SS_{AC} is required, expand the quantity $(a - 1)(b + 1)(c - 1)$. We then have

$$(abc + ac + b + 1) - (a + c + ab + bc)$$

the contrast previously arrived at by the odd-even rule. For the SS_A, expand $(a - 1)(b + 1)(c + 1)$ and obtain

$$(a + ab + ac + abc) - (b + c + bc + 1)$$

The approach is simply to insert a minus sign only within those parentheses containing the letters appearing in the designation of the effect of interest.

In Section 5.7 sums of squares for a $2 \times 2 \times 3$ design were computed.

The D, R, and DR effects are all on one df. Applying our single df approach to the cell totals of Table 5–16, we obtain

$$SS_D = \frac{(1,951 - 1,627)^2}{120}$$

$$= 874.80$$

$$SS_R = \frac{(1,718 - 1,860)^2}{120}$$

$$= 168.03$$

$$SS_{DR} = \frac{(955 + 864 - 996 - 763)^2}{120}$$

$$= 30.00$$

These are exactly the results obtained in Section 5.7.

5.11 CONCLUDING REMARKS

The completely randomized designs discussed in Chapters 4 and 5 have several advantages. The analysis of the data is simpler than for most other designs. For any given number of measurements, the error df will be larger for these designs than for comparable designs. The requirements of the underlying model are most easily met by completely randomized designs, and violations of the assumptions embedded in the model are least likely to affect inferences derived from the F ratio. These designs share one main deficit. Since the within-cells variability that forms the error term is a function of individual differences, the efficiency of this design is relatively low. Other designs, which allow the experiment to remove from the error term variability due to individual differences, will generally yield a more precise estimate of population effects. A completely randomized approach should be considered whenever subjects are reasonably homogeneous with regard to the variable being measured; whenever a large n is available, compensating to some extent for the variability of measurements; or whenever the available n is so small that the loss in degrees of freedom that always accompanies more efficient designs yields a considerable loss in power. It should also be noted that there are many experimental situations in which it is impossible to do anything other than assign different subjects to different levels of the variables. This is self-evident in the case where the independent variable is personality type or training technique. It may also be true where much time is needed to obtain a measure from the subject and it is therefore preferable to obtain only one measure from each subject.

EXERCISES

5.1 Plot *all* main and interaction effects. Assuming errorless data, which effects are significant?

	A_1			A_2			A_3		
	B_1	B_2	B_3	B_1	B_2	B_3	B_1	B_2	B_3
C_1	22	12	14	19	6	8	16	5	6
C_2	18	8	10	21	8	10	18	7	8
C_3	14	4	6	20	7	9	23	12	13

5.2 Suppose that in the design of Exercise 5.1, we have the following cell frequencies:

	A_1			A_2			A_3		
	B_1	B_2	B_3	B_1	B_2	B_3	B_1	B_2	B_3
C_1	10	10	10	5	5	5	15	15	15
C_2	10	10	10	5	5	5	15	15	15
C_3	4	4	4	2	2	2	6	6	6

Assuming that the entries in Exercise 5.1 are cell totals, compute sums of squares for all main and interaction effects.

5.3 Prove that $\sum_j n_{j.}(\bar{Y}_{.j.} - \bar{Y}_{...}) = 0$.

5.4 Ninety-six children are subjects in a study of perceptual discrimination. Half of the children are six years old, half are nine years old. Half of the subjects are tested with two-dimensional objects, half with three-dimensional objects. Half are required to discriminate on the basis of shape, half on the basis of color. Thus, there are eight groups differing with respect to age (A; 6 and 9), dimensions (D; 2 or 3), and relevant cue (C; shape and color). We have the following hypotheses:
1. Nine-year olds will make fewer errors than 6-year olds
2. On the average, three-dimensional objects will be easier than two-dimensional objects, but
3. The difference between two and three-dimensional objects will be more marked for 6-year olds than for 9-year olds, and
4. The difference will hold for shape but not for color.
What effects should be significant?

5.5 Consider a two-factor design with n subjects in each cell. One possible model is

$$Y_{ijk} = \mu + \alpha_j + \alpha_k + (\alpha\beta)_{jk} + \varepsilon_{ijk}$$

where

$$\alpha_j = \mu_j - \mu$$

$$\beta_k = \mu_k - \mu$$

$$(\alpha\beta)_{jk} = \mu_{jk} - \mu_j - \mu_k + \mu$$

$$\varepsilon_{ijk} = Y_{ijk} - \mu_{jk}$$

Then the analysis of variance table is

SV	df	EMS
A	$a - 1$	$\sigma_e^2 + nb\theta_a^2$
B	$b - 1$	$\sigma_e^2 + na\theta_b^2$
AB	$(a - 1)(b - 1)$	$\sigma_e^2 + n\theta_{ab}^2$
S/AB	$ab(n - 1)$	σ_e^2

Note:
$$\theta_{AB}^2 = \frac{\sum_j \sum_k (\alpha\beta)_{jk}^2}{(a - 1)(b - 1)}$$

An alternative model is

$$Y_{ijk} = \mu + \alpha_j + \beta_k + \varepsilon_{ijk}$$

where

$$\varepsilon_{ijk} = Y_{ijk} - \mu_j - \mu_k + \mu$$

Assuming the second model to be true, and

$$\hat{\varepsilon}_{ijk} = Y_{ijk} - \bar{Y}_j - \bar{Y}_k + \bar{Y}_{...},$$

prove that

$$E\left(\sum_i \sum_j \sum_k \hat{\varepsilon}_{ijk}^2\right) = (abn - a - b + 1)\sigma_e^2$$

If the second model is valid, what might be the advantage in applying it? (*Hint:* How does the analysis of variance table change?) What problem arises if the second model is assumed but is not valid? (i.e., $\theta_{AB}^2 \neq 0$)

5.6 *abcen* children are divided into *abce* groups according to age (*A*), level of *B*, level of *C*, and experimenter (*E*). Only the main and interaction effects of *B* and *C* are of interest. To save computational time, it is decided to pull out one *SV* containing all terms that include *A*, *E*, or both. The *SV*s are therefore

B
C
BC
Pooled A, E effects
Within-cells error

(a) Give the *SS* and *df* formulas for the pooled term.
(b) Suppose we analyzed the data as having the following *SV*s:

B
C
BC
Residual

What are the *Residual df*? Why might the first analysis be preferred?

5.7 We have the following cell totals with eight subjects in a cell.

		D_1		D_2	
		C_1	C_2	C_1	C_2
	B_1	14	22	31	18
A_1	B_2	12	34	33	21
	B_3	26	24	43	19
		$\overline{52}$	$\overline{80}$	$\overline{107}$	$\overline{58}$
	B_1	42	46	20	41
A_2	B_2	15	18	25	30
	B_3	27	17	15	44
		$\overline{84}$	$\overline{81}$	$\overline{60}$	$\overline{115}$

Set up the single *df* computations for
(a) SS_D
(b) SS_{AC}
(c) SS_{ACD}

REFERENCES

Kempthorne, O. *The Design and Analysis of Experiments*, New York: Wiley, 1952.

Overall, J. E. and Spiegel, D. K., "Concerning least squares analysis of experimental data," *Psychological Bulletin*, 72: 311–322 (1969).

Peng, K. C. *The Design and Analysis of Scientific Experiments*, Reading: Addison-Wesley, 1967.

Snedecon, G. W. and Cochran, W. G. *Statistical Methods*, (6th ed.), Ames: Iowa State University, 1967.

6

Designs Using a Concomitant Variable

6.1 TREATMENTS × BLOCKS: INTRODUCTION

It was noted in preceding chapters that the major disadvantage of completely randomized designs is their relative inefficiency. The variability among subjects within groups, the error term against which the variability among treatment means is tested, is generally large. Much of this error variance can be attributed to individual differences in factors which contribute to performance. Even people treated alike will differ in their scores because of differences in such factors as attitude, previous experience, and intelligence. If the contribution of such individual difference variables could somehow be removed from the data matrix, the error variance would be reduced and it would be easier to detect the effects of the independent variable. In this chapter we consider one procedure for doing this, for removing some of the error variance attributable to individual differences.

In the design under consideration, subjects are divided into b blocks on the basis of their scores on a concomitant variable, a measure thought to be highly correlated with the dependent variable. For example, in a study in which some measure of paired-associate learning is of interest, the available population of subjects might be divided into blocks on the basis of intelligence test scores or even on the basis of scores in a paired-associate task other than the one to be used in the experiment. The simplest way to accomplish the distribution of subjects among blocks is to rank order them on the basis of the concomitant score. Assume that we have abn subjects. Then the highest scoring an subjects will be assigned to one block (e.g., B_1), the next an subjects will be assigned to B_2, and so on. The an subjects within each block are then randomly assigned to the levels of A, with the sole restriction that there be an equal number at each level. The result of the procedure that we have described is a two-factor (A, B) design with n subjects in each of the ab cells. The an subjects in each block are considered to be a random sample from an infinitely large population defined by the range of concomitant scores for that block. The members of this population have been randomly assigned to the a levels of A.

The treatments × blocks design has several advantages over the completely randomized, one-factor design. First, treatment groups are roughly matched for at least one measure that should affect performance. Second, since the design is essentially a two-factor design, the treatments × blocks interaction effects may be investigated. This means that we may consider such a question as, Are differences in the effects of different training methods greater at one level of intelligence than at another? Third, and most important, the treatments × blocks design will generally be much more efficient than a one-factor design involving the same total number of dependent measures at each treatment level. To see why this is so, we next turn to a detailed analysis of the efficiency of the treatments × blocks design relative to that of the completely randomized design.

6.2 RELATIVE EFFICIENCY

This section has several purposes. We will prove the contention that the treatments × blocks design is usually more efficient than the completely randomized design. As a by-product of the proof, an estimate of the ratios of error variances will be obtained for the two designs. If it is known how much more efficient the treatments × blocks design has been than the completely randomized design would have been, there is a basis for judging whether it would be worthwhile to establish blocks in subsequent related experiments. Finally, it should be noted that although a specific derivation will be given for the relative efficiency of the treatments × blocks and the completely randomized one-factor designs, the procedure to be presented can be readily generalized for investigation of relative efficiencies of various other designs.

Table 6–1 contains the information necessary for a statement of the relative efficiency of the completely randomized and treatments × blocks designs. Instead of the usual σ_e^2, $\sigma_{e_{c.r.}}^2$ and $\sigma_{e_{t \times b}}^2$ are used to distinguish between the error variances. Sums of squares formulas have been omitted, since those for the one-factor case were previously presented in Chapter 4 and those for the treatments × blocks design are the same as those for the two-factor design of Chapter 5. For the completely randomized design, bn scores are assumed at each level of A so that both designs are based on a total of abn scores.

The expected total sum of squares for the completely randomized design is

(6.1) $ESS_{\text{tot}_{c.r.}} = (a - 1)(\sigma_{e_{c.r.}}^2 + bn\theta_A^2) + a(bn - 1)\sigma_{e_{c.r.}}^2$

TABLE 6–1 *Expectations for two designs*

Completely Randomized

SV	df	EMS
A	$a - 1$	$\sigma^2_{e_{c.r.}} + bn\theta^2_A$
S/A	$a(bn - 1)$	$\sigma^2_{e_{c.r.}}$

Treatments × Blocks

SV	df	EMS
A	$a - 1$	$\sigma^2_{e_{t \times b}} + bn\theta^2_A$
B	$b - 1$	$\sigma^2_{e_{t \times b}} + an\theta^2_B$
AB	$(a - 1)(b - 1)$	$\sigma^2_{e_{t \times b}} + n\theta^2_{AB}$
S/AB	$ab(n - 1)$	$\sigma^2_{e_{t \times b}}$

and the expected total sum of squares for the treatments × blocks design is

$$(6.2) \quad ESS_{\text{tot}_{t \times b}} = (a - 1)(\sigma^2_{e_{t \times b}} + bn\theta^2_A) + (b - 1)(\sigma^2_{e_{t \times b}} + an\theta^2_B)$$
$$+ (a - 1)(b - 1)(\sigma^2_{e_{t \times b}} + n\theta^2_{AB}) + ab(n - 1)\sigma^2_{e_{t \times b}}$$

Since both designs involve a groups of bn subjects, it is reasonable to assume that $ESS_{\text{tot}_{c.r.}} = ESS_{\text{tot}_{t \times b}}$. Then, setting the right-hand side of Equation (6.1) equal to the right-hand side of Equation (6.2), canceling $bn(a - 1)\theta^2_A$, and simplifying, we have

$$(abn - 1)\sigma^2_{e_{c.r.}} = (abn - ab + a - 1)\sigma^2_{e_{t \times b}} + (b - 1)(\sigma^2_{e_{t \times b}} + an\theta^2_B)$$
$$(6.3) \qquad\qquad + (a - 1)(b - 1)(\sigma^2_{e_{t \times b}} + n\theta^2_{AB})$$
$$= (abn - ab + a - 1)\sigma^2_{e_{t \times b}} + ESS_B + ESS_{AB}$$

and

$$(6.4) \qquad \sigma^2_{e_{c.r.}} = \sigma^2_{e_{t \times b}} + \frac{1}{abn - 1}[(ESS_B + ESS_{AB}) - a(b - 1)\sigma^2_{e_{t \times b}}]$$

We now have the population error variance for the completely randomized design as a function of population parameters estimated from data obtained with the treatments × blocks design. But relative efficiency involves comparisons of mean squares rather than expected mean squares. Therefore, Equation (6.4) is replaced by a similar statement in which sample statistics are substituted for population parameters:

$$(6.5) \qquad MS_{S/A} = MS_{S/AB} + \frac{1}{abn - 1}[(SS_B + SS_{AB}) - a(b - 1)MS_{S/AB}]$$

By Equation (6.5), the data from a treatments × blocks design may be used to estimate what the error variance would have been if the completely randomized design had been used. Such information should aid in selecting designs for future experiments in an area.

The efficiency of the treatments × blocks design relative to that of the completely randomized design is defined as

$$R.E. = \frac{MS_{S/A}}{MS_{S/AB}}$$

$MS_{S/A}$ is therefore replaced by the right side of Equation (6.5), resulting in

(6.6) $$R.E. = 1 - \frac{a(b-1)}{abn-1} + \frac{SS_B + SS_{AB}}{(abn-1)MS_{S/AB}}$$

From Equation (6.6) it is seen that R.E. will be greater than 100 percent (the treatments × blocks design will be more efficient than the completely randomized) whenever

$$\frac{SS_B + SS_{AB}}{(abn-1)MS_{S/AB}} > \frac{a(b-1)}{abn-1}$$

or, multiplying both sides by

$$\frac{(abn-1)}{a(b-1)}$$

whenever

$$\frac{(SS_B + SS_{AB})/a(b-1)}{MS_{S/AB}} > 1$$

The quantity $SS_B + SS_{AB}$ is distributed on

$$(b-1) + (a-1)(b-1) = a(b-1)$$

df. Therefore

$$\frac{SS_B + SS_{AB}}{a(b-1)}$$

is a mean square and

$$\frac{(SS_B + SS_{AB})/a(b-1)}{MS_{S/AB}}$$

is essentially an F ratio. Thus, R.E. will be greater than 1 whenever an F test of the combined B and AB effects is greater than 1 (not necessarily significant, merely greater than 1) and will increase as the variability due to either B or AB increases. The condition that this F be greater than 1 will have high probability whenever the concomitant and dependent variables are correlated in the population. For example, if intelligence and paired-associate scores are correlated, high-intelligence subjects should have higher paired-associate scores than low-intelligence subjects. This implies differences among the blocks based on level of intelligence and will generally be reflected in large values of SS_B and will result in high relative efficiency.

While $MS_{S/A}$ will generally be larger than $MS_{S/AB}$, the former is distributed on $abn - a$ df while the latter is distributed on $abn - ab$ df (assuming abn measures for both designs). Thus, the completely randomized design is less efficient than the treatments × blocks, but its greater number of error df may result in a more powerful F ratio. Fisher (1952) has proposed an adjustment to account for the discrepancies in df. He suggests that relative efficiency may be defined as

(6.7) $$\text{R.E.} = \frac{(df_{S/AB} + 1)(df_{S/A} + 3)}{(df_{S/AB} + 3)(df_{S/A} + 1)} \cdot \frac{MS_{S/A}}{MS_{S/AB}}$$

For example, if $a = 2$ and N (the total number of observations) $= 8$, for the completely randomized design $df_{S/A} = 2(4 - 1) = 6$, and for the treatments × blocks design with $b = 2$, $df_{S/AB} = 2[2(2 - 1)] = 4$. Then the adjusted relative efficiency is

$$\text{R.E.} = \left(\frac{5}{7}\right)\left(\frac{9}{7}\right)\frac{MS_{S/A}}{MS_{S/AB}}$$

$$= (.92)\frac{MS_{S/A}}{MS_{S/AB}}$$

6.3 SELECTING THE OPTIMAL NUMBER OF BLOCKS

The number of blocks, b, influences experimental results in two ways. On the one hand, as b increases, the number of error df decreases (assuming that the total number of observations is held constant), resulting in reduced power of the F test. On the other hand, increasing the number of blocks results in a reduction of error variance. The variability within a block × treatment combination should be smaller when there are three blocks (e.g., high, medium, and low intelligence) than when there are two blocks (high and low intelligence). Because of these opposed effects of increased block number, an optimal block number exists. There is some level of b such that lesser and greater values result in less precise tests of treatment effects. This optimal value of b changes as a function of a (number of treatment levels), n, and ρ (the correlation coefficient for the population of concomitant and dependent measures). In this section we will attempt to clarify the relationships among these factors. In doing so, we will provide information of practical value in designing treatments × blocks experiments, as well as the basis for a better understanding of the way in which the design operates.

Some measure of the adequacy of the design is required in order to decide upon the value of b for various experiments. Such a measure should also clearly reflect the influences of a, n, and ρ. Since the influence of ρ is not clear in the R.E. formula, and since R.E. must be computed relative to another design, we seek some other index of error variability for the present purpose. The most generally accepted measure is I_a, the *apparent imprecision* of the design, the formula for which is

$$(6.8) \qquad I_a = \frac{df_{S/AB} + 3}{df_{S/AB} + 1} I_t$$

$$= \left[\frac{ab(n-1) + 3}{ab(n-1) + 1}\right] \left\{ \frac{1 - \rho^2[1 - (\bar{\sigma}_x^2/\sigma_X^2)]}{1 - \rho^2} \right\}$$

where σ_X^2 is the variance of concomitant measures for the jth treatment population, and $\bar{\sigma}_x^2$ is computed by obtaining the variance of concomitant measures for each block in the jth treatment population and then averaging over the b blocks; ρ is the correlation between X and Y measures in the jth treatment population. I_t is referred to as the true imprecision. I_a is the preferred measure, since Fisher's correction, $(df_{S/AB} + 3)/(df_{S/AB} + 1)$, adjusts for the loss of df (and therefore power) due to estimation of B and AB effects. To provide some idea of how we arrive at Equation (6.8), as well as a better feeling for the meaning of imprecision, we consider the sample-to-sample fluctuation in the difference between two treatment group means. We denote this variance of the difference between the means as $\sigma_{\bar{Y}_j - \bar{Y}_{j'}}^2$. An expression for aver $\sigma_{\bar{Y}_j - \bar{Y}_{j'}}^2$ ("aver" refers to the average variance over the b blocks) can be derived if it is assumed that

(a) within the population from which each block is sampled, Y (the dependent measure) is a linear function of X (the concomitant measure) and the slope of this function is the same for all b populations

(b) the population variability of the Ys about the best fitting straight line is the same for all b populations.

The appropriate expression is

$$(6.9) \qquad \text{aver } \sigma_{\bar{Y}_j - \bar{Y}_{j'}}^2 = \frac{2\sigma_Y^2}{n}\left[1 - \rho^2\left(1 - \frac{\bar{\sigma}_x^2}{\sigma_X^2}\right)\right]$$

where σ_Y^2 is the variance of dependent measures for the jth treatment population; σ_X^2, $\bar{\sigma}_x^2$, and ρ have been defined above. It is assumed that these population values are the same for all a treatment populations.

It is evident from Equation (6.9) that as the number of observations (n) is increased, the average sampling error decreases. The same relationship holds for ρ and sampling error. Note that when ρ is zero,

$$\text{aver } \sigma^2_{\bar{Y}_j - \bar{Y}_{j'}} = \frac{2\sigma^2_{\bar{Y}}}{n}$$

When ρ is 1,

$$\text{aver } \sigma^2_{\bar{Y}_j - \bar{Y}_{j'}} = \left(\frac{2\sigma^2_{\bar{Y}}}{n}\right)\left(\frac{\bar{\sigma}^2_x}{\sigma^2_X}\right)$$

Since the average variability over the blocks in a population will be less than the total variability in the population, $\bar{\sigma}^2_x/\sigma^2_X < 1$, and the sampling error when ρ is 1 will be some fraction of the sampling error when ρ is zero.

While the number of blocks, b, does not explicitly appear in Equation (6.9), its influence can be understood by considering $\bar{\sigma}^2_x$. If there is only one block, we actually have a completely randomized design, $\bar{\sigma}^2_x$ is identical to σ^2_X, and

$$\text{aver } \sigma^2_{\bar{Y}_j - \bar{Y}_{j'}} = \frac{2\sigma^2_{\bar{Y}}}{n}$$

As the number of blocks increases, the range of concomitant scores within each block must necessarily decrease, and $\bar{\sigma}^2_x$ becomes progressively smaller. Therefore, when the number of blocks is very large, $\bar{\sigma}^2_x$ is near zero and the average error is approximately

(6.10) $$\min \sigma^2_{\bar{Y}_j - \bar{Y}_{j'}} = \frac{2\sigma^2_{\bar{Y}}}{n}(1 - \rho^2)$$

Cox has proposed that the ratio of the average to the theoretically minimum sampling error (i.e., $\min \sigma^2_{\bar{Y}_j - \bar{Y}_{j'}}$) be utilized as an index of the precision of the design. Specifically, for the treatments × blocks design we have the *true imprecision*, I_t, where

(6.11) $$I_t = \frac{\text{aver } \sigma^2_{\bar{Y}_j - \bar{Y}_{j'}}}{\min \sigma^2_{\bar{Y}_j - \bar{Y}_{j'}}}$$

$$= \frac{1 - \rho^2[1 - (\bar{\sigma}^2_x/\sigma^2_X)]}{1 - \rho^2}$$

Multiplying by Fisher's adjustment, we have Equation (6.8), which defines I_a.

Feldt has used I_a as the basis for determining the optimal number of blocks to be included in an experiment. That number of blocks resulting in a lower value of I_a than any other number of blocks is considered the optimal number. Table 6–2 has been reproduced from Feldt's article. Data available from previous research or from pilot studies may be used to estimate ρ. With this estimate and knowledge of a and N (the total number of scores), the experimenter has a sound basis for deciding upon the number of blocks to be included in the experiment.*

The relationships among the optimal value of b and the values of a, ρ, and N may be summarized as follows:

(a) As ρ increases, the optimal value of b also increases. The increased loss in error df due to increase in b is more than compensated for by the improvement in precision due to increase in ρ.

(b) As N increases, the optimal value of b increases. The increased loss in error df due to increases in b has less effect on precision as the total number of observations increases.

(c) As a decreases, the optimal value of b increases. The increased loss in error df due to increases in b is compensated for by the reduced loss due to df_{AB} as a decreases.

TABLE 6–2 *Optimal values of b for selected experimental conditions, assuming levels defined by equal proportions of the population*

				N			
ρ	a	20	30	50	70	100	150
.2	2	2	3	4	5	7	9
	5	1	2	2	3	4	6
.4	2	3	4	6	9	13	17
	5	2	3	4	5	7	10
.6	2	4	6	9	13	17	25**
	5	2	3	5	7	9	14
.8	2	5*	7*	12*	17	23	25**
	5	2*	3*	5*	7*	10*	15*

* Limit imposed by the requirement $N \geq ab$
** Slight improvement possible with more than 25 levels

This table is reproduced from L. S. Feldt, "A Comparison of the Precision of Three Experimental Designs Employing a Concomitant Variable," *Psychometrika*, 23:335–354 (1958), by permission of the author and editor.

The imprecision for a completely randomized design is easily obtained if we note that this is actually a treatments × blocks design with only one block. Consequently $\bar{\sigma}_x^2 = \sigma_X^2$, and substituting in Equation (6.8),

(6.12)
$$I_a = \left[\frac{df_{S/A} + 3}{df_{S/A} + 1} \right] \frac{1}{1 - \rho^2}$$

*The entries in Table 6–2 have been rounded to the nearest integer.

TABLE 6–3 *Ratio I_a for completely randomized design to I_a for treatment $\times$ blocks design*

		N					
ρ	a	20	30	50	70	100	150
.2	2	1.015	1.023	1.029	1.034	1.036	1.038
	5	1.000	1.009	1.021	1.027	1.031	1.035
.4	2	1.117	1.141	1.161	1.170	1.176	1.181
	5	1.060	1.097	1.136	1.153	1.165	1.174
.6	2	1.388	1.443	1.494	1.516	1.531	1.541
	5	1.235	1.340	1.431	1.470	1.500	1.523
.8	2	2.196	2.378	2.543	2.615	2.666	2.699
	5	1.608	1.943	2.276	2.427	2.543	2.629

Based on Table 4 from L. S. Feldt, "A Comparison of the Precision of Three Experimental Designs Employing a Concomitant Variable," *Psychometrika*, 23:335–354 (1958), by permission of the author and editor.

On the basis of Equations (6.8) and (6.12), Table 6–3 was derived; this presents the ratio of values of I_a for the completely randomized design to values of I_a for the treatments $\times$ blocks design (assuming that the optimal value of b is used). The tabled values are consistently greater than 1, indicating that the completely randomized design is more imprecise, or less precise. The advantage of the treatments $\times$ blocks design increases as ρ and N increase and as a decreases. To put it another way, the effort involved in assigning subjects to blocks is most worthwhile when the optimal number of blocks is large.

In using Table 6–2 to design experiments, it is helpful to note that the experimenter can readily use values of ρ, N, and a other than those in the table. For example, suppose $\rho = .6$, $N = 30$, and $a = 4$. The value of optimal b for $a = 4$ is not tabled, but can be approximated by linear interpolation. When $\rho = .6$, $N = 30$, and $a = 5$, the value of b is 3. When $a = 2$, $b = 6$. Since $a = 4$ is $\frac{1}{3}$ the distance between $a = 5$ and $a = 2$, we take as the required value of b the number falling $\frac{1}{3}$ of the distance between 3 and 6. The optimal value chosen for the experiment is therefore 4. We may interpolate among various values of N in exactly the same manner.

If the value of ρ which has been estimated for an experiment does not appear in Table 6–2, we may linearly interpolate between values of ρ^2. For example, suppose that $N = 150$, $\rho = .3$, and $a = 2$. The square of ρ is .09, which falls 5/12 of the distance between $.2^2 (= .04)$ and $.4^2 (= .16)$. When $N = 150$, $\rho = .2$, and $a = 2$, the optimal value of b is 9; when $\rho = .4$, $b = 17$. The level of b for our example should therefore fall at the value 5/12 of the distance between 9 and 17. To the nearest integer, the result is 12.

6.4 EXTENSIONS OF THE TREATMENTS × BLOCKS DESIGN

The discussion has thus far been limited to the case in which only one treatment variable is under investigation. Extension to experiments in which several treatment variables are of interest is straightforward. For example, if one wants to test hypotheses about the single and joint effects of two variables, A and C, the model and the analysis is the same as that for the three-factor design of Chapter 5, blocks (B) being the third factor. Subjects are divided into blocks on the basis of the control variable; within each block, there is random assignment of subjects to the ac treatment combinations. The derivation of relative efficiency follows directly from that of Section 6.2. I_t, the true imprecision, will not be influenced by the number of treatment variables. However, apparent imprecision will be increased if additional treatments are investigated and N is held constant. To see why this is so, note that the adjustment factor which transforms I_t into I_a is $(df_{error} + 3)/(df_{error} + 1)$. If, for example, $N = 80$, $a = 2$, and $b = 4$,

$$df_{error} = 80 - 1 - (2 - 1) - (4 - 1) - (2 - 1)(4 - 1)$$
$$= 72$$

and

$$I_a = \frac{75}{73} I_t$$

$$= 1.03 I_t$$

If five levels of a variable, C, are added to the design, we would then have

$$df_{error} = 72 - (5 - 1) - (2 - 1)(5 - 1) - (4 - 1)(5 - 1)$$
$$- (2 - 1)(4 - 1)(5 - 1)$$
$$= 40$$

In this case,

$$I_a = \frac{43}{41} I_t$$

$$= 1.09 I_t$$

When the error df are considered, the design involving more treatment variables is slightly less precise.

Tables similar to Table 6–2 have not been designed for the multivariable cases. However, because of the reduction in precision just noted, values of b smaller than those presented in Table 6–2 should be used whenever there is more than one treatment variable. As N becomes larger, the loss of error df, occasioned by the inclusion of additional variables, has a correspondingly smaller effect on I_a. Therefore, the tabled values become better approximations to optimal b.

6.5 THE EXTREME GROUPS DESIGN

A design related to the treatments × blocks design is one in which only two extreme blocks of subjects are used. For example, the eyelid conditioning rates of subjects who score high on the Taylor Manifest Anxiety Scale are compared with those for low-scoring subjects, or the gambling behavior of subjects scoring high on the MMPI Psychopathic Deviant Scale is compared with the gambling behavior of subjects who score low on the scale. At the outset, the distinction between this design and the treatments × blocks should be made clear. In the treatments × blocks design, subjects are divided into blocks on the assumption that the block, or concomitant, variable is correlated with the dependent variable. The purpose of the design is to improve the precision of the test of some treatment effect. In the extreme groups design, the blocks variable is the variable whose effect is of interest. Rather than assume a correlation between concomitant and dependent variable, the experimenter investigates whether such a correlation exists. Do high anxious subjects give more conditioned responses than low anxious subjects? Do those subjects who score high on a psychopathic deviant scale take more risks in a laboratory gambling task than those who score low on the scale?

There are two major questions to be resolved in relation to the extreme groups design. First, what percentage of the available population should define the extreme groups? Should the number of conditioned responses be compared for the top and bottom 10 percent on the anxiety scale? The top and bottom 25 percent? The top and bottom halves? Second, when should the extreme groups approach be discarded in favor of sampling from all levels of the concomitant measure? To clarify this second question, consider the following example. Anxiety measures are available on 200 subjects. It is not feasible to obtain eyelid conditioning scores from more than 100 of the original group. These 100 could be the highest and the lowest 50 on the anxiety scale, in which case a *t* test comparing their mean conditioning scores would be used to assess the relationship between anxiety and conditioning. Alternatively, the 100 subjects to be conditioned could be randomly sampled from the original group of 200, with an equal probability of representation for all anxiety scores. In this instance a correlation coefficient would be computed and its deviation from zero tested by a *t* statistic. Which test of relationship is more powerful?

We first consider the question of the optimal size of the extreme groups. The criterion of optimality is the power of the *t* test comparing the means of the two groups. Feldt has shown that when ρ is .10, the power of the *t* test is greatest if each extreme group consists of 27 percent of the population tested on the concomitant measure. As the correlation between the concomitant and dependent measures increases, the optimal percentage decreases, but not markedly. When ρ is .80, power is greatest if each extreme group contains 23.3 percent of the population. Obviously, we never know ρ; the purpose of the experiment is to determine whether ρ is different from zero. However, it seems reasonable to conclude that if we use the top and bottom fourths of concomitant measures, we will not depart far from optimality, regardless of the true value of ρ. For example, the Taylor Manifest Anxiety Scale might be administered to a class of 100 elementary psychology students. Fifty of these, the 25 scoring highest and the 25 scoring lowest, would then be asked to serve as subjects in an eyelid conditioning experiment. Mean number of conditioned responses would be compared for the two extreme groups.

We next consider the question of whether it is better to use the extreme groups approach, or to sample randomly from the population of test takers. The answer to this depends on two factors: the percentage of the original test takers who are available for the experimental session and the correlation between the two measures. Before attempting to answer the question, we need some ground rules. Suppose that 100 students have been tested on the Taylor scale and we can run some maximum percentage on the eyelid conditioning apparatus. If the percentage is above 50, the rule will be that for purposes of comparing the extreme groups and random sampling approaches, only 50 percent will be used for the extreme groups approach. For example, it may be undesirable in terms of the expenditure of time and effort to run more than 60 of our population of 100 in the conditioning phases. Since, for the extreme groups approach, two groups, each consisting of 25 subjects, are optimal, this is all we will use. The power of the t test between means for this optimal grouping will be compared with a t test of the correlation coefficient for a random sample of 60 subjects. If less than 50 percent of the population is available for the conditioning phase, the available percentage is divided in half to determine the size of the extreme groups.

We may briefly summarize Feldt's findings with regard to this comparison of two approaches. If both concomitant and dependent measures are available on 80 percent or more of the population, the random sampling—correlational approach is more powerful. If the dependent measure can be obtained from only 75 percent of the subjects originally tested, the extreme groups approach should be used if $\rho < .492$; otherwise 75 percent should be sampled at random from the population, tested in the second phase of the experiment, and a correlation coefficient then computed. As the percentage of subjects for whom both measures are available decreases, a higher value of ρ is required if the correlational approach is to be used. Thus, if no more than 50 percent of the subjects given the Taylor scale are to be tested in the conditioning phase, the extreme groups approach should be used whenever $\rho < .847$.

Let us attempt to reduce this discussion to some simple recommendations. We again note that the value of ρ is never known; the purpose of the experiment is to get some information about it. However, we may be able to obtain a rough estimate from other research. In the case of the Taylor scale and conditioned eyelid response, correlations between other anxiety measures and other forms of conditioning may be found in the literature. Such correlations tend to be low, rarely exceeding .2. In general, correlations greater than .6 are rarely observed in the psychological literature. Therefore, unless both concomitant and dependent measures are obtainable from at least 80 percent of the population, the extreme groups approach should generally be used. There will be some rare occasions when the value of ρ is surprisingly high, and the correlational approach would therefore have proved more powerful. However, if ρ is very large, even a less powerful approach should be sufficiently sensitive. If more than 80 percent of the population is available for both phases of the experiment, then the total available proportion should be randomly sampled from the population and a correlation coefficient computed.

Feldt points out that his results depend upon the assumptions that the distribution of the concomitant and dependent measures is bivariate normal and that the relationship between the two variables is linear. If the latter assumption is incorrect, the extreme groups approach is at worst misleading, at best relatively uninformative. Consider Figure 6–1 to understand why the

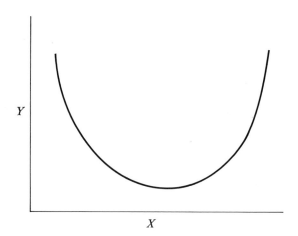

FIGURE 6–1 *Example of a nonlinear relationship between Y and X*

extreme groups approach might be misleading. If the extreme groups are tested, we will surely conclude that they do not differ with respect to Y, the dependent measure. This is true, but completely misses the important fact that there is a relationship between X and Y, that the extreme groups differ from a middle group. Even if the relationship between X and Y is monotonic (i.e., Y consistently increases or consistently decreases as X increases), if it is not linear, the extreme groups approach results in a loss of information.

Whenever there is reason to suspect a nonlinear relationship, it is better to sample the available proportion randomly, regardless of its size, from the total population from which the concomitant measure was taken. In this way, information about the shape of the function relating X and Y can be obtained.

EXERCISES

6.1 Assume that we are interested in the main and joint effects of two variables, A and B. Derive the efficiency of an $A \times B \times$ blocks design relative to that of a two-factor (A,B) completely randomized design.

6.2 Analyze the following set of data obtained with an $A \times$ blocks design

	A_1	A_2	A_3
B_1	1	16	3
	2	10	1
B_2	4	13	3
	3	12	3
B_3	6	7	9
	8	10	5
B_4	10	7	5
	8	7	12

Estimate the efficiency relative to a completely randomized design. What is the primary reason that the $A \times$ blocks design is more efficient?

6.3 Several previous studies have yielded estimates of the correlation between an intelligence test and paired-associate learning ranging from .25 to .35. Given four levels of A and a total of 80 available subjects, what value of b would yield optimal precision?

6.4 Assume that X and Y are correlated so that a reliable estimate of ρ is .6. Assume that 36 Ss are available for an experiment and that there are three levels of A. We divide the subjects into 6 blocks on the basis of the X score. The Anova table is:

SV	df	MS
A	2	100
B	5	160
AB	10	22
S/AB	18	16

If we wanted to reject H_0 at the 1-percent level when $\mu_3 = \mu_2 + 4$ and $\mu_2 = \mu_1 + 4$, or when the spread is greater:
(a) What estimated power do we have? (Note: for the $T \times B$ design, Eq. (4.41) should read

$$\phi^2 = \frac{bn \sum_j (\mu_j - \mu)^2 / a}{\sigma_e^2}$$

(b) What power would we expect to have if we had used a completely randomized design instead?

REFERENCES

Cox, D. R., "The use of a concomitant variable in selecting an experimental design," *Biometrika*, 44: 150–158 (1957).

Feldt, L. S., "A comparison of the precision of three experimental designs employing a concomitant variable," *Psychometrika*, 23: 335–354 (1958).

Feldt, L. S., "The use of extreme groups to test for the presence of a relationship," *Psychometrika*, 26: 307–316 (1961).

Fisher, R. A. *Statistical Methods for Research Workers*, 12th ed., London: Oliver & Boyd, 1952.

Repeated Measurements Designs

7.1 INTRODUCTION

In Chapters 4 and 5 error variance was separated from the variance among treatment effects. Within the completely randomized designs, error variance was a large complex contributed to by individual differences and errors of measurement. In Chapter 6 one way of reducing the error variance was considered. It was shown that some of the variability due to individual differences could be removed if subjects were divided into blocks on the basis of a concomitant variable correlated with the dependent variable. In this chapter consideration will be given to designs in which error variability due to individual differences is further reduced, in fact completely removed.

The one-factor repeated measurements design entails a test of each subject under each of the a levels of the treatment variable A. The order of presentation of the A_j is randomized independently for each subject. The design may be laid out as a two-factor design where subjects constitute the second factor, as has been done in Table 7–1. The important point is that subjects may be treated like any main effect in a two-factor design, the variability associated with the effect can be isolated, and as a consequence the error variance in this design is not inflated by variability due to individual

TABLE 7–1 *Data matrix for a one-factor repeated measurements design*

	A_1	$\cdots$	A_2	$\cdots$	A_j	$\cdots$	A_a
S_1	Y_{11}		Y_{12}		Y_{1j}		Y_{1a}
S_2	Y_{21}		Y_{22}		Y_{2j}		Y_{2a}
$\vdots$							
S_i	Y_{i1}		Y_{i2}		Y_{ij}		Y_{ia}
$\vdots$							
S_n	Y_{n1}		Y_{n2}		Y_{nj}		Y_{na}

differences. The design will generally be more efficient than even the treatments × block design, since variability due to individual differences will be more effectively separated from error variance.

Increased precision is not the sole reason for the use of the repeated measurements design. Both the completely randomized and treatments × block designs require more subjects than does the repeated measurements design to achieve the same power of the F test. Therefore, the use of the repeated measurements design should be considered whenever a limited number of subjects are available for long periods of time. This will frequently be the case in small clinics, military research installations, and industrial settings. It will less often be a factor in the choice of a design in large universities.

The repeated measurements design is the natural one to select when one is concerned with performance trends over time. For example, if one wants to measure the course of dark adaptation over time, the most efficient use of subjects requires that each subject be tested at all points in time that are of interest. In this instance, time is the treatment variable A.

Although the repeated measurements design does not involve any new computational problems, two other issues arise that have not been considered in the context of previous designs. For the first time, the levels of one of our variables, S (subjects), may be considered a random sample from a population of levels. In Section 7.2 some implications of random, as opposed to fixed, effects will be considered. There is also the strong possibility that scores for two treatments will be correlated, since both sets of scores are obtained from the same individuals. The implication of this violation of the usual independence assumption will also be dealt with in Section 7.2.

7.2 MODELS FOR THE ONE-FACTOR REPEATED MEASUREMENTS DESIGN

7.2.1 The additive model. Consider a group of n subjects, each of whom is tested once under each level of A, the treatment variable. The order of exposure of the subject to the treatment levels is random, and the randomization is carried out independently for each subject. In order to develop a model for this experiment, it is assumed that the n subjects are a random sample from an infinite population of subjects. Furthermore, the observed score for the ith subject under treatment A_j is viewed as being randomly sampled from an infinite population of independent measurements on the ith subject under A_j. We first consider the additive model:

(7.1) $$Y_{ij} = \mu + \eta_i + \alpha_j + \varepsilon_{ij}$$

where Y_{ij} is the score of subject i under A_j; $\eta_i = \mu_i - \mu$ (μ_i is the population mean of the a scores for the ith subject); $\alpha_j = \mu_j - \mu$ (μ_j is the mean of the scores of an infinite population of subjects tested under A_j); and ε_{ij} is the deviation of Y_{ij} from μ that is not accounted for by the treatment effect (α_j)

or by individual differences (η_i). Equivalently, ε_{ij} may be viewed as equal to $Y_{ij} - \mu_{ij}$, where μ_{ij} is the expected value of the hypothetical population of measurements on the ith subject under A_j.

Ordinarily, the a levels of the treatment variable are arbitrarily selected by the experimenter. It is as if there were only a levels in the population and they were all represented in the study. If this is the case, $\sum_{j=1}^{a} \alpha_j = 0$, since the sum of all deviations of treatment means about their mean, μ, must be zero. As was seen in Chapter 4, this fact is used in deriving the expected mean squares. A second implication of the fact that the effects of A are fixed (i.e., its levels are arbitrarily chosen) is that inferences about treatment effects should be limited to those treatment levels included in the experiment. To put it another way, $\theta_A^2 \left[= \sum_j (\mu_j - \mu)^2/(a - 1) \right]$ is a measure of the variability of only those a treatment effects included in the population and, strictly speaking, our null hypothesis is that θ_A^2 equals zero. Of source, in practice we expect the experimenter to use his brains as well as his F ratios to draw inferences. An investigator working with a set of a drugs may well extend his conclusions to other drugs of related chemical composition. Such conclusions may well be as correct as those based on the statistical analysis and are desirable whenever the investigator's substantive knowledge permits extrapolation beyond the treatment levels employed. Nevertheless, it is wise to recognize that conclusions about the effects of the treatments arbitrarily included in the experiment are based on different grounds than are those about the effects of related treatments not included in the experiment.

When subjects are considered, the situation is somewhat different. The most reasonable assumption about the subjects is that they are a random sample from an infinite population. Admittedly, it would be hard to defend the usual procedures of subject selection as truly random, and it is sometimes difficult to characterize precisely the population sampled. However, it is clear that the n subjects have not been arbitrarily selected because of a desire to compare n qualitatively or quantitatively differing characteristics. The random sampling assumption best seems to describe the true state of affairs.

It is further assumed that the η_i are distributed normally in the population sampled and that this population has mean zero and variance σ_S^2; i.e., $E(\eta_i) = 0$ and $\text{var}(\eta_i) = \sigma_S^2$. Note that $\sum_{i=1}^{n} \eta_i \neq 0$, since the η_i sampled in any one experiment do not exhaust the population of such deviation scores. In further contrast to fixed effect variables, conclusions about the variability of the η_i are conclusions about the variance of the population from which the $n\eta_i$ have been sampled; the conclusions are not restricted to the particular sample of subjects in the experiment.

As usual, the ε_{ij} are assumed to be random samples from a normally distributed treatment populations, each with mean zero and variance σ_e^2. The η, α, and ε are assumed to be uncorrelated.

Table 7–2, which summarizes the analysis of variance for the repeated measurements design, is a consequence of Equation (7.1) and the sampling and distributional assumptions that have been presented.

TABLE 7–2 *Analysis of variance for the additive model*

SV	df	SS	MS	EMS	F
S	$n-1$	$\dfrac{\sum_i(\sum_j Y_{ij})^2}{a} - C$	$\dfrac{SS_S}{n-1}$	$\sigma_e^2 + a\sigma_S^2$	$\dfrac{MS_S}{MS_{SA}}$
A	$a-1$	$\dfrac{\sum_j(\sum_i Y_{ij})^2}{n} - C$	$\dfrac{SS_A}{a-1}$	$\sigma_e^2 + n\theta_A^2$	$\dfrac{MS_A}{MS_{SA}}$
SA	$(n-1)(a-1)$	$\sum_i\sum_j Y_{ij}^2 - C - SS_S - SS_A$	$\dfrac{SS_{SA}}{(n-1)(a-1)}$	σ_e^2	

We consider the SV first. Equation (7.1) suggests that the total variance should have three components: A, S, and error. The error source has been labeled SA as a reminder that the computations involved are those for an interaction. To obtain the error sums of squares, the variability among cell means (in this design the mean is based on one score) is adjusted for the contributions of A and S variability. Despite the form of the computations, it is important to note that under the additive model, MS_{SA} estimates the population error variance and nothing more; no interaction is assumed to exist in the population.

The computations for both SS_A and SS_B follow directly from those for the two-factor designs of Chapter 5. There is one index of summation less in the present case because there is only one score in each cell of this "two-factor" design. The df are also straightforward; nothing new need be added to the discussions of Chapters 4 and 5.

The F tests for both A and S follow the usual rule: a denominator is required such that, under H_0, the ratio of mean squares = 1. Clearly, MS_{SA} will be appropriate for both F tests.

Missing data. Suppose that one score is missing; subject i has failed to appear to be tested under A_j or the recording apparatus was not working in one session. Call the missing score X_{ij}. From Equation (7.1),

$$E(X_{ij}) = \mu + \eta_i + \alpha_j$$

Our best estimate of μ is

$$\hat{\mu} = \frac{\text{grand total}}{an}$$

In this case, the grand total is $T_{..} + X_{ij}$ where $T_{..}$ is the sum of the $an - 1$ scores that were obtained. Similarly, we estimate

$$\hat{\eta}_i = \hat{\mu}_i - \hat{\mu}$$

where

$$\hat{\mu}_i = \frac{T_i + X_{ij}}{a}, \; T_i = \text{obtained total for } S_i$$

and

$$\hat{\alpha}_j = \hat{\mu}_j - \hat{\mu}$$

where

$$\hat{\mu}_j = \frac{T_j + X_{ij}}{n}, \; T_j = \text{obtained total for } A_j$$

then

$$\hat{X}_{ij} = \frac{T_{..} + \hat{X}_{ij}}{an} + \left(\frac{T_i + \hat{X}_{ij}}{a} - \frac{T_{..} + \hat{X}_{ij}}{an} \right) + \left(\frac{T_j + X_{ij}}{n} - \frac{T_{..} - \hat{X}_{ij}}{an} \right)$$

from which we obtain our estimate,

(7.2)
$$\hat{X}_{ij} = \frac{nT_i + aT_j - T_{..}}{(n - 1)(a - 1)}$$

Suppose we have several missing values, $X_{ij}^{(1)}$, $X_{ij}^{(2)}$, etc. Guess at all except $X_{ij}^{(1)}$, which is approximated by Equation (7.2), using guesses of $X_{ij}^{(2)}$, $X_{ij}^{(3)}, \ldots$, as if they were part of the data. With our approximation of $X_{ij}^{(1)}$ and guesses of $X_{ij}^{(3)}$, $X_{ij}^{(4)}, \ldots$, we may now apply Equation (7.2) to approximate $X_{ij}^{(2)}$. We continue this process, repeating the cycle with first-round approximations of X_{ij}, until two successive cycles exhibit little change. One final point should be noted: a df is lost from the error term df for each value estimated.

7.2.2 The nonadditive model. The assumption that η_i and α_j contribute in an additive manner to the "true" (errorless) score of the ith subject in treatment A_j permits the simple analysis just presented in the preceding section. Unfortunately, data are rarely so obliging as to conform to Equation (7.1). More often than not, the variability among subjects' scores will be a function of the particular treatment under observation. This means that an interaction of subject and treatment level contributes to the score. Such a state of affairs requires a revision of our model to include an interaction term in the population as a contribution to the score, Y_{ij}. The equation that expresses this nonadditive model is

$$(7.3) \qquad Y_{ij} = \mu + \eta_i + \alpha_j + (\eta\alpha)_{ij} + \varepsilon_{ij}$$

where $(\eta\alpha)_{ij}$ is the interaction effect of the ith subject and the jth level of A and is assumed to be a random variable sampled from a normally distributed population with mean zero and variance σ_{SA}^2. Turning to Table 7–3, we see that sources of variance, df, and sums of squares are the same under the nonadditive model as they were under the additive model; merely assuming separate interaction and error components in the population does not enable us to compute them separately. To put it another way, regardless of the model, SS_A, SS_S, and SS_{SA} add to SS_{tot}. However, the *EMS* have changed. The interaction variance σ_{SA}^2, now contributes to the expectations for A and SA. Of particular interest is the $E(MS_A)$, which is the first instance we have encountered of an *EMS* containing a component other than σ_e^2 and the null hypothesis term. In view of this development, it would seem worthwhile to discuss further the *EMS* for the nonadditive model.

First consider $\overline{Y}_{i.}$, the mean of the a scores obtained from the ith subject. Summing over j and dividing by a in Equation (7.3), we have

$$(7.4) \qquad \overline{Y}_{i.} = \mu + \frac{\sum_j \eta_i}{a} + \frac{\sum_j \alpha_j}{a} + \frac{\sum_j (\eta\alpha)_{ij}}{a} + \frac{\sum_j \varepsilon_{ij}}{a}$$

Since A is a fixed-effects variable, $\sum_j \alpha_j = 0$. Furthermore, $\sum_j (\eta\alpha)_{ij}$ represents the sum of *all* interaction effects for the ith subject and therefore will also be zero. Thus, Equation (7.4) may be replaced by the simpler result:

$$(7.5) \qquad \overline{Y}_{i.} = \mu + \eta_i + \frac{\sum_j \varepsilon_{ij}}{a}$$

Clearly, the variability among the means of the n subjects in the experiment does not involve variability among interaction effects, since such effects do not contribute to any subject's mean performance. This is reflected in $E(MS_S)$ in Table 7–3.

Now consider $\bar{Y}_{.j}$, the mean of the n scores obtained under treatment A_j. Again turning to Equation (7.3), we sum over i and divide by n to obtain

$$(7.6) \qquad \bar{Y}_{.j} = \mu + \alpha_j + \frac{\sum_i \eta_i}{n} + \frac{\sum_i (\eta\alpha)_{ij}}{n} + \frac{\sum_i \varepsilon_{ij}}{n}$$

The term, $\sum_i \eta_i/n$, is not a function of j and therefore individual differences will not contribute to the variability among the $\bar{Y}_{.j}$. However, $\sum_i (\eta\alpha)_{ij}$ neither equals zero (as $\sum_j (\eta\alpha)_{ij}/n$ did) nor is constant over levels of j (as $\sum_i \eta_i/n$ is). It does not represent the sum of all population interaction effects obtainable under A_j but rather a random sample of size n from that infinitely large set of values. Furthermore, the composition of the sample of n interaction effects will vary over treatment levels (as the subscript j suggests). The consequence of this argument is that the SS_A will reflect variability among interaction effects, as the $E(MS_A)$ in Table 7–3 indicates.

A numerical example may help to further clarify why interaction effects contribute to variability among subject means but not to variability among treatment level means. Table 7–4 contains data for a population consisting of four subjects. Assume that two subjects, S_1 and S_2, are sampled for an "experiment." We "test" both subjects at all three levels of A. Thus, "subjects" is a random effects variable; while A is a fixed effects variable. Check the calculations for $S(\eta_i)$, $A(\alpha_j)$, and $SA[(\eta\alpha)_{ij}]$ effects. Note that the mean interaction effect is zero for each subject and, therefore, interaction variability does not contribute to the variability among subject means. On the other hand, calculate the mean interaction effect for each level of A, limiting the calculations to the two subjects included in the experiment. We have

$$(\eta\alpha)_{.1} = \tfrac{1}{2}(-1 + 3) = 1$$

$$(\eta\alpha)_{.2} = \tfrac{1}{2}(1 + 2) = 1.5$$

$$(\eta\alpha)_{.3} = \tfrac{1}{2}(0 - 5) = -2.5$$

Clearly, the variability among treatment means is based partly upon variability among interaction effects.

7.2.3 The test of the subject effect. In the preceding sections two alternative models for the analysis of variance were presented for the repeated measurements design. Though neither the partitioning of the total variability nor the computations for the various sums of squares differ under the two models, the choice of models does have implications for our inferences. The most

TABLE 7-3 *Analysis of variance for the nonadditive model*

SV	df	SS	MS	EMS	F
S	$n - 1$	$\dfrac{\sum_i(\sum_j Y_{ij})^2}{a} - C$	$\dfrac{SS_S}{n-1}$	$\sigma_e^2 + a\sigma_S^2$	
A	$a - 1$	$\dfrac{\sum_j(\sum_i Y_{ij})^2}{n} - C$	$\dfrac{SS_A}{a-1}$	$\sigma_e^2 + \sigma_{SA}^2 + n\theta_A^2$	$\dfrac{MS_A}{MS_{SA}}$
SA	$(n-1)(a-1)$	$\sum_i\sum_j Y_{ij}^2 - C - SS_S - SS_A$	$\dfrac{SS_{SA}}{(n-1)(a-1)}$	$\sigma_e^2 + \sigma_{SA}^2$	

TABLE 7–4 *Data for a population of four subjects*

	A_1		A_2		A_3			
	Y_{i1}	$(\eta\alpha)_{i1}$	Y_{i2}	$(\eta\alpha)_{i2}$	Y_{i3}	$(\eta\alpha)_{i3}$	μ_i	η_i
S_1	2	-1	6	1	10	0	6	$-.5$
S_2	4	3	5	2	3	-5	4	-2.5
S_3	5	-1	8	0	14	1	9	2.5
S_4	3	-1	3	-3	15	4	7	.5
μ_j	3.5		5.5		10.5		$\mu = 6.5$	
α_j	-3		-1		4			

obvious implication is that nonadditivity permits only a negatively biased test of the subject effect. By negatively biased, we mean that $E(MS_S)/E(MS_{SA}) < 1$ when H_0 is true. Returning to Table 7–3, we find that

$$\frac{E(MS_S)}{E(MS_{SA})} = \frac{\sigma_e^2 + a\sigma_S^2}{\sigma_e^2 + \sigma_{SA}^2}$$

and if H_0 is true ($\sigma_S^2 = 0$), the ratio is clearly less than 1. This tendency for too many Type II errors to result is not a particularly important consequence of nonadditivity, since subject variability is usually large enough to be detected by even the most conservative test. In fact, if there were not strong a priori evidence for significant subject variability, the use of the design would be of questionable value since there would be no basis for expecting precision to be better than with any other design.

7.2.4 Additivity, nonadditivity, and covariance. A more important reason for distinguishing between the additive and nonadditive models is the problem of heterogeneity of covariance. The covariance of two groups of scores is the average cross-product of deviations about the group mean; in the repeated measurements design, the covariance for treatments j and j' is

(7.7) $$\text{cov}\,(Y_{ij}Y_{ij'}) = \frac{\sum_i (Y_{ij} - \overline{Y}_{.j})(Y_{ij'} - \overline{Y}_{.j'})}{n - 1}$$

Note that this is the numerator of the correlation coefficient for the two sets of scores; we could also express the covariance by

(7.8) $$\text{cov}\,(Y_{ij}Y_{ij'}) = r_{j,j'}S_jS_{j'}$$

where $r_{j,j'}$ is the correlation of scores for A_j and $A_{j'}$ and S_j is the standard deviation of scores at A_j.

The importance of the covariance lies in the fact that homogeneity of the treatment population covariances is required for the ratio MS_A/MS_{SA}

to be distributed as F. By homogeneity of the treatment population covariances we mean that $E(Y_{ij} - \mu_{.j})(Y_{ij'} - \mu_{.j'})$ is constant for all values of j and j' $(j \neq j')$. This condition will always be met when the additive model holds; that is, whenever all population interaction effects are zero. We can see this by considering a "population" of scores in which the interaction effects are zero, as in Table 7–5. The covariance is always the same for any two

TABLE 7–5 *A population of scores with all interaction effects zero*

	A_1	A_2	A_3
S_1	3	6	5
S_2	8	11	10
S_3	6	9	8
S_4	9	12	11

treatment levels. A closer look indicates that the covariances are not only all equal to each other, but equal to the variance at any level of j as well. This can be demonstrated by substituting the right hand side of Equation (7.1) for Y_{ij} and $Y_{ij'}$ in the expression for the population covariance. Assuming that the η_i and ε_{ij} are independently distributed and substituting,

$$E(Y_{ij} - \mu_{ij})(Y_{ij'} - \mu_{ij'}) = E(\eta_i^2) = \sigma_S^2$$

If the nonadditive model is correct, heterogeneity of covariance may occur. Table 7–6 presents an example for such a situation. Without any calculations, it is clear that the covariances for the first and second, and for the second and third, columns are negative while those for the first and third columns are positive.

TABLE 7–6 *A nonadditive population of scores*

	A_1	A_2	A_3
S_1	3	12	5
S_2	6	11	8
S_3	8	9	10
S_4	9	6	11

Assuming that the appropriate model for our data is the nonadditive model, we are faced with two questions. What is the direction of the bias in the F test of treatments? What can be done about the bias? In answer to the first question, Box (1954) has shown that the bias is generally positive and that too many Type I errors are made in testing treatment effects when the covariances are heterogeneous. The consequences of heterogeneity of co-variance are greater than those for heterogeneity of variance.

We could approach the problem of bias by first testing for homogeneity of covariances and variances. Such a test is extremely tedious. It involves the calculation of $[a(a - 1)]/2$ correlation coefficients and the inversion of an $a \times a$ matrix containing the sample covariances and variances; the last operation is unfamiliar to most psychologists. For those who are interested, the test is available in a 1950 paper by Box. The present author prefers an alternative that stems from Box's 1954 paper. Box has shown that the statistic MS_A/MS_{SA} is distributed as F but on $(a - 1)\lambda$ and $(a - 1)(n - 1)\lambda$ df. If the homogeneity of variance and the homogeneity of covariance assumptions are met by the data, $\lambda = 1$ and we have the usual F test. As heterogeneity of variances and covariances increases, λ approaches a lower bound of $1/(a - 1)$. Therefore, for an extreme violation of the homogeneity assumptions, F would be distributed on 1 and $(n - 1)$ df. In view of these comments upon the F distribution, it is suggested that the F statistics be first assessed against the F required for significance on 1 and $(n - 1)$ df. In other words, first assume the worst possible degree of heterogeneity of variances and co-variances and consequently set λ equal to $1/(a - 1)$. If λ is actually larger than this value; that is, if the variances and covariances are more homo-geneous than has been assumed, we have thrown away df and are using a negatively biased F test. Thus, if the obtained F is significant with this generally conservative approach, we can have reasonable faith in our rejec-tion of the null hypothesis.

What if the conservative F test just outlined is not significant? It may mean that the null hypothesis is true. On the other hand, nonsignificance may merely indicate that our homogeneity assumptions are not as badly violated as the conservative test implies. To check this last possibility, now assess the obtained F against that required for significance on $a - 1$ and $(n - 1)(a - 1)$ df. In other words, assume homogeneity of variances and covariances and consequently set λ equal to 1. If this positively biased F test is not significant, it is reasonably sure that the null hypothesis should be accepted.

It is possible for the two F tests to yield contradictory results. The F assessed on 1 and $(n - 1)$ df might not be significant, while that assessed on $a - 1$ and $(n - 1)(a - 1)$ df might be significant. Since the first F may be negatively biased and the second F may be positively biased, a clear-cut decision about the null hypothesis is difficult to achieve. However, a statement about the approximate level of significance may generally be made. For example, the negatively biased F may not be significant at the 5-percent level but may be significant at the 10-percent level, while the positively biased F is significant at the 5-percent level, leading to the conclusion that the true significance level is somewhere between 5 percent and 10 percent. If this sort of approximation is not sufficient, an exact F test can be computed (Rao,

1952). A slightly simpler procedure, which yields a reasonable approximation to the nominal α level (Collier, Baker, Mandeville, and Hayes, 1967; Stoloff, 1967), involves computing an estimate of λ, the adjustment for *df*.

The calculation of λ requires values of the variances for each level of A and covariances for each pair of levels of A. The appropriate equation is

$$(7.9) \qquad \hat{\lambda} = a^2(\bar{V}_{jj} - \bar{V}_{..})^2 / \left[(a-1)\left(\sum_{j=1}^{a} \sum_{j'=1}^{a} V_{jj'}^2 - 2a \sum_{j=1}^{a} \bar{V}_j^2 + a^2\bar{V}_{..}^2 \right) \right]$$

where $\bar{V}_{jj}$ is the average of the a variances; that is

$$\bar{V}_{jj} = \sum_{j=1}^{a} \left(\frac{\sum_{i=1}^{n}(Y_{ij} - \bar{Y}_{.j})^2}{a(n-1)} \right)$$

V_{ij} is the covariance for A_j and $A_{j'}$ as computed by Equation (7.7), or by the raw score formula given by Equation (2.20) (note that $\sum_{j=1}\sum_{j'=1}$ is the sum of squared covariances *and* variances), $\bar{V}_j$ is an average of the variance and $a - 1$ covariances associated with treatment A_j; that is

$$\bar{V}_j = \frac{1}{a}(V_{j1} + V_{j2} + \cdots + V_{jj} + \cdots + V_{ja})$$

and $V_{..}$ is the average of the a^2 variances and covariances or, equivalently

$$\bar{V}_{..} = \frac{1}{a} \sum_{j=1}^{a} \bar{V}_j$$

TABLE 7–7 *A variance-covariance matrix*

	A_1	A_2	A_3	A_4	A_5	A_6	$\sum_i V_{ij}$	$\bar{V}_j$
A_1	3.100	.101	−.279	−.083	−.009	1.557	4.387	.731
A_2	.101	5.780	1.013	−.114	−1.014	.039	5.805	.968
A_3	−.279	1.013	5.560	1.039	1.366	−.169	8.530	1.422
A_4	−.083	−.114	1.039	5.600	3.080	.258	9.780	1.630
A_5	−.009	−1.014	1.366	3.080	6.820	.222	10.465	1.744
A_6	1.557	.039	−.169	.258	.222	5.170	7.077	1.179
								7.674

$\bar{V}_{jj}$ is the average of the variances, the diagonal entries = 5.34
$\bar{V}_{..}$ is the average of all 36 entries = 1.28
$(\bar{V}_{ij} - \bar{V}_{..})^2 = 16.48$
$\sum_j \sum_k V_{jk}^2 = 212.86$
$2a\sum_j \bar{V}_j = 126.98$
$a^2\bar{V}_{..}^2 = 58.91$
$\lambda = [(6^2)(16.48)]/[(6-1)(212.86 - 126.98 + 58.91)] = .82$

To illustrate the computations, we use a data set from an article by Greenhouse and Geisser (1959). The variance-covariance matrix is presented in Table 7–7, together with summary statistics; note that the matrix must be symmetric with $V_{jj'} = V_{j'j}$ for all j and j'. Assume $n = 10$. Rounding the adjusted *df* to the nearest integer, we use 4 *df* and 37 *df* in testing the *A* effects in Table 7–7.

7.2.5 Additivity, nonadditivity and efficiency. Even if the covariance problem does not exist, an efficiency problem may. If the data are best described by a nonadditive model, efficiency will be less than in the additive case, and the resulting *F* test will be less powerful. In other words, if the null hypothesis is false, $E(MS_A)/E(MS_{SA})$ where

$$\frac{E(MS_A)}{E(MS_{SA})} = \frac{\sigma_e^2 + n\theta_A^2}{\sigma_e^2}$$

will be greater than $E(MS_A)/E(MS_{SA})$ under the nonadditive model where

$$\frac{E(MS_A)}{E(MS_{SA})} = \frac{\sigma_e^2 + \sigma_{SA}^2 + n\theta_A^2}{\sigma_e^2 + \sigma_{SA}^2}$$

To see why this is so, consider the following illustrative ratios:

$$\frac{15}{5}, \quad \frac{20}{10}, \quad \frac{50}{40}, \quad \frac{100}{90}$$

As we add the same constant to the numerator and denominator of any one ratio, we obtain a smaller ratio. Thus,

$$\frac{15 + 5}{5 + 5} < \frac{15}{5} \quad \text{and} \quad \frac{50 + 50}{40 + 50} < \frac{50}{40}$$

Values of $E(MS_A)/E(MS_{SA})$ for the two models are like any two of the illustrative ratios. If the nonadditive model is correct, the constant σ_{SA}^2 has essentially been added to both numerator and denominator of the value of $E(MS_A)/E(MS_{SA})$ for additive data; thus the size of the expected *F* ratio has been decreased.

We will next provide a numerical example that should further make our point about efficiency and, at the same time, illustrate the calculations involved in analyzing data from the one-factor repeated measurements design. Table 7–8 contains response time scores to which we will apply the analysis of variance. Since the *df* for *subjects* are $n - 1$,

TABLE 7–8 *Response time data*

	A_1	A_2	A_3	$\sum_j Y_{ij}$
S_1	1.7	1.9	2.0	5.6
S_2	4.4	4.5	5.7	14.6
S_3	6.6	7.4	10.5	24.5
$\sum_i Y_{ij} =$ 12.7		13.8	18.2	$\sum_i \sum_j Y_{ij} =$ 44.7

$$SS_S = \frac{\sum_{i=1}^{3}(\sum_{j=1}^{3} Y_{ij})^2}{3} - C$$

$$= \frac{(5.6)^2 + (14.6)^2 + (24.5)^2}{3} - \frac{(44.7)^2}{9}$$

$$= 59.58$$

For the treatment effects, we have

$$SS_A = \frac{\sum_{j=1}^{3}(\sum_{i=1}^{3} Y_{ij})^2}{3} - C$$

$$= \frac{(12.7)^2 + (13.8)^2 + (18.2)^2}{3} - \frac{(44.7)^2}{9}$$

$$= 5.65$$

For the error variability, we have

$$SS_{SA} = \sum_i \sum_j Y_{ij}^2 - C - SS_S - SS_A$$

$$= 291.17 - \frac{(44.7)^2}{9} - 59.58 - 5.65$$

$$= 3.93$$

To test the treatment effects, we compute

$$F = \frac{5.65/2}{3.93/4}$$

$$= 2.87$$

which is not significant at the 5-percent level.

Most investigators would agree that response speed, the reciprocal of response time, is a perfectly meaningful measure. Instead of considering the time per response, the number of responses per unit of time is noted. Speed measurements have been obtained by taking the reciprocals of the data in

TABLE 7–9 *Response speed data obtained by taking reciprocals of the scores in Table 7–8*

	A_1	A_2	A_3	$\sum_j Y_{ij}$
S_1	.589	.526	.500	1.615
S_2	.227	.222	.175	.624
S_3	.152	.135	.095	.382
$\sum_i Y_{ij} =$	.968	.883	.770	$\sum_i \sum_j Y_{ij} = 2.621$

Table 7–8; the results are presented in Table 7–9. The analysis of variance is executed as before, but to help prevent misplaced decimal points, the precaution was taken of multiplying all the entries of Table 7–9 by 1,000. (Note that this will not affect the F ratios.) Then,

$$SS_S = \frac{(1{,}615)^2 + (624)^2 + (382)^2}{3} - \frac{(2{,}621)^2}{9}$$

$$= 284{,}548.3$$

and

$$SS_A = \frac{(968)^2 + (883)^2 + (770)^2}{3} - \frac{(2{,}621)^2}{9}$$

$$= 6{,}577.6$$

and

$$SS_{SA} = 1{,}055{,}389 - \frac{(2{,}621)^2}{9} - 284{,}548.3 - 6{,}577.6$$

$$= 969.7$$

The test of treatment effects yields

$$F = \frac{6{,}577.6/2}{969.7/4}$$

$$= 13.56$$

a result significant at the .01 level.

Why do such diverse inferences result from the two sets of data which, after all, are merely transforms of each other? The answer lies in a closer examination of the data in the two tables. Returning to Table 7–7, we note that the spread among scores at A_3 is considerably greater than that at A_1. In Table 7–9, the variability among subjects seems less affected by the level

FIGURE 7-1 *A plot of the data in Tables 7-8 and 7-9*

of A. The point is more clearly made in Figure 7-1, in which both data sets are plotted, that of Table 7-8 against the left-hand axis and that of Table 7-9 against the right-hand axis. The response time data more clearly contain interaction effects. As a result of this nonadditivity, efficiency is lower and the F test is less powerful.

7.2.6 A test for nonadditivity. Several implications of the distinction between the additive and nonadditive models for repeated measurements designs have been indicated. At best, data that conform to the nonadditive model may result in a less efficient F test of the treatment effect; at worst, the F test may actually be biased. There is another, nonstatistical, consideration. Additive data may be more parsimoniously described than nonadditive data. For example, again consider the data of Figure 7-1. The experimenter who presents a mathematical model of response time as a function of level of A must estimate a different regression coefficient for each subject; each curve has a different slope. On the other hand, looking at the response speed data, only the slope-intercepts differ among subjects. One regression coefficient describes the inclination of all three curves.

In view of the potential advantages of working with additive data, two questions arise. How can one determine which model is appropriate for any given data set? Assuming that the nonadditive model is appropriate, how should the experimenter handle this data? The answers to the question generally involve computational labor beyond the usual analysis of variance and are not always satisfactory. Nevertheless, they seem to merit consideration.

The choice between data models reduces to a test of the null hypothesis that σ_{SA}^2 is zero. If the null hypothesis is rejected, the nonadditive model is appropriate; otherwise, the additive model applies. Tukey (1949) has proposed a single *df* test that involves the further analysis of the interaction sum of squares into two components. One of these represents nonadditivity and is distributed on one *df*. The remaining interaction variability provides the error term for the nonadditivity test and is distributed on $(n - 1)(a - 1) - 1$ *df*.

As an example of the calculations, the Tukey test will be carried out on the response time data originally presented in Table 7–8, and reproduced in Table 7–10. Note that the deviations of each column and row mean from the grand mean have also been tabled. This is the first step in calculating the *F* statistic for additivity. Next obtain the cross-product terms, $\sum_j Y_{ij}(\bar{Y}_{.j} - \bar{Y}_{..})$. These are

$$(-.7)(1.7) + (-.4)(1.9) + (1.1)(2.0) = .25$$

$$(-.7)(4.4) + (-.4)(4.5) + (1.1)(5.7) = 1.39$$

$$(-.7)(6.6) + (-.4)(7.4) + (1.1)(10.5) = 3.97$$

TABLE 7–10 *Data for Tukey's single df test*

	A_1	A_2	A_3	$\bar{Y}_{i.}$	$\bar{Y}_{i.} - \bar{Y}_{..}$	$\sum_j Y_{ij}(\bar{Y}_{.j} - \bar{Y}_{..})$
S_1	1.7	1.9	2.0	1.9	-3.1	.25
S_2	4.4	4.5	5.7	4.9	$-.1$	1.39
S_3	6.6	7.4	10.5	8.2	3.2	3.97
$\bar{Y}_{.j}$	4.3	4.6	6.1	$\bar{Y}_{..} = 5.0$		
$\bar{Y}_{.j} - \bar{Y}_{..}$	$-.7$	$-.4$	1.1			

These are in turn multiplied by the row deviations:

$$\sum_i \sum_j Y_{ij}(\bar{Y}_{.j} - \bar{Y}_{..})(\bar{Y}_{i.} - \bar{Y}_{..}) = (.25)(-3.1) + (1.39)(-.1) + (3.97)(3.2)$$

$$= 11.8$$

The formula for Tukey's single *df* sum of squares is

(7.10) $$SS_{nonadd} = \frac{[\sum_i \sum_j Y_{ij}(\bar{Y}_{.j} - \bar{Y}_{..})(\bar{Y}_{i.} - \bar{Y}_{..})]^2}{\sum_j(\bar{Y}_{.j} - \bar{Y}_{..})^2 \sum_i(\bar{Y}_{i.} - \bar{Y}_{..})^2}$$

For our example, we have

$$SS_{nonadd} = \frac{(11.8)^2}{(1.8)(19.9)}$$

$$= 3.89$$

Subtracting the above from the SS_{SA} for this data computed previously, we have

$$SS_{bal} = 3.93 - 3.89$$

$$= .04$$

The F ratio is

$$F = \frac{MS_{nonadd}}{MS_{bal}}$$

$$= \frac{3.89}{.04/3}$$

$$= 292$$

a result which, even on one and three *df*, is clearly significant.

What do the SS_{nonadd} represent? Some feeling for the answer to this question may be obtained by considering an experiment in which each of n subjects is tested on each of a successive days. We would expect an individual subject's scores to be correlated with the daily averages, the $\overline{Y}_{.j}$; for example, if the group means show improvement over days, then, for any individual subject, we expect the values of Y_{ij} to increase over days. We would have an SA interaction if the rate at which Y_{ij} changed as a function of $\overline{Y}_{.j}$ varied for individual subjects, for then the curves for subjects would not be parallel. Further suppose that an individual's rate of change over values of $\overline{Y}_{.j}$ was correlated to his average performance. For example, the subject with the highest average performance over days might show little change over days because he was already performing well on Day 1, while other subjects with lower values of $\overline{Y}_{i.}$ might show more rapid improvement relative to the changes in $\overline{Y}_{.j}$. It is this type of interaction, this correlation between a subject's average performance and the rate at which his performance changes relative to the changes in the group performance, to which Tukey's test is sensitive.

Some further appreciation of how the test works may be obtained by looking at the cross-products column in Table 7–10. Note that the cross-products (.25, 1.39, 3.97) increase as a function of the row means (1.9, 4.9, 8.2). The nonadditivity sum of squares depends upon the slope of the function relating the cross-products and the means. In contrast, consider the perfectly additive data set of Table 7–11. Note that there is no change in the cross-products as a function of changes in the row means. The SS_{nonadd} will be zero for this data set.

Tukey's test will not be sensitive to all interactions. Clearly, there are interaction effects in Table 7–12, but the SS_{nonadd} equals zero since $[(1)(4) + (0)(0) + (-1)(4)]^2$ equals zero. Why doesn't the Tukey test work here? The answer lies in the cross-products column. Note that the entries

TABLE 7–11 *An additive data set*

	A_1	A_2	A_3	$\bar{Y}_{i.}$	$\bar{Y}_{i.} - \bar{Y}_{..}$	$\sum\limits_{j} Y_{ij}(\bar{Y}_{.j} - \bar{Y}_{..})$
S_1	3	5	1	3	-2	8
S_2	4	6	2	4	-1	8
S_3	8	10	6	8	3	8
$\bar{Y}_{.j}$	5	7	3	$\bar{Y}_{..} = 5$		
$\bar{Y}_{.j} - \bar{Y}_{..}$	0	2	-2			

decrease and then increase as the row means increase. In other words, the slope of the straight line that best relates cross-products to row means is zero. Figure 7–2 contains the plot of cross-products against row means for the

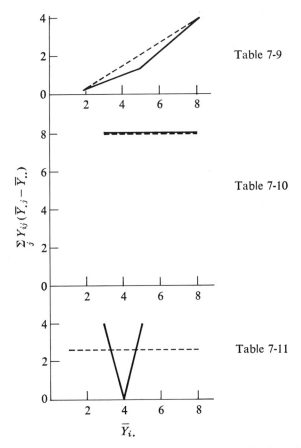

FIGURE 7–2 *A plot of cross-products against row means for three data sets*

TABLE 7–12 *Nonadditive data not sensitive to Tukey's F test*

	A_1	A_2	A_3	$\bar{Y}_{i.}$	$\bar{Y}_{i.} - \bar{Y}_{..}$	$\sum_j Y_{ij}(\bar{Y}_{.j} - \bar{Y}_{..})$
S_1	4	7	4	5	1	4
S_2	4	4	4	4	0	0
S_3	2	5	2	3	−1	4
$\bar{Y}_{.j}$	10/3	16/3	10/3	$\bar{Y}_{..} = 4$		
$\bar{Y}_{.j} - \bar{Y}_{..}$	−2/3	4/3	−2/3			

data sets of Tables 7–10 to 7–12. It is the slope of the dashed line, the best fitting straight line, which is of interest. If Tukey's test is to be sensitive to interaction effects, the slope of the dashed line must deviate from a slope of zero; the greater the departure from a horizontal line, the greater will be the SS_{nonadd}. Fortunately, the nature of most interaction effects is such that the cross-product row-mean relationship will have a linear component and Tukey's test will therefore be serviceable.

7.2.7 Transformations of the data. We turn next to the problem of dealing with data when the evidence suggests that the nonadditive model is appropriate. The example of Tables 7–10 and 7–11 suggest one solution. It was possible to transform the original data to a data set that appears to be well described by an additive model. The choice of an appropriate transformation is not always this simple. Generally, one might first try the transformations recommended in Chapter 4 for heterogeneity of variance; however, the additive transformation is not necessarily the transformation that stabilizes the variances. As a further rough guide, Tukey indicates that when the sign of the sum of cross-products (11.8, in our example) is positive, a square root transform may be appropriate; when the sign is negative, raising each score to a power (e.g., squaring or cubing) or taking the logarithm of each score may work.

7.3 MORE THAN ONE TREATMENT VARIABLE

Although the discussion has thus far been limited to two-dimensional designs involving subjects and a single treatment variable, there is no reason why several treatment variables cannot be investigated. As before, *n* subjects would be randomly sampled, and each one would be tested under all combinations of all treatment variables. Thus, if there were the variables *A* and *B*, each subject would be tested *ab* times, once under each treatment combination, the order of the *ab* presentations being independently randomized for each subject. Our previous comments about the choice between models and the implications of this choice still hold. The computations are straightforward. For a $S \times A \times B$ design, proceed as with a three-factor design (see Chapter 5) with one score in each cell. The analysis for this case is presented in Table 7–13. A nonadditive model, with A and B having fixed effects and subjects having random effects, has been assumed. The composition of each EMS can be summarized by noting that it contains σ_e^2 plus a null hypothesis term. If the effect does not involve S (i.e., if A, B, or AB is being considered), the EMS will also contain the interaction of the effect with S. Thus, for the AB effect, we immediately have $\sigma_e^2 + n\theta_{AB}^2$. Since S is not part of the effect, σ_{ABS}^2 is also included in the EMS. Coefficients, such as the n by which σ_{AB}^2 is multiplied, are determined as in the past. If a factor (including S) does not appear in the subscript, then the number of levels of the factor appear in the coefficient. More general rules will be developed in Chapter 8. However, those just stated will suffice for any simple repeated measurements design involving S and any number of fixed effect variables.

7.4 CONCLUDING REMARKS

The psychological literature abounds with examples of designs involving repeated measurements—both the relatively simple designs of this chapter and the more complex variations to be considered in subsequent chapters. The reasons for this are not difficult to discern. The precision of the design is potentially far better than that of the designs considered before. Furthermore, the design is natural when the supply of subjects is limited relative to the number of treatment combinations to be studied, or when the experimenter's goal is to collect data on some performance measure plotted as a function of time. However, in using these designs the experimenter must be aware of the potential problems. In particular, we have been concerned with the possible consequences of population interactions between subjects and treatments. We have suggested alternative F tests when heterogeneity of covariance is suspected, an F test to detect nonadditivity, and possible ways of transforming the data to an additive scale. We have also indicated that these techniques will not always suffice; adjusting df when heterogeneity is suspected may still not permit a clear inference about the null hypothesis, Tukey's test will not detect all departures from additivity, and appropriate transformations will not always be found. Nevertheless, these techniques do provide a starting point for coping with some of the problems that may result from the use of the repeated measurements design. More important, awareness of the potential problems is the first requirement for deciding whether to use the design and for an intelligent evaluation of summary statistics such as the F ratio.

TABLE 7-13 *Analysis of variance for a two-factor repeated measurements design*

SV	df	SS	EMS	F
Total	$abn - 1$	$\sum_i \sum_j \sum_k Y_{ijk}^2 - C$		
A	$a - 1$	$\dfrac{\sum_j (\sum_i \sum_k Y_{ijk})^2}{nb} - C$	$\sigma_e^2 + b\sigma_{AS}^2 + bn\theta_A^2$	$\dfrac{MS_A}{MS_{AS}}$
B	$b - 1$	$\dfrac{\sum_k (\sum_i \sum_j Y_{ijk})^2}{na} - C$	$\sigma_e^2 + a\sigma_{BS}^2 + an\theta_B^2$	$\dfrac{MS_B}{MS_{BS}}$
S	$n - 1$	$\dfrac{\sum_i (\sum_j \sum_k Y_{ijk})^2}{ab} - C$	$\sigma_e^2 + ab\sigma_S^2$	
AB	$(a - 1)(b - 1)$	$\dfrac{\sum_j \sum_k (\sum_i Y_{ijk})^2}{n} - C - SS_A - SS_B$	$\sigma_e^2 + \sigma_{ABS}^2 + n\theta_{AB}^2$	$\dfrac{MS_{AB}}{MS_{ABS}}$
AS	$(a - 1)(n - 1)$	$\dfrac{\sum_j \sum_i (\sum_k Y_{ijk})^2}{b} - C - SS_A - SS_S$	$\sigma_e^2 + b\sigma_{AS}^2$	
BS	$(b - 1)(n - 1)$	$\dfrac{\sum_k \sum_i (\sum_j Y_{ijk})^2}{a} - C - SS_B - SS_S$	$\sigma_e^2 + a\sigma_{BS}^2$	
ABS	$(a - 1)(b - 1)(n - 1)$	$\sum_i \sum_j \sum_k Y_{ijk}^2 - C - SS_A - \cdots - SS_{BS}$	$\sigma_e^2 + \sigma_{ABS}^2$	

Where the independent variable is something other than time or trial number, it is important that great care be taken to randomize the order of presentation of treatments independently for each subject. In part, this is done to guard against the confounding of time and treatments. What inference can be drawn in the extreme case in which A_1 is always presented first, A_2 always second, and so on? In part, proper randomization of the order of presentation is important because it should serve to minimize heterogeneity of covariance. Scores for treatments close together in time should be more highly correlated than those for treatments further apart in time. By randomizing the treatment presentations independently for each subject, each pair of treatments is given an equal opportunity to appear any given distance apart in time.

It is also helpful to provide sufficient time between presentations of treatments to minimize "carry-over" effects. For example, if rats are being tested in a Skinner box under each of several drugs, time between testings should be sufficient to allow the effects of the last drug to wear off. Even if the different orders of presentation balance out so that treatments and trials are not confounded, carry-over effects, if present, will result in increased variability among orders of presentation and thus will reduce the efficiency of the design.

The design considered in this chapter is the simplest possible design involving repeated measurements. In subsequent chapters complications will be introduced, such as having subjects randomly distributed over levels of a second treatment variable and systematically counterbalancing the orders of presentation of treatments. The importance of this chapter lies in the fact that it provides a relatively simple context in which to consider a number of problems peculiar to all repeated measurements designs—*EMS* under nonadditivity, implications of nonadditivity, the use of transformations, the implications of heterogeneity of covariance. With some understanding of these aspects of repeated measurements designs, it will be possible to concentrate more on the actual data analysis in subsequent chapters.

EXERCISES

7.1 The repeated measurements design is really a special case of the randomized blocks design. Suppose that in the design of Chapter 6, the blocks were chosen at random; they might be social groups, litters of rats, etc. On the basis of the discussion of the present chapter, how would the analysis of the treatment × blocks experiment change?

7.2 Apply Tukey's single *df* test to the data of Table 7–8. How do the results compare with those for Table 7–7?

7.3 Consider the following summary statistics for three sets of data. Which data sets should be sensitive to the Tukey test?

		$\bar{Y}_{i.} - \bar{Y}_{..}$	$\sum_j Y_{ij}(\bar{Y}_{.j} - \bar{Y}_{.1})$
Set 1	S_1	-3	2
	S_2	-1	4
	S_3	1	4
	S_4	3	2
Set 2	S_1	-3	2
	S_2	-1	6
	S_3	1	2
	S_4	3	6
Set 3	S_1	-3	2
	S_2	-1	5
	S_3	1	4
	S_4	3	7

7.4 Prove that $\sum_j(\eta\alpha)_{ij} = 0$ where α is a fixed-effect variable and η is a random-effects variable.

7.5 Compute the *df* adjustment for heterogeneity of covariance for the following data set.

	A_1	A_2	A_3
S_1	12	14	20
S_2	9	6	18
S_3	10	11	23
S_4	8	10	20

7.6 What components of variance would you expect to contribute to the various *EMS* in a three-factor design if A and B were random-effects variables and C had a fixed effect? Model your approach on the reasoning in Section 7.2.2.

REFERENCES

Box, G. E. P., "Problems in the analysis of growth and wear curves," *Biometrics*, 6: 362–389 (1950).

Box, G. E. P., "Some theories on quadratic forms applied in the study of analysis of variance problems: II. Effects of inequality of variance and covariance between errors in the two-way classification," *Annals of Mathematical Statistics*, 25: 484–498 (1954).

Collier, R. O., Baker, F. D., Mandeville, G. K., and Hayes, T. F., "Estimates of test size for several test procedures based on conventional variance ratios in the repeated measure design," *Psychometrika*, 32: 339–353 (1967).

Rao, C. R. *Advanced Statistical Methods in Biometric Research*. Canada: Wiley, 1952.

Stolof, P. H. An empirical evaluation of the effects of violating the assumption of homogeneity of covariance for the repeated measures design of the analysis of variance. University of Maryland, Technical Report TR–66–28, NSG–398, May, 1966.

Tukey, J. W., "One degree of freedom for nonadditivity," *Biometrics*, 5:232–242 (1949).

8

Mixed Designs: Between- and Within-Subjects Variability

8.1 INTRODUCTION

Thus far, two types of designs have been discussed, those in which different treatments involve different subjects and those in which all subjects are tested under each treatment. The most prevalent design in the psychological literature is a combination of these two approaches. For example, n subjects might be tested at A_1, n other subjects tested at A_2, and so on, until an subjects have been accounted for. The subjects will have been randomly assigned to the a treatments, and thus the design will appear to be a completely randomized one-factor design. However, each of the an subjects are also tested at each of the b levels of the independent variable, B, the order of presentation of the b treatments being randomized independently for each subject, as in the repeated measurements design. We will generally refer to A as a *between-subjects* variable and B as a *within-subjects* variable. The data matrix for this mixed design is presented in Table 8–1.

It is not difficult to determine why the mixed designs are so frequently used in psychological research. One reason is that psychologists are often interested in comparing group performances over time or trials. For example, in a free operant situation, different subjects may be tested at different percentages of reinforcement for several successive minutes. Percentage is the A variable and blocks of time is the B variable. As another example, one might measure the signal brightness required for detection at various times after entering a darkened room, different groups of subjects having been exposed to different illuminations prior to the dark-adaptation test. In experiments such as these, the interest lies in comparing the average performance over time for the various levels of A (the A main effect); in determining whether, averaging over all subjects, performance is modified over time (the B main effect); and in determining whether the shapes and slopes of the performance curves are similar for the a groups (the AB interaction effect).

TABLE 8–1 *Data matrix for a mixed design, one between- and one within-subjects variable*

		B_1	B_2	$\cdots$	B_k	$\cdots$	B_b
A_1	S_{11}	Y_{111}	Y_{112}		Y_{11k}		Y_{11b}
	S_{21}	Y_{211}	Y_{212}		Y_{21k}		Y_{21b}
	$\vdots$						
	S_{i1}	Y_{i11}	Y_{i12}		Y_{i1k}		Y_{i1b}
	$\vdots$						
	S_{n1}	Y_{n11}	Y_{n12}		Y_{n1k}		Y_{n1b}
$\vdots$							
A_j	S_{1j}	Y_{1j1}	Y_{1j2}		Y_{1jk}		Y_{1jb}
	S_{2j}	Y_{2j2}	Y_{2j2}		Y_{2jk}		Y_{2jb}
	$\vdots$						
	S_{ij}	Y_{ij1}	Y_{ij2}		Y_{ijk}		Y_{ijb}
	$\vdots$						
	S_{nj}	Y_{nj1}	Y_{nj2}		Y_{njk}		Y_{njb}
$\vdots$							
A_a	S_{1a}	Y_{1a1}	Y_{1a2}		Y_{1ak}		Y_{1ab}
	S_{2a}	Y_{2a1}	Y_{2a2}		Y_{2ak}		Y_{2ab}
	$\vdots$						
	S_{ia}	Y_{ia1}	Y_{ia2}		Y_{iak}		Y_{iab}
	$\vdots$						
	S_{na}	Y_{na1}	Y_{na2}		Y_{nak}		Y_{nab}

A second frequent use of the mixed designs is in psychometric research. Arts and Science, Engineering, and Education majors might be tested on each of several measures in some standard battery of tests. One usually wants to know whether the average performance over the tests is the same for all groups and whether the group profiles are of the same shape, profiles being the bar graphs showing group performance on each measure. Thus, if Arts and Science majors perform better on a verbal aptitude measure and engineers perform better on a quantitative aptitude measure, the profiles will not be parallel. This variability in the simple effects for each measure will be reflected in the magnitude of the AB interaction. At this point it might also be noted that unless the b measures are on the same scale, the profile analysis (test of interaction) is difficult to interpret. Generally, people working with these measures have used Z or T scores; data from different levels of B are then comparable in the sense that the b populations may be considered to have been drawn from the same parent population.

The mixed designs are also appropriate whenever B and AB effects are of greater interest than A effects, since the B and AB effects will usually be tested against a smaller error term with more *df* than will the *A* effects. There will also be instances in which the mixed design is required by the nature of the independent variables; one variable is clearly a between-subjects variable while the second seems to meet the requirements of the repeated measurements model. Variables that will generally be between-subjects variables are those that will entail carry-over effects (e.g., method of training) and individual characteristics (e.g., age, personality characteristic as evidenced by position on some scale). On the other hand, stage of practice is naturally a within-subjects variable.

8.2 ONE BETWEEN- AND ONE WITHIN-SUBJECTS VARIABLE

8.2.1 The analysis of variance model. In past chapters, we have begun with a detailed statement of the model, relating individual scores to parameters of the population from which they have been sampled. In the present instance, we have a somewhat more complicated design. Consequently, it will be easier to begin by partitioning the variance of the scores in Table 8–1 in a way that appears reasonable in light of our experience with the designs of Chapters 4 through 7.

One approach is to structure the design so that it looks more familiar. For example, we might ignore the variable *A*, in which case we would have the simple repeated measurements design of Chapter 7, with *an* subjects and *b* levels of the treatment variable. By analogy to Table 7–2, we would have

SV	*df*
S	$an - 1$
B	$b - 1$
S × *B*	$(an - 1)(b - 1)$

which yields the required total of $abn - 1$ *df*. A second look at the data matrix of Table 8–1 suggests that the above analysis is somewhat unrealistic; it clearly neglects the variable *A*, which is presumably of more than passing interest. Since the three terms above account for all the possible *df*, the *A* source of variance cannot be an additional term but rather must be a component of one of the above terms. Table 8–1 suggests the answer: the *an* subjects may differ because they are at *a* different levels of *A*. Thus the SS_S should be partitioned into two components, an *A* and an *S/A* term on $a - 1$ and $a(n - 1)$ *df* respectively, just as in our treatment of the completely randomized design in Chapter 4. We now have

SV	*df*
S	$an - 1$
A	$a - 1$
S/A	$a(n - 1)$
B	$b - 1$
SB	$(an - 1)(b - 1)$

Returning to Table 8–1, we see that the design could be viewed as a two-factor (A and B) design, with n scores in each cell; our analysis thus far, however, has failed to take into account any possible AB interaction. The necessary $(a - 1)(b - 1)$ df can only be extracted from the SB term. We note that

$$(an - 1)(b - 1) = (a - 1)(b - 1) + a(n - 1)(b - 1)$$

This suggests that

$$SS_{SB} = SS_{AB} + SS_{SB/A}$$

Table 8–2 contains the complete analysis. The indented terms—A, S/A, B, AB, and SB/A—are the ones of interest; the *Between S* and *Within S* terms represent intermediate steps designed to ease the calculations.

The foregoing discussion of the partitioning of the total variability should facilitate our understanding of the model that will now be presented. We assume that the a groups of n subjects are random samples from a corresponding treatment populations consisting of infinite numbers of subjects. This is the exact approach of Chapters 4 and 5. As in Chapter 7, we assume that Y_{ijk}, the observed score for the ijth subject (i.e., the ith subject at the jth level of A) under treatment B_k is randomly sampled for an infinite population of independent measurements taken on that subject under B_k; the expected value of this hypothetical population is μ_{ijk}. Other relevant parameters are μ_{ij}, the expected value of the total population of scores for the ijth subject; μ_j, the expected value of the total population of scores under treatment A_j; μ_k, the expected value for the total population of scores under B_k; μ_{jk}, the expected value for the total population of scores under A_j and B_k. The structural model is

(8.1) $$Y_{ijk} = \mu + \alpha_j + \eta_{i/j} + \beta_k + (\alpha\beta)_{jk} + (\eta\beta)_{ik/j} + \varepsilon_{ijk}$$

where α, β_k, and $(\alpha\beta)_{jk}$ are defined as in previous chapters, and $\eta_{i/j} = \mu_{ij} - \mu_j$, a measure of the unique contribution of the ijth subject;

$$\eta_{ik/j} = (\mu_{ijk} - \mu) - (\mu_k - \mu) - (\mu_{ij} - \mu) - (\alpha\beta)_{jk}$$

$$= \mu_{ijk} - \mu_{ij} - \mu_{jk} + \mu_j,$$ a measure of the interaction effect associated with the ijth subject and kth level of B, and adjusted for the interaction of A_j and B_k;

$\varepsilon_{ijk} = Y_{ijk} - \mu_{ijk}$, the error of measurement.

Assuming that A and B are variables having fixed effects and that subjects are randomly sampled from a large population of subjects, several statements about the population parameters follow. With regard to fixed effects, $\sum\alpha_j = 0$, $\sum\beta_k = 0$, and $\sum(\alpha\beta)_{jk} = 0$. It is further assumed that the $\eta_{i/j}$, $(\eta\beta)_{ik/j}$, and ε_{ijk} are all randomly sampled from normally distributed populations with mean zero and variances $\sigma_{S/A}^2$, $\sigma_{SB/A}^2$, and σ_e^2, respectively.

TABLE 8–2 Analysis of variance for the mixed design, one between- and one within-subjects variable

SV	df	SS	EMS	F
Total	$anb - 1$	$\sum_i^n \sum_j^a \sum_k^b Y_{ijk}^2 - C$		
Between S	$an - 1$	$\dfrac{\sum_i^n \sum_j^a (\sum_k^b Y_{ijk})^2}{b} - C$		
A	$a - 1$	$\dfrac{\sum_j^a (\sum_i^n \sum_k^b Y_{ijk})^2}{nb} - C$	$\sigma_e^2 + b\sigma_{S/A}^2 + nb\theta_A^2$	$\dfrac{MS_A}{MS_{S/A}}$
S/A	$a(n - 1)$	$SS_{B.S} - SS_A$	$\sigma_e^2 + b\sigma_{S/A}^2$	
Within S	$an(b - 1)$	$SS_{tot} - SS_{B.S.}$		
B	$b - 1$	$\dfrac{\sum_k^b (\sum_i^n \sum_j^a Y_{ijk})^2}{na} - C$	$\sigma_e^2 + \sigma_{SB/A}^2 + na\theta_B^2$	$\dfrac{MS_B}{MS_{SB/A}}$
AB	$(a - 1)(b - 1)$	$\dfrac{\sum_j^a \sum_k^b (\sum_i^n Y_{ijk})^2}{n} - C - SS_A - SS_B$	$\sigma_e^2 + \sigma_{SB/A}^2 + n\theta_{AB}^2$	$\dfrac{MS_{AB}}{MS_{SB/A}}$
SB/A	$a(n - 1)(b - 1)$	$SS_{w.s} - SS_B - SS_{AB}$	$\sigma_e^2 + \sigma_{SB/A}^2$	

8.2.2 The analysis of variance. Table 8–2 presents the *SV*, *df*, *SS*, *EMS*, and *F* (*MS* have been omitted; as usual, they are simply ratios of *SS* to *df*). The *SV* are consistent with Equation (8.1), with one source for each term in the equation except ε_{ijk}. The omission of an independent error term is consistent with the analyses of Chapter 7. Since there is only a single score in each combination of *S*, *A*, and *B*, no within-cells variability exists, and the sources of Table 8–2 therefore account for the total variability in the data matrix. For convenience, we have grouped the sources into two sets, those that account for the variability between subjects and those that account for the variability within subjects. Within the first set, we have *A* and *S/A*, corresponding to α_j and $\eta_{i/j}$ effects. Within the second set of terms, we have *B*, *AB*, and *SB/A*, corresponding to β_k, $(\alpha\beta)_{jk}$, and $(\eta\beta)_{ik/j}$. The nature of the correspondence will become clearer shortly when we turn to the *EMS*.

Scanning the *SV*, the reader may wonder why there is no *SA* term present. The answer lies in the distinction between *crossing* and *nesting*. When data are obtained for all combinations of two variables, the variables are said to cross. In this case, an interaction sum of squares may be computed for the two variables, since the question, Is the difference in the effects of *A* a function of the level of *B*?, is a meaningful one. Subjects and *B*, and *A* and *B*, cross in the design under discussion.

Scores cannot be obtained for all combinations of subjects and levels of *A*, since any given subject appears in combination with only one level of *A*. The question of interaction is meaningless. Consider asking whether the difference between Subject 1 and Subject 2 is greater at A_1 than at A_2. In our design, either both subjects appear only at A_1, or both appear only at A_2, or one appears only at A_1 while the other appears only at A_2. In any of these cases, the question posed above has no answer. What we have instead of crossing is the nesting of subjects within levels of *A*; i.e., there are *n* subjects at A_1, *n* others at A_2, and so on.

How do we interpret the nested terms? It is as if the analysis of variance were carried out separately at each level of *A* and the resulting sums of squares were then pooled. Thus, $SS_{S/A}$ could be obtained by computing the variability for each set of *n* subjects' means about their mean and summing the *a* terms so obtained. Equivalently, it is that variability among subjects that remains after we remove the variability due to the fact that subjects are at different levels of *A*. Similarly, we could obtain $SS_{SB/A}$ by computing an *SB* interaction within each level of *A* and then summing the resulting *a* terms. Or we can view $SS_{SB/A}$ as the interaction variability for subjects and *B* that remains after we remove the contribution due to the interaction variability of *A* and *B*.

Next turn to the *df* column of Table 8–2. No discussion of the entries for *A*, *B*, or *AB* seems necessary; their rationale has been previously considered. The *df* for the *between-subjects* source reflect the variability of *an* means about the grand mean. The $df_{S/A}$ may be calculated as a residual:

$$df_{S/A} = df_{B.S} - df_A$$
$$a(n - 1) = (an - 1) - (a - 1)$$

or we may note that, at each level of A, the variability of subject means about their mean is based on $n - 1$ df. Pooling over levels of A gives the appropriate result.

The *within-subjects* variability is obtained by computing for each subject the variability of his b measures about their mean and then pooling over subjects. Therefore, we have $b - 1$ df for each subject, and since there are an subjects, pooling results in $an(b - 1)$ df. Alternatively,

$$df_{W.S} = df_{tot} - df_{B.S}$$
$$an(b - 1) = (abn - 1) - (an - 1)$$

The $df_{SB/A}$ may be computed in several ways. It is a residual from the *within-subjects* variability:

$$df_{SB/A} = df_{W.S} - df_B - df_{AB}$$
$$= an(b - 1) - (b - 1) - (a - 1)(b - 1)$$
$$= a(n - 1)(b - 1)$$

The $df_{SB/A}$ reflects the difference between the overall interaction of subjects and levels of B (disregarding the presence of the variable A) and the AB interaction:

$$df_{SB/A} = df_{SB} - df_{AB}$$
$$= (an - 1)(b - 1) - (a - 1)(b - 1)$$
$$= a(n - 1)(b - 1)$$

The $df_{SB/A}$ reflects the SB variability at each level of A, pooled over levels of A:

$$df_{SB/A} = (n - 1)(b - 1) + \cdots + (n - 1)(b - 1)$$
$$= a(n - 1)(b - 1)$$

The computational formulas for sums of squares present no difficulties. They follow the same logic just developed for df. In those cases where several approaches to the calculation of df (e.g., SB/A) have been presented, the simplest one has been used as a basis for computing sums of squares. An alternative calculation for $SS_{SB/A}$ will be suggested to prepare the reader for subsequent designs in which similar nested terms cannot always be calculated as simple residual quantities. The alternative approach rests upon the expansion of df proposed earlier in this book. We begin with

$$df_{SB/A} = a(n - 1)(b - 1)$$
$$= anb - an - ab + a$$

We require a squared quantity for each df; therefore, we have

$$\sum_i^n \sum_j^a \sum_k^b (\quad)^2 - \sum_i^n \sum_j^a (\quad)^2 - \sum_j^a \sum_k^b (\quad)^2 + \sum_j^a (\quad)^2$$

Since all indices of summation must be represented, we now have

$$\sum_i^n \sum_j^a \sum_k^b (Y_{ijk})^2 - \sum_i^n \sum_j^a \left(\sum_k^b Y_{ijk}\right)^2 - \sum_j^a \sum_k^b \left(\sum_i^n Y_{ijk}\right)^2 + \sum_j^a \left(\sum_i^n \sum_k^b Y_{ijk}\right)^2$$

Since each squared quantity must be divided by the number of scores summed prior to squaring, the final result is

$$(8.2) \quad S_{SB/A} = \sum_i^n \sum_j^a \sum_k^b Y_{ijk}^2 - \frac{\sum_i^n \sum_j^a (\sum_k^b Y_{ijk})^2}{b} - \frac{\sum_j^a \sum_k^b (\sum_i^n Y_{ijk})^2}{n}$$
$$+ \frac{\sum_j^a (\sum_i^n \sum_k^b Y_{ijk})^2}{nb}$$

Note that each of the four component terms has been computed previously as part of some other sum of squares quantity. For example, $\sum_i^n \sum_j^a (\sum_k^b Y_{ijk})^2/b$ is calculated in obtaining $SS_{B.S}$. Where do the other three terms appear among other sum of squares entries?

The rationale for the *EMS* is essentially that of Section 7.2.2 for the nonadditive repeated measurements model. However, the designs are now becoming sufficiently complex so that some rules for generating *EMS* would seem helpful. The following will apply to designs involving fixed effects, random effects, or both. The major stipulation in the use of these rules is that an infinite number of levels are assumed in populations from which a random sample is obtained. This assumption will be approximately correct in most psychological research.

RULES OF THUMB FOR GENERATING *EMS*

RULE 1. Decide for each independent variable (*including subjects*) whether it is fixed or random. Assign a letter to designate each variable. Assign another letter to be used as a coefficient that represents the number of levels of each variable. In the example of Table 8–2, the variables are designated A, B, and S; the coefficients are a, b, and n; A and B are fixed-effect variables while S is random.

RULE 2. List σ_e^2, as part of each *EMS*.

RULE 3. For each *EMS* list the null hypothesis component; i.e., the component corresponding directly to the *SV* under consideration. Thus we add $nb\theta_A^2$ to the *EMS* for the *A* line, $b\sigma_{S/A}^2$ to the *EMS* for the *S/A* line. Note that a component consists of three parts:

(1) a coefficient representing the number of scores at each level of the effect (e.g., *nb* scores at each level of *A*, or *b* scores for each subject),
(2) a σ^2 or θ^2, depending upon whether the effect is assumed to be random or fixed [σ^2 is the variance of the population of effects; e.g., $\sigma_{S/A}^2 = E(\eta_{i/j}^2)$; $\theta_A^2 = \sum_j \alpha_j^2/(a - 1)$], and
(3) as subscripts, those letters that designate the effect under consideration.

RULE 4. Now add to each *EMS* all components whose subscripts contain *all* of the letters designating the *SV* in question. For example, since the subscript *SB/A* contains the letters *S* and *A*, add $\sigma_{SB/A}^2$ to the *EMS* for the *S/A* line (this is later deleted according to Rule 6).

RULE 5. Next, examine the components for each *SV*. If a slash appears in the subscript, define only the letters to the left of the slash as "essential." If there are several slashes (as in the next chapter), only the letters prior to the leftmost slash are essential. If there is no slash in the subscript, all letters are considered essential.

RULE 6. Among the essential letters, ignore any that are necessary to designate the *SV*. For example, if the source is *A*, when considering $n\theta_{AB}^2$, ignore the *A*. If the source is *S/A*, in considering the $\sigma_{SB/A}^2$ component, *S* and *B* are essential subscripts and *S* is to be ignored. If *any* of the remaining essential letters designate fixed variables, delete the entire component from the *EMS*. Thus, in the preceding examples, since *B* represents a fixed variable, $n\theta_{AB}^2$ does not contribute to the *EMS* for *A* and $\sigma_{SB/A}^2$ does not contribute to the *EMS* for *S/A*.

In Table 8–3 are listed the *EMS* as they would appear following the execution of Rules 1 to 4. The underlined components are those deleted on

TABLE 8–3 EMS *for the mixed design, one between- and one within-subjects variable*

SV	EMS			
A	$\sigma_e^2 + nb\theta_A^2$		$+ \underline{n\theta_{AB}^2} + b\sigma_{S/A}^2$	$+ \underline{\sigma_{SB/A}^2}$
S/A	σ_e^2		$+ b\sigma_{S/A}^2$	$+ \underline{\sigma_{SB/A}^2}$
B	σ_e^2	$+ na\theta_B^2 + \underline{n\theta_{AB}^2}$		$+ \sigma_{SB/A}^2$
AB	σ_e^2		$+ n\theta_{AB}^2$	$+ \sigma_{SB/A}^2$
SB/A	σ_e^2			$+ \sigma_{SB/A}^2$

the basis of Rule 6. If the underlined components are erased, we have the results in the *EMS* column of Table 8–2.

With the *EMS* available, the *F* tests readily follow. As always, error terms are required such that the ratio of *EMS* equals 1 when the null hypothesis is true. Such error terms are possible for tests of *A*, *B*, and *AB* effects. One word of caution is necessary with regard to the tests of *B* and *AB* effects; heterogeneity of covariance can positively bias the test, as in the simpler repeated measurements designs of Chapter 7. We will consider this problem in more detail in Section 8.4 after presenting several more mixed designs.

8.2.3 A numerical example. An illustrative set of data is presented in Table 8–4. Running only two subjects at each level of *A* is not generally recommended; it is hoped that the reader will use this example as a model for computations rather than for experimentation.

Prior to the actual analysis of variance, subtotals are obtained for each subject, for each level of *A*, for each level of *B*, and for each *AB* combination. Since all of these quantities are eventually used in the analysis, it is wise to have them available from the start.

TABLE 8–4 *Data for a numerical example, one between- and one within-subjects variable*

		B_1	B_2	B_3	$\sum_k Y_{ijk}$
A_1	S_{11}	7	1	7	15
	S_{21}	9	2	10	21
	$\sum_i Y_{i1k} = 16$		3	17	$\sum_i \sum_k Y_{i1k} = 36$
A_2	S_{12}	11	6	7	24
	S_{22}	16	14	9	39
	$\sum_i Y_{i2k} = 27$		20	16	$\sum_i \sum_k Y_{i2k} = 63$
	$\sum_i \sum_j Y_{ijk} = 43$		23	33	$\sum_i \sum_j \sum_k Y_{ijk} = 99$

We first calculate the correction term. As usual, this is the squared total of all scores divided by the total number of observations. In this case,

$$C = \frac{(99)^2}{12}$$

$$= 816.75$$

The SS_{tot} is obtained by squaring each score, summing the squared scores and subtracting C. Thus, we have

$$SS_{tot} = \sum_i^n \sum_j^a \sum_k^b Y_{ijk} - C$$

$$= (7)^2 + (1)^2 + \cdots + (14)^2 + (9)^2 - C$$

$$= 1{,}023.00 - 816.75$$

$$= 206.25$$

We next turn to the first major component of the SS_{tot}, the $SS_{B.S}$:

$$SS_{B.S} = \frac{\sum_i^n \sum_j^a (\sum_k^b Y_{ijk})^2}{b} - C$$

$$= \frac{(15)^2 + (21)^2 + (24)^2 + (39)^2}{3} - C$$

$$= \frac{2{,}763}{3} - 816.75$$

$$= 104.25$$

The $SS_{B.S}$ are then partitioned:

$$SS_A = \frac{\sum_j^a (\sum_i^n \sum_k^b Y_{ijk})^2}{nb} - C$$

$$= \frac{(36)^2 + (63)^2}{6} - C$$

$$= \frac{5{,}265}{6} - 816.75$$

$$= 60.75$$

and

$$SS_{S/A} = SS_{B.S} - SS_A$$

$$= 104.25 - 60.75$$

$$= 43.50$$

The second major component of the SS_{tot} is the $SS_{W.S}$. This is simply the difference between the SS_{tot} and $SS_{B.S}$. Therefore,

$$SS_{W.S} = SS_{tot} - SS_{B.S}$$

$$= 206.25 - 104.25$$

$$= 102.00$$

The within-subjects variability is now analyzed into its components:

$$SS_B = \frac{\sum_k^b (\sum_i^n \sum_j^a Y_{ijk})^2}{na} - C$$

$$= \frac{(43)^2 + (23)^2 + (33)^2}{4} - C$$

$$= \frac{3,467}{4} - 816.75$$

$$= 50.00$$

and

$$SS_{AB} = \frac{\sum_j^a \sum_k^b (\sum_i^n Y_{ijk})^2}{n} - C - SS_A - SS_B$$

$$= \frac{(16)^2 + (3)^2 + \cdots + (16)^2}{2} - C - SS_A - SS_B$$

$$= \frac{1,939}{2} - 816.75 - 60.75 - 50.00$$

$$= 42.00$$

and

$$SS_{SB/A} = SS_{W.S} - SS_B - SS_{AB}$$

$$= 102.00 - 50.00 - 42.00$$

$$= 10.00$$

Table 8–5 summarizes the analysis. Assuming that our α level is .05, both the B and AB effects are significant. To instill some meaning into this

TABLE 8–5 *Analysis of variance for the data of Table 8–4*

SV	df	SS	MS	F
Total	11	206.25		
Between S	3	104.25		
A	1	60.75	60.75	2.79
S/A	2	43.50	21.75	
Within S	8	102.00		
B	2	50.00	25.00	10.00*
AB	2	42.00	21.00	8.40*
SB/A	4	10.00	2.50	
				*p < .05

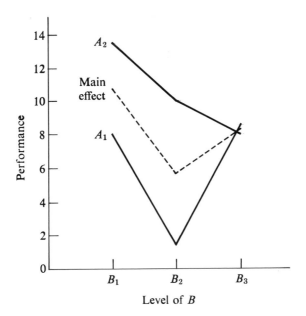

FIGURE 8–1 *A plot of cell means of Table* 8–4

last statement, we turn to Figure 8–1, which contains a plot of the six cell means. In addition, the dashed line represents the main effect of B. The source of the significant B effect is due to the roughly V-shaped function obtained when the average performance is plotted against the levels of B. If high scores are desirable, it would seem that B_1 is the preferred treatment. It is tempting to go somewhat further and more precisely conclude that B_1 is better than both B_2 and B_3, and that B_3 is better than B_2; however, our F test only permits the inference that at least one of these three contrasts involves a difference in the population. Presumably, it is safe to assume that B_1 and B_2 differ in their effects, since the difference in their means is greatest. In Chapter 13 the set of three contrasts will be evaluated as well as others that are not quite so obvious.

It has just been noted that inferences about the population effects of B are limited until further comparisons can be made within pairs of means. Our conclusions about B are now further qualified by noting that the AB interaction is significant. The simple effects of B at each level of A are not identical to the main effect. The source of the AB interaction is not difficult to detect in Figure 8–1. The B means, when plotted at A_1, form a symmetrical V-shaped function; when the means are plotted for the A_2 data, the function decreases monotonically as the level of B increases.

8.3 ADDITIONAL MIXED DESIGNS

Countless variations of the design in the preceding section have appeared in the experimental journals. They should present no insurmountable difficulties. The developments follow those of Section 8.2; to illustrate this, two additional mixed designs will be dealt with in the present section.

8.3.1 Two between- and one within-subjects variables. Table 8–6 presents the data matrix for the design to be considered next. A random sample from a

TABLE 8–6 *Data matrix for a mixed design, two between- and one within-subjects variables*

			C_1	$\cdots$	C_m	$\cdots$	C_c
		S_{111}	Y_{1111}		Y_{111m}		Y_{111c}
		$\vdots$					
	A_1	S_{i11}	Y_{i111}		Y_{i11m}		Y_{i11c}
		$\vdots$					
		S_{n11}	Y_{n111}		Y_{n11m}		Y_{n11c}
	$\vdots$						
B_1	A_j	$\vdots$ S_{ij1} $\vdots$	Y_{ij11}		Y_{ij1m}		Y_{ij1c}
	$\vdots$						
	A_a	$\vdots$ S_{ia1} $\vdots$	Y_{ia11}		Y_{ia1m}		Y_{ia1c}
$\vdots$							
B_k	A_j	$\vdots$ S_{ijk} $\vdots$	Y_{ijk1}		Y_{ijkm}		Y_{ijkc}
$\vdots$							
B_b	A_j	$\vdots$ S_{ijb} $\vdots$	Y_{ijb1}		Y_{ijbm}		Y_{ijbc}

large population of subjects has been randomly distributed among the *ab* combinations of levels of the variables *A* and *B*, with the restriction that there may be *n* subjects in each combination. Each subject is tested at *c* levels of the variable *C*, the order of presentation of the *c* treatments being independently randomized for each subject. The notation used in Table 8–6, and throughout this section, will be as follows:

$$i \quad 1, 2, \ldots, n$$
$$j \quad 1, 2, \ldots, a$$
$$k \quad 1, 2, \ldots, b$$
$$m \quad 1, 2, \ldots, c$$

Development of the model. As in Section 8.2.1, we initially view the design as consisting of two factors: abn subjects and c levels of the within-subject variable C. This suggests

SV	df
Between S	$abn - 1$
C	$c - 1$
Between S $\times$ C	$(abn - 1)(c - 1)$

which accounts for the total df, $abcn - 1$. We next consider the main and interaction effects involving A and B, the between-subject variables. These sources are extracted from the between subject variability, yielding

SV	df
Between S	$abn - 1$
A	$a - 1$
B	$b - 1$
AB	$(a - 1)(b - 1)$
S/AB	$ab(n - 1)$
C	$c - 1$
Between S $\times$ C	$(abn - 1)(c - 1)$

This breakdown is still incomplete; it is clear from Table 8–6 that A and B, as well as subjects, cross with C and that such interaction terms must be taken into account. If we consider the interaction of each of the four between-subject terms with C, we obtain four additional terms: AC, BC, ABC, and SC/AB. The corresponding df are

$$(a - 1)(c - 1) \quad (b - 1)(c - 1) \quad (a - 1)(b - 1)(c - 1)$$

and

$$ab(n - 1)(c - 1)$$

which together add to $(abn - 1)(c - 1)$, the df for the *Between S* $\times$ C term. Replacing that term by its four components, we obtain the final result, Table 8–7.

A model consistent with the above is provided by

$$(8.3) \qquad Y_{ijkm} = \mu + \alpha_j + \beta_k + (\alpha\beta)_{jk} + \eta_{i/jk} + \gamma_m + (\alpha\gamma)_{jm} + (\beta\gamma)_{km}$$
$$+ (\alpha\beta\gamma)_{jkm} + (\eta\gamma)_{im/jk} + \varepsilon_{ijkm}$$

TABLE 8-7 *Analysis of variance for the mixed design, two between- and one within-subjects variables*

SV	df	SS	EMS	F
Total	$abnc - 1$	$\sum_i^n \sum_j^a \sum_k^b \sum_m^c Y_{ijkm}^2 - C$		
Between S	$abn - 1$	$\dfrac{\sum_i^n \sum_j^a \sum_k^b (\sum_m^c Y_{ijkm})^2}{c} - C$		
A	$a - 1$	$\dfrac{\sum_j^a (\sum_i^n \sum_k^b \sum_m^c Y_{ijkm})^2}{nbc} - C$	$\sigma_e^2 + c\sigma_{S/AB}^2 + nbc\theta_A^2$	$\dfrac{MS_A}{MS_{S/AB}}$
B	$b - 1$	$\dfrac{\sum_k^b (\sum_i^n \sum_j^a \sum_m^c Y_{ijkm})^2}{nac} - C$	$\sigma_e^2 + c\sigma_{S/AB}^2 + nac\theta_B^2$	$\dfrac{MS_B}{MS_{S/AB}}$
AB	$(a-1)(b-1)$	$\dfrac{\sum_j^a \sum_k^b (\sum_i^n \sum_m^c Y_{ijkm})^2}{nc} - C - SS_A - SS_B$	$\sigma_e^2 + c\sigma_{S/AB}^2 + nc\theta_{AB}^2$	$\dfrac{MS_{AB}}{MS_{S/AB}}$
S/AB	$ab(n-1)$	$SS_{B.S} - SS_A - SS_B - SS_{AB}$	$\sigma_e^2 + c\sigma_{S/AB}^2$	
Within S	$abn(c-1)$	$SS_{tot} - SS_{B.S}$		
C	$(c-1)$	$\dfrac{\sum_m^c (\sum_i^n \sum_j^a \sum_k^b Y_{ijkm})^2}{nab} - C$	$\sigma_e^2 + \sigma_{SC/AB}^2 + nba\theta_C^2$	$\dfrac{MS_C}{MS_{SC/AB}}$
AC	$(a-1)(c-1)$	$\dfrac{\sum_j^a \sum_m^c (\sum_i^n \sum_k^b Y_{ijkm})^2}{nb} - C - SS_A - SS_C$	$\sigma_e^2 + \sigma_{SC/AB}^2 + nb\theta_{AC}^2$	$\dfrac{MS_{AC}}{MS_{SC/AB}}$
BC	$(b-1)(c-1)$	$\dfrac{\sum_k^b \sum_m^c (\sum_i^n \sum_j^a Y_{ijkm})^2}{na} - C - SS_B - SS_C$	$\sigma_e^2 + \sigma_{SC/AB}^2 + na\theta_{BC}^2$	$\dfrac{MS_{BC}}{MS_{SC/AB}}$
ABC	$(a-1)(b-1)(c-1)$	$\dfrac{\sum_j^a \sum_k^b \sum_m^c (\sum_i^n Y_{ijkm})^2}{n} - C - SS_{AB} - SS_{AC}$ $- SS_{BC} - SS_{AC} - SS_B - SS_C$	$\sigma_e^2 + \sigma_{SC/AB}^2 + n\theta_{ABC}^2$	$\dfrac{MS_{ABC}}{MS_{SC/AB}}$
SC/AB	$ab(n-1)(c-1)$	$SS_{w.s} - SS_C - SS_{AC} - SS_{BC} - SS_{ABC}$	$\sigma_e^2 + \sigma_{SC/AB}^2$	

where α_j, β_k, and γ_m are the main effects associated with A_j, B_k, and C_m, their interaction effects are defined as usual, and $\eta_{i/jk}$ is the unique contribution associated with the ith subject in A_jB_k, $(\eta\gamma)_{im/jk}$ is the nested subject $\times C$ interaction effect, and ε_{ijkm} is the error of measurement associated with sampling Y_{ijkm}. It is assumed that A, B, and C are fixed-effect variables; this, as usual, implies that the sum of effects over the appropriate indices is zero. The ε_{ijkm}, $\eta_{i/jk}$, and $(\eta\gamma)_{im/jk}$ effects are randomly sampled from infinite populations of such effects. The population distributions are normal, with zero expected value and with variances σ_e^2, $\sigma_{S/AB}^2$, and $\sigma_{SC/AB}^2$, respectively.

The analysis of variance. The analysis of variance for the two between- and one within-subjects design is summarized in Table 8–6. The *SV* are consistent with Equation (8.3). The *df* are derived as in Section 8.2.2. The df_{tot}, as usual, are one less than the total number of scores, which is *abcn* in the present case. This quantity is partitioned into two parts, the $df_{\text{B.S}}$, which is one less than the total number of subjects, and the $df_{\text{w.s}}$, which is the *df* for each subject $(c - 1)$ pooled over the number of subjects. The *df* for all treatment main and interaction effects are identical to those presented for previous designs, and the *df* for the two nested effects can be readily obtained as residuals. Computational formulas for sums of squares all follow the precedents set in the preceding section and in preceding chapters. The results of applying the rules of thumb for generating *EMS* are similar to those achieved in Section 8.2; we again obtain two error terms, one for between-subjects and one for within-subjects effects.

A numerical example. An analysis of variance on the data of Table 8–8 will demonstrate the computational process. We first compute

$$SS_{\text{tot}} = \sum_i^n \sum_j^a \sum_k^b \sum_m^c Y_{ijkm}^2 - C$$

$$= (4)^2 + (8)^2 + \cdots + (6)^2 + (8)^2 - \frac{(169)^2}{24}$$

$$= 1{,}403.000 - 1{,}190.042$$

$$= 212.958$$

The total variability among subjects is

$$SS_{\text{B.S}} = \frac{\sum_i^n \sum_j^a \sum_k^b (\sum_m^c Y_{ijkm})^2}{c} - C$$

$$= \frac{(22)^2 + (27)^2 + \cdots + (15)^2}{3} - \frac{(169)^2}{24}$$

$$= 3{,}721.000 - 1{,}190.042$$

$$= 50.291$$

TABLE 8–8 *Data for a numerical example, two between- and one within-subjects design*

			C_1	C_2	C_3	$\sum_m Y_{ijkm}$
A_1	B_1	S_{111}	4	8	10	22
		S_{211}	6	9	12	27
		$\sum_i Y_{i11m} = 10$		17	22	$\sum_i \sum_m Y_{i11m} = 49$
	B_2	S_{112}	3	7	11	21
		S_{212}	5	11	12	28
		$\sum_i Y_{i12m} = 8$		18	23	$\sum_i \sum_m Y_{i12m} = 49$
		$\sum_i \sum_k Y_{i1km} = 18$		35	45	$\sum_i \sum_k \sum_m Y_{i1km} = 98$
A_2	B_1	S_{121}	4	6	9	19
		S_{221}	5	8	8	21
		$\sum_i Y_{i21m} = 9$		14	17	$\sum_i \sum_m Y_{i21m} = 40$
	B_2	S_{122}	4	3	9	16
		S_{222}	1	6	8	15
		$\sum_i Y_{i22m} = 5$		9	17	$\sum_i \sum_m Y_{i22m} = 31$
		$\sum_i \sum_k Y_{i2km} = 14$		23	34	$\sum_i \sum_k \sum_m Y_{i2km} = 71$
		$\sum_i \sum_j \sum_k Y_{ijkm} = 32$		58	79	$\sum_i \sum_j \sum_k \sum_m Y_{ijkm} = 169$

Subtotals for *BC* Cells

	C_1	C_2	C_3	$\sum_i \sum_j \sum_m Y_{ijkm}$
B_1	19	31	39	89
B_2	13	27	40	80
$\sum_i \sum_j \sum_k Y_{ijkm} = 32$	58	79		$\sum_i \sum_j \sum_k \sum_w Y_{ijkm} = 169$

This variability may be partitioned into several components:

$$SS_A = \frac{\sum_j^a (\sum_i^n \sum_k^b \sum_m^c Y_{ijkm})^2}{nbc} - C$$

$$= \frac{(98)^2 + (71)^2}{12} - \frac{(169)^2}{24}$$

$$= 1{,}220.417 - 1{,}190.042$$

$$= 30.375$$

$$SS_B = \frac{\sum_k^b (\sum_i^n \sum_j^a \sum_m^c Y_{ijkm})^2}{nac} - C$$

$$= \frac{(89)^2 + (80)^2}{12} - \frac{(169)^2}{24}$$

$$= 1{,}193.417 - 1{,}190.042$$

$$= 3.375$$

$$SS_{AB} = \frac{\sum_j^a \sum_k^b (\sum_i^n \sum_m^c Y_{ijkm})^2}{nc} - C - SS_A - SS_B$$

$$= \frac{(49)^2 + (49)^2 + (40)^2 + (31)^2}{6} - \frac{(169)^2}{24} - SS_A - SS_B$$

$$= 1{,}227.167 - 1{,}190.042 - 30.375 - 3.375$$

$$= 3.375$$

$$SS_{S/AB} = SS_{B.S} - SS_A - SS_B - SS_{AB}$$

$$= 50.291 - 30.375 - 3.375 - 3.375$$

$$= 13.166$$

Note that the A, B, and AB terms may be swiftly computed by the single df approach first presented in Section 5.10. Thus,

$$SS_A = \frac{(98 - 71)^2}{24} \qquad\qquad SS_B = \frac{(89 - 80)^2}{24}$$

$$= 30.375 \qquad\qquad\qquad\qquad = 3.375$$

and

$$SS_{AB} = \frac{(49 - 49 - 40 - 31)^2}{24}$$

$$= 3.375$$

The within-subjects variability may be obtained as the difference between the total and the between-subjects variability:

$$SS_{W.S} = SS_{tot} - SS_{B.S}$$
$$= 212.958 - 50.291$$
$$= 162.667$$

This term is next partitioned into its components. We have

$$SS_C = \frac{\sum_m^c(\sum_i^n \sum_j^a \sum_k^b Y_{ijkm})^2}{nab} - C$$

$$= \frac{(32)^2 + (58)^2 + (79)^2}{8} - \frac{(169)^2}{24}$$

$$= 1,328.625 - 1,190.042$$

$$= 138.583$$

$$SS_{AC} = \frac{\sum_j^a \sum_m^c(\sum_i^n \sum_k^b Y_{ijkm})^2}{nb} - C - SS_A - SS_C$$

$$= \frac{(18)^2 + (35)^2 + \cdots + (34)^2}{4} - \frac{(169)^2}{24} - SS_A - SS_C$$

$$= 1,363.750 - 1,190.042 - 30.375 - 138.583$$

$$= 4.750$$

$$SS_{BC} = \frac{\sum_k^b \sum_m^c(\sum_i^n \sum_j^a Y_{ijkm})^2}{ma} - C - SS_B - SS_C$$

$$= \frac{(19)^2 + (31)^2 + \cdots + (40)^2}{4} - \frac{(169)^2}{24} - SS_B - SS_C$$

$$= 1,335.250 - 1,190.042 - 3.375 - 138.583$$

$$= 3.250$$

$$SS_{ABC} = \frac{\sum_j^a \sum_k^b \sum_m^c(\sum_i^n Y_{ijkm})^2}{n} - C - SS_{AB} - SS_{AC} - SS_{BC} - SS_A$$
$$- SS_B - SS_C$$

$$= \frac{(10)^2 + (17)^2 + \cdots + (9)^2 + (17)^2}{2} - \frac{(169)^2}{24} - SS_A - SS_B$$
$$- SS_C - SS_{AB} - SS_{AC} - SS_{BC}$$

$$= 1,375.500 - 1,190.042 - 30.375 - 3.375 - 138.583 - 3.375$$
$$- 4.750 - 3.250$$

$$= 1.750$$

and

$$SS_{SC/AB} = SS_{w.s} - SS_C - SS_{AC} - SS_{BC} - SS_{ABC}$$
$$= 162.667 - 138.583 - 4.750 - 3.250 - 1.750$$
$$= 14.334$$

The final results of the analysis are summarized in Table 8–9. Clearly, only the A and C main effects are significant; the population mean for the A_1 treatment is higher than for the A_2 treatment, and the mean performance is also an increasing function of the level of C.

TABLE 8–9 *Analysis of variance for the data of Table 8–8*

SV	df	SS	MS	F
Total	23	212.958		
Between S	7	50.291		
A	1	30.375	30.375	9.228*
B	1	3.375	3.375	1.025
AB	1	3.375	3.375	1.025
S/AB	4	13.166	3.292	
Within S	16	162.667		
C	2	138.583	69.292	38.674*
AC	2	4.750	2.375	1.325
BC	2	3.250	1.625	.907
ABC	2	1.750	.875	.488
SC/AB	8	14.334	1.792	
				* p < .01

8.3.2 One between- and two within-subjects variables. This design is of interest because it requires consideration of a more complex analysis of within-subjects variability than any previously discussed. The design is quite common; indeed, three and even four within-subjects variables are frequently manipulated. As an example of one experiment using this design, consider a study of paired-associate learning in which the subject must learn the 16 responses that are correct for the 16 stimuli presented to him. Trials to criterion (perhaps two errorless runs) is the dependent variable. The stimuli and responses are nonsense syllables. Half of the stimuli have high association values, the other half have low association values; similarly, the responses are divided between high and low association syllables. Thus, four pairs are high-high ($B_1 C_1$), four are low-high ($B_2 C_1$), four are high-low ($B_1 C_2$), and four are low-low ($B_2 C_2$). Since each S must learn all 16 pairs, stimulus association value (B) and response association value (C) are both within-subjects variables.

TABLE 8–10 *Data matrix for a mixed design, one between- and two within-subjects variables*

	B_1			...	B_k			...	B_b		
	C_1	C_m	C_c	...	C_1	C_m	C_c	...	C_1	C_m	C_c
A_1											
S_{11}	Y_{1111}	Y_{111m}	Y_{111c}		Y_{11k1}	Y_{11km}	Y_{11kc}		Y_{11b1}	Y_{11bm}	Y_{11bc}
$\vdots$											
S_{i1}	Y_{i111}	Y_{i11c}	Y_{i11c}		Y_{i1k1}	Y_{i1km}	Y_{i1kc}		Y_{i1b1}	Y_{i1bm}	Y_{i1bc}
$\vdots$											
S_{n1}	Y_{n111}	Y_{n11m}	Y_{n11c}		Y_{n1k1}	Y_{n1km}	Y_{n1kc}		Y_{n1b1}	Y_{n1bm}	Y_{n1bc}
...											
A_j											
$\vdots$ S_{ij} $\vdots$	Y_{ij11}	Y_{ij1m}	Y_{ij1c}		Y_{ijk1}	Y_{ijkm}	Y_{ijkc}		Y_{ijb1}	Y_{ijbm}	Y_{ijbc}
...											
A_a											
$\vdots$ S_{ia} $\vdots$	Y_{ia11}	Y_{ia1m}	Y_{ia1c}		Y_{iak1}	Y_{iakm}	Y_{iakc}		Y_{iab1}	Y_{iabm}	Y_{iabc}

Table 8–10 presents a data matrix for the class of designs under discussion. A random sample from a large population of subjects has been randomly distributed among the a levels of the between-subjects variable, A, with the restriction that there are exactly n subjects at each level. The within-subjects variables are B and C, and there are bc scores obtained from each subject. The order of presentation of the bc treatment combinations is randomized independently for each subject. The indices, i, j, k, and m are used in reference to the levels of subjects, A, B, and C, respectively.

Development of the model. As in the preceding sections of this chapter, we use the developments of Chapter 7 as a point of departure. The design may initially be viewed as an $S \times B \times C$ design with an subjects. Following the pattern set by Table 7–13, we have

SV	df
Between S	$an - 1$
B	$b - 1$
Between S $\times$ B	$(an - 1)(b - 1)$
C	$c - 1$
Between S $\times$ C	$(an - 1)(c - 1)$
BC	$(b - 1)(c - 1)$
Between S $\times$ BC	$(an - 1)(b - 1)(c - 1)$

These terms account for the total of $abcn - 1$ df. This intermediate analysis clearly neglects possible effects due to A and its interactions with the within-subject variables, B and C. The further breakdown of the above table to obtain the remaining terms is straightforward. As in the past,

$$SS_{\text{B.S}} = SS_A + SS_{S/A}$$

Now consider the crossing of A and S/A with B; this yields AB and SB/A on $(a - 1)(b - 1)$ and $a(n - 1)(b - 1)$ df, which together account for $(an - 1)(b - 1)$ df. Therefore, the *Between S $\times$ B* term above may be replaced by two terms, AB and SB/A. Crossing A and S/A with C yields the two components of *Between S $\times$ C*, AC and SC/A; crossing A and S/A with BC yields the two components of *Between S $\times$ BC*, ABC and SBC/A. The complete set of SV are presented in Table 8–11.

Equation (8.4) presents a model that provides for the general set of effects suggested by the above discussion:

$$(8.4) \qquad Y_{ijkm} = \mu + \alpha_j + \eta_{i/j} + \beta_k + \gamma_m + (\beta\gamma)_{km} + (\alpha\beta)_{jk} + (\alpha\gamma)_{jm}$$
$$+ (\alpha\beta\gamma)_{jkm} + (\eta\beta)_{ik/j} + (\eta\gamma)_{im/j} + (\eta\beta\gamma)_{ikm/j} + \varepsilon_{ijkm}$$

As usual, specific assumptions about the population parameters are required in order to derive *EMS* and in order that the ratios of mean squares testing the null hypotheses are distributed as F. It is assumed that the effects of the variables A, B, and C are fixed; the levels of these variables have been

TABLE 8-11 *Analysis of variance for the mixed design, one between- and two within-subjects variables*

SV	df	SS	EMS	F
Total	$anbc - 1$	$\sum_i^n \sum_j^a \sum_k^b \sum_m^c Y_{ijkm}^2 - C$		
Between S	$an - 1$	$\dfrac{\sum_i^n \sum_j^a (\sum_k^b \sum_m^c Y_{ijkm})^2}{bc} - C$		
A	$a - 1$	$\dfrac{\sum_j^a (\sum_i^n \sum_k^b \sum_m^c Y_{ijkm})^2}{nbc} - C$	$\sigma_e^2 + bc\sigma_{S/A}^2 + nbc\theta_A^2$	$\dfrac{MS_A}{MS_{S/A}}$
S/A	$a(n - 1)$	$SS_{B.s} - SS_A$	$\sigma_e^2 + bc\sigma_{S/A}^2$	
Within S	$an(bc - 1)$	$SS_{tot} - SS_{B.s}$		
B	$b - 1$	$\dfrac{\sum_k^b (\sum_i^n \sum_j^a \sum_m^c Y_{ijkm})^2}{nac} - C$	$\sigma_e^2 + c\sigma_{SB/A}^2 + nac\theta_B^2$	$\dfrac{MS_B}{MS_{SB/A}}$
AB	$(a - 1)(b - 1)$	$\dfrac{\sum_j^a \sum_k^b (\sum_i^n \sum_m^c Y_{ijkm})^2}{nc} - C - SS_A - SS_B$	$\sigma_e^2 + c\sigma_{SB/A}^2 + nc\theta_{AB}^2$	$\dfrac{MS_{AB}}{MS_{SB/A}}$
SB/A	$a(n - 1)(b - 1)$	$\dfrac{\sum_i^n \sum_j^a \sum_k^b (\sum_m^c Y_{ijkm})^2}{c} - \dfrac{\sum_j^a \sum_k^b (\sum_i^n \sum_m^c Y_{ijkm})^2}{nc} - \dfrac{\sum_i^n \sum_j^a (\sum_k^b \sum_m^c Y_{ijkm})^2}{bc} + \dfrac{\sum_j^a (\sum_i^n \sum_k^b \sum_m^c Y_{ijkm})^2}{nbc}$	$\sigma_e^2 + c\sigma_{SB/A}^2$	

TABLE 8-11 (continued)

Source	df	SS	E(MS)	F
C	$c-1$	$\dfrac{\sum_m^c(\sum_i^n\sum_j^a\sum_k^b Y_{ijkm})^2}{nab} - C$	$\sigma_e^2 + b\sigma_{SC/A}^2 + nab\theta_C^2$	$\dfrac{MS_C}{MS_{SC/A}}$
AC	$(a-1)(c-1)$	$\dfrac{\sum_j^a\sum_m^c(\sum_i^n\sum_k^b Y_{ijkm})^2}{nb} - C - SS_A - SS_C$	$\sigma_e^2 + b\sigma_{SC/A}^2 + nb\theta_{AC}^2$	$\dfrac{MS_{AC}}{MS_{SC/A}}$
SC/A	$a(n-1)(c-1)$	$\dfrac{\sum_i^n\sum_j^a\sum_m^c(\sum_k^b Y_{ijkm})^2}{b} - \dfrac{\sum_i^n\sum_j^a(\sum_k^b\sum_m^c Y_{ijkm})^2}{bc} + \dfrac{\sum_j^a(\sum_i^n\sum_k^b\sum_m^c Y_{ijkm})^2}{nbc}$	$\sigma_e^2 + b\sigma_{SC/A}^2$	
BC	$(b-1)(c-1)$	$\dfrac{\sum_k^b\sum_m^c(\sum_i^n\sum_j^a Y_{ijkm})^2}{na} - C - SS_B - SS_C$	$\sigma_e^2 + \sigma_{SBC/A}^2 + na\theta_{BC}^2$	$\dfrac{MS_{BC}}{MS_{SBC/A}}$
ABC	$(a-1)(b-1)(c-1)$	$\dfrac{\sum_j^a\sum_k^b\sum_m^c(\sum_i^n Y_{ijkm})^2}{n} - C - SS_{AB} - SS_{AC}$ $- SS_{BC} - SS_A - SS_B - SS_C$	$\sigma_e^2 + \sigma_{SBC/A}^2 + n\theta_{BC}$	$\dfrac{MS_{ABC}}{MS_{SBC/A}}$
SBC/A	$a(n-1)(b-1)(c-1)$	$SS_{W.s} - SS_B - SS_{AB} - SS_{SB/A} - SS_C$ $- SS_{AC} - SS_{SC/A} - SS_{BC} - SS_{ABC}$	$\sigma_e^2 + \sigma_{SBC/A}^2$	

arbitrarily chosen. Consequently, the fixed effect population components (e.g., θ_A^2) are defined exactly as in Section 8.3.1, p. 204). It is assumed that subjects have been randomly sampled from a large population of subjects. Consequently, $\eta_{i/j}$, $(\eta\beta)_{ik/j}$, $(\eta\beta\gamma)_{ikm/j}$, as well as ε_{ijkm}, are assumed to be random samples from normally distributed populations with mean zero and variances $\sigma_{S/A}^2$, $\sigma_{SB/A}^2$, $\sigma_{SC/A}^2$, and σ_e^2, respectively.

The analysis of variance. The SV of Table 8–11 parallel the terms in Equation (8.4), with the qualification that there is no separate error term corresponding to ε_{ijkm}. In the df column, only the nested subject × treatment interactions warrant comment. The previous section on the model provides the clue to these quantities. For example, the SB/A effect is the difference between the *Between S × B (SB)* effect and the AB effect. It follows that $df_{SB/A} = df_{SB} - df_{AB}$, or

$$a(n - 1)(b - 1) = (an - 1)(b - 1) - (a - 1)(b - 1)$$

An alternative way of arriving at the above result is to view the nested interaction as a pool of a SB interactions, each computed at a different level of A. Then there are $(n - 1)(b - 1)$ df at each level of A, and the pooled result is $a(n - 1)(b - 1)$ df.

The SS calculations are also familiar, with the possible exception of the nested interaction terms. Either of the two approaches indicated for df can be used to calculate SS. For example, we could calculate an SS_{SB} as we would for a simple repeated measurements design and then subtract SS_{AB}, thereby obtaining $SS_{SB/A}$. As an alternative, the overall design could be treated as a two-factor repeated measurements designs. An SB interaction sum of squares could be computed for each of the a sets of data, and these would then be pooled. My preference for a computational formula is again based on the isomorphism of df and SS; this formula will ordinarily be the easiest to generate and the quickest to execute. For example, in the case of the SB/A term, we expand the df, obtaining

$$a(n - 1)(b - 1) = abn - an - ab + a$$

Remembering that there must be one squared term for each df, we have

$$\sum_j^a \sum_k^b \sum_i^n (\quad)^2 - \sum_j^a \sum_i^n (\quad)^2 - \sum_j^a \sum_k^b (\quad)^2 + \sum_j^a (\quad)^2$$

Indices not appearing outside the parentheses must appear within. Therefore, we obtain

$$\sum_j^a \sum_k^b \sum_i^n \left(\sum_m^c Y_{ijkm} \right)^2 - \sum_j^a \sum_i^n \left(\sum_k^b \sum_m^c Y_{ijkm} \right)^2 - \sum_j^a \sum_k^b \left(\sum_i^n \sum_m^c Y_{ijkm} \right)^2$$

$$+ \sum_j^a \left(\sum_i^n \sum_k^b \sum_m^c Y_{ijkm} \right)^2$$

Finally, we divide by the number of scores being summed prior to squaring, yielding

$$SS_{SB/A} = \frac{\sum_j^a \sum_k^b \sum_i^n (\sum_m^c Y_{ijkm})^2}{c} - \frac{\sum_j^a \sum_i^n (\sum_k^b \sum_m^c Y_{ijkm})^2}{bc}$$

$$- \frac{\sum_j^a \sum_k^b (\sum_i^n \sum_m^c Y_{ijkm})^2}{nc} + \frac{\sum_j^a (\sum_i^n \sum_k^b \sum_m^c Y_{ijkm})^2}{nbc}$$

Note that of the four terms comprising the computational formula for $SS_{SB/A}$, three have been previously calculated in the analysis of variance table. The quantity $\sum_i^n \sum_j^a (\sum_k^b \sum_m^c Y_{ijkm})^2 / bc$ is computed in order to obtain $SS_{B.S}$; $\sum_j^a \sum_k^b (\sum_i^n \sum_m^c Y_{ijkm})^2 / nc$ is computed as part of the calculations for SS_{AB}; and $\sum_j^a (\sum_i^n \sum_k^b \sum_m^c Y_{ijkm})^2 / nbc$ is part of the calculation for SS_A. Thus, only one term need be calculated in order to obtain the $SS_{SB/A}$.

The *EMS* follow from the rules of thumb presented earlier in this chapter. The reader would be well advised to check through each line to ensure that he understands the application of the rules. Once the *EMS* have been generated, the *F* tests present no problem. Each main or interaction effect is tested against the nested effect that follows most cloesly in the table.

A numerical example. Table 8–12 contains data for an illustrative analysis. In addition to the data, subtotals are presented that will be of use during the analysis. We begin by computing the total sum of squares:

$$SS_{tot} = \sum_i \sum_j \sum_k \sum_m Y_{ijkm}^2 - C$$

$$= (3)^2 + (5)^2 + \cdots + (11)^2 - \frac{(412)^2}{48}$$

$$= 3{,}912.000 - 3{,}536.333$$

$$= 375.667$$

For the variability among subjects, we have

$$SS_{B.S} = \frac{\sum_i \sum_j (\sum_k \sum_m Y_{ijkm})^2}{bc} - C$$

$$= \frac{(54)^2 + (56)^2 + \cdots + (80)^2 + (88)^2}{8} - \frac{(412)^2}{48}$$

$$= 3{,}647.000 - 3{,}536.333$$

$$= 110.667$$

TABLE 8-12 Data and sub-totals for an analysis of variance, one between- and two within-subjects variables

| | | B_1 | | | | | B_2 | | | | | |
		C_1	C_2	C_3	C_4	$\sum_m Y_{ij1m}$	C_1	C_2	C_3	C_4	$\sum_m Y_{ij2m}$	$\sum_k\sum_m Y_{ijkm}$
A_1	S_{11}	3	5	7	8	23	7	6	9	9	31	54
	S_{21}	2	5	6	8	21	8	5	10	12	35	56
	$\sum_i Y_{i11m} =$	5	10	13	16	$\sum_i\sum_m Y_{i11m} = 44$	$\sum_i Y_{i12m} = 15$	11	19	21	$\sum_i\sum_m Y_{i12m} = 66$	$\sum_i\sum_k\sum_m Y_{i1km} = 110$
A_2	S_{12}	8	9	9	11	37	6	7	7	9	29	66
	S_{22}	7	7	9	10	33	7	8	9	11	35	68
	$\sum_i Y_{i21m} = 15$		16	18	21	$\sum_i\sum_m Y_{i21m} = 70$	$\sum_i Y_{i22m} = 13$	15	16	20	$\sum_i\sum_m Y_{i22m} = 64$	$\sum_i\sum_k\sum_m Y_{i2km} = 134$
A_3	S_{13}	10	12	13	15	50	5	7	8	10	30	80
	S_{23}	11	11	14	16	52	8	8	9	11	36	88
	$\sum_i Y_{i31m} = 21$		23	27	31	$\sum_i\sum_m Y_{i31m} = 102$	$\sum_i Y_{i32m} = 13$	15	17	21	$\sum_i\sum_m Y_{i32m} = 66$	$\sum_i\sum_k\sum_m Y_{i3km} = 168$
	$\sum_i\sum_j Y_{ij1m} = 41$		49	58	68	$\sum_i\sum_j\sum_m Y_{ij1m} = 216$	$\sum_i\sum_j Y_{ij2m} = 41$	41	52	62	$\sum_i\sum_j\sum_m Y_{ij2m} = 196$	$\sum_i\sum_j\sum_k\sum_m Y_{ijkm} = 412$

TABLE 8-12 (continued)

Subjects × C Subtotals $\left(\sum_k Y_{ijkm}\right)$

		C_1	C_2	C_3	C_4	$\sum_k \sum_m Y_{ijkm}$
A_1	S_{11}	10	11	16	17	54
	S_{21}	10	10	16	20	56
	$\sum_i \sum_k Y_{i1km} =$	20	21	32	37	$\sum_i \sum_k \sum_m Y_{i1km} = 110$
A_2	S_{12}	14	16	16	20	66
	S_{22}	14	15	18	21	68
	$\sum_i \sum_k Y_{i2km} =$	28	31	34	41	$\sum_i \sum_k \sum_m Y_{i2km} = 134$
A_3	S_{13}	15	19	21	25	80
	S_{23}	19	19	23	27	88
	$\sum_i \sum_k Y_{i3km} =$	34	38	44	52	$\sum_i \sum_k \sum_m Y_{i3km} = 168$
	$\sum_i \sum_j \sum_k Y_{ijkm} =$	82	90	110	130	$\sum_i \sum_j \sum_k \sum_m = 412$

This term has two components

$$SS_A = \frac{\sum_j (\sum_i \sum_k \sum_m Y_{ijkm})^2}{nbc} - C$$

$$= \frac{(110)^2 + (134)^2 + (168)^2}{16} - \frac{(412)^2}{48}$$

$$= 3{,}642.500 - 3{,}536.333$$

$$= 106.167$$

and

$$SS_{B/A} = SS_{\text{B.S}} - SS_A$$

$$= 110.667 - 106.167$$

$$= 4.500$$

We turn next to the within-subjects variability:

$$SS_{\text{W.S}} = SS_{\text{tot}} - SS_{\text{B.S}}$$

$$= 375.667 - 110.667$$

$$= 265.000$$

For the main effect of B, we can use the single df formula, thus saving some computational effort:

$$SS_B = \frac{(216 - 196)^2}{48}$$

$$= 8.333$$

For the AB interaction, we have

$$SS_{AB} = \frac{\sum_j \sum_k (\sum_i \sum_m Y_{ijkm})^2}{nc} - C - SS_A - SS_B$$

$$= \frac{(44)^2 + (70)^2 + \cdots + (66)^2}{8} - \frac{(412)^2}{48} - 106.167 - 8.333$$

$$= 3{,}756.000 - 3{,}536.333 - 106.167 - 8.333$$

$$= 105.167$$

The error variability for the previous two terms is

$$SS_{SB/A} = \frac{\sum_i \sum_j \sum_k (\sum_m Y_{ijkm})^2}{c} - \frac{\sum_i \sum_j (\sum_k \sum_m Y_{ijkm})^2}{bc}$$

$$- \frac{\sum_j \sum_k (\sum_i \sum_m Y_{ijkm})^2}{nc} + \frac{\sum_j (\sum_i \sum_k \sum_m Y_{ijkm})^2}{nbc}$$

$$= \frac{(23)^2 + (31)^2 + \cdots + (52)^2 + (36)^2}{4} - 3{,}647.000$$

$$- 3{,}756.000 + 3{,}642.500$$

$$= 9.500$$

For the C main effect, we have

$$SS_C = \frac{\sum_m (\sum_i \sum_j \sum_k Y_{ijkm})^2}{abn} - C$$

$$= \frac{(82)^2 + (90)^2 + (110)^2 + (130)^2}{12} - \frac{(412)^2}{48}$$

$$= 3{,}652.000 - 3{,}536.333$$

$$= 115.667$$

For the AC interaction, we have

$$SS_{AC} = \frac{\sum_j \sum_m (\sum_i \sum_k Y_{ijkm})^2}{nb} - C - SS_A - SS_C$$

$$= \frac{(20)^2 + (21)^2 + \cdots + (44)^2 + (52)^2}{4} - \frac{(412)^2}{48} - 106.167 - 115.667$$

$$= 3{,}764.000 - 3{,}536.333 - 106.167 - 115.667$$

$$= 5.833$$

The error variability for the previous two terms is

$$SS_{SC/A} = \frac{\sum_i \sum_j \sum_m (\sum_k Y_{ijkm})^2}{b} - \frac{\sum_i \sum_j (\sum_k \sum_m Y_{ijkm})^2}{bc}$$

$$- \frac{\sum_j \sum_m (\sum_i \sum_k Y_{ijkm})^2}{nb} + \frac{\sum_j (\sum_i \sum_k \sum_m Y_{ijkm})^2}{nbc}$$

$$= \frac{(10)^2 + (11)^2 + \cdots + (23)^2 + (27)^2}{2} - 3{,}647.000$$

$$- 3{,}764.000 + 3{,}642.500$$

$$= 5.500$$

The BC interaction sum of squares is obtained by

$$SS_{BC} = \frac{\sum_k \sum_m (\sum_i \sum_j Y_{ijkm})^2}{na} - C - SS_B - SS_C$$

$$= \frac{(41)^2 + (49)^2 + \cdots + (52)^2 + (62)^2}{6} - \frac{(412)^2}{48} - 8.333 - 115.667$$

$$= 3{,}663.333 - 3{,}536.333 - 8.333 - 115.667$$

$$= 3.000$$

We have next

$$SS_{ABC} = \frac{\sum_j \sum_k \sum_m (\sum_i Y_{ijkm})^2}{n} - C - SS_A - SS_B - SS_C - SS_{AB}$$
$$- SS_{AC} - SS_{BC}$$

$$= \frac{(5)^2 + (10)^2 + \cdots + (17)^2 + (21)^2}{2} - C - SS_A - SS_B - SS_C$$
$$- SS_{AB} - SS_{AC} - SS_{BC}$$

$$= 3{,}889.000 - 3{,}536.333 - 106.167 - 8.333 - 115.667 - 105.167$$
$$- 5.833 - 3.000$$

$$= 8.500$$

The error variability for the two previous terms may be taken as a residual from the $SS_{\text{w.s}}$:

$$SS_{SBC/A} = SS_{\text{w.s}} - SS_B - SS_{AB} - SS_{SB/A} - SS_C - SS_{AC} - SS_{SC/A}$$
$$- SS_{BC} - SS_{ABC}$$

$$= 265.000 - 8.333 - 105.167 - 9.500 - 115.667 - 5.833 - 5.500$$
$$- 3.333 - 8.167$$

$$= 3.500$$

The final results of the analysis are summarized in Table 8–13.

TABLE 8–13 *Analysis of variance for the data of Table* 8–12

SV	df	SS	MS	F
Total	47	375.667		
Between S	5	110.667		
A	2	106.167	53.083	35.389**
S/A	3	4.500	1.500	
Within S	42	265.000		
B	1	8.333	8.333	2.631
AB	2	105.167	52.583	16.603*
SB/A	3	9.500	3.167	
C	3	115.667	38.556	63.103**
AC	6	5.833	.972	1.591
SC/A	9	5.500	.611	
BC	3	3.000	1.000	2.571
ABC	6	8.500	1.417	3.630
SBC/A	9	3.500	.389	

$$* \ p < .025$$
$$** \ p < .01$$

8.4 CONCLUDING REMARKS

The mixed design is a compromise between the simplicity of the completely randomized design and the high relative efficiency of the simple repeated measurements design. In testing the null hypothesis that $\theta_2^A = 0$ where A is a between-subjects variable, precision is usually low; this part of the design is essentially completely randomized. By introducing one or more within-subjects variables, fewer subjects are needed than would be required if all variables were between subjects. A potentially precise test of the within-subjects effects is provided, since the error terms are not inflated by individual differences. But this saving of subjects and increase in precision costs something. As in the simpler repeated measurements designs of Chapter 7, the possibility of subject × treatment interactions is introduced, with the consequent possibilities of lowered efficiency (if the interaction with subjects is large) and heterogeneity of covariance. If the population covariances are not similar for all possible pairs of levels of within-subjects variables, only the F tests of within-subjects effects will be positively biased. For levels of the between-subjects variables, the expected covariances should all be zero if randomization has been properly carried out.

Tukey's single df test, described in Chapter 7, can be extended to test for nonadditivity in the mixed design. For example, in a one between- and one within-subjects design, a sum of squares for nonadditivity could be computed at each level of A. These a quantities could then be pooled, yielding a sum of squares for nonadditivity on a df that would be tested against the balance of the SB/A interaction. Transformations to additivity may also be sought following the guidelines of Chapter 7. If additivity is achieved, the nested interaction effects are dropped from the model and such terms as $\sigma^2_{SB/A}$ disappear from the EMS. The covariances will be homogeneous, and the tests of within-subjects effects will generally be efficient and unbiased.

In the absence of strong evidence to the contrary, most investigators assume that the nonadditive model presented in this chapter is appropriate. Therefore, it becomes important to be aware of the possibility of hetero-geneity of covariance and the consequent positive bias of the tests of within-subjects effects. Conservative F tests parallel to those presented in Section 7.2.4 are available. For example, consider the one between- and one within-subjects design of the preceding section. The statistics MS_B and MS_{AB} are distributed on $(b-1)\lambda$ and $(a-1)(b-1)\lambda$ df and their common error term, $MS_{SB/A}$, is distributed on $a(n-1)(b-1)\lambda$ df; λ lies between $1/(b-1)$ (maximum heterogeneity of variance and covariance) and 1 (homogeneous variances and covariances). The conservative test of the B effect would therefore employ 1 and $a(n-1)$ df, while a similar test of the AB interaction would employ $a-1$ and $a(n-1)$ df.

If the conservative test is not significant while the usual test on the full df is, the investigator may wish to directly estimate the adjustment, λ. Equation (7.8) is still appropriate but the variances and covariances are slightly modified for the mixed design. To obtain λ, we require a $b \times b$ variance-covariance matrix whose elements are obtained by averaging over the levels of the between-subjects variable A. Thus we have

$$(8.5) \qquad V_{kk'} = \frac{\sum_j \sum_i (Y_{ijk} - \bar{Y}_{..k})(Y_{ijk'} - \bar{Y}_{..k'})}{a(n-1)}$$

For a more detailed discussion of the covariance problem and of nonadditivity in general, the reader is referred back to Chapter 7.

EXERCISES

8.1 Carry out the appropriate analysis of variance on the data.

		B_1C_1	B_1C_2	B_1C_3	B_2C_1	B_2C_2	B_2C_3
	S_{11}	18	8	7	14	9	4
A_1	S_{21}	20	10	13	11	6	10
	S_{31}	23	12	14	13	12	3
	S_{12}	16	5	7	10	15	5
A_2	S_{22}	13	8	4	16	11	6
	S_{32}	18	6	10	17	11	4
	S_{13}	14	6	8	11	3	1
A_3	S_{23}	20	10	8	7	9	2
	S_{33}	13	7	7	13	5	6

8.2 We have *abc* combinations of the variables *A*, *B*, and *C*. Assuming that the levels of all three are randomly sampled, find the *EMS*. Are there any special problems that occur? How would the *EMS* and *F* tests be affected if it were known that $\sigma_{AB}^2 = 0$?

8.3 We have a mixed design with *A* (between *S*s) and *B* (within *S*s). The data is punched on cards and we wish to analyze it on the computer. Unfortunately, the only available program is for a 2-factor (crossing) anova. Show how each *SS* in the mixed anova would be obtained if you are allowed to put the data through the computer any number of times. You are allowed to combine (add or subtract) the results of the several computer runs on a desk calculator.

9

Hierarchical Designs

9.1 INTRODUCTION

In Chapter 8 an examination was made of the models and analyses that are appropriate when subjects are nested within levels of the variable A and are crossed with the levels of a second variable, B. In this chapter the paradigm is extended to designs in which there are several levels of nests. Thus, subjects may be nested within levels of a variable G, while the levels of G are in turn nested within the levels of A. Such designs are often labeled "hierarchical" in reference to the hierarchy of variables that is typical of them. The statistical models and computations are straightforward extensions of the developments of the preceding chapter.

Hierarchical designs have an important place in both psychological and educational research. To cite one example, consider a group dynamics experiment designed for the purpose of studying the effects of stress upon attitude change in the members of four-man conference teams. We might have several conference groups under a high stress and an equal number under a low stress condition; the resultant design might be characterized as *subjects within conference groups within levels of stress.* A similar example might be taken from educational psychology. Several first-grade classes are taught reading under one method while an equal number are taught under a second method; all students are tested at the end of the term. The design might be characterized as *subjects within classes within methods.* Still another example might be taken from the animal laboratory. Different methods of rearing rats might be compared, each method being applied to several different litters. We might characterize this design as *subjects within litters within methods of rearing.* In all of these examples it is reasonable to assume that the total variability among subjects has three potential sources: subjects' scores may differ because of

(a) Treatment effects. The variable A (e.g., stress, method) is a potential source.

(b) Group effects. Differences in the composition of groups (e.g., conference groups, school classes) may contribute to the variability in the data.

(c) Residual individual differences. Even the scores of subjects within the same group may vary due to such factors as attitude or ability.

The primary new aspect of the hierarchical design is the assumption that an individual's score is in part influenced by the social unit of which he is a member. Even though the same experimental treatment is applied to both, two individuals in different social groups (or school classes or litters) will differ, not merely because they are different individuals, but also because they are subject to interactions with different sets of individuals and occurrences. Once we recognize that some experiments involve such group effects, it becomes necessary to consider those effects within the statistical model.

Hierarchical designs do not necessarily involve social units, such as groups or school classes. For example, we might have four difficult and four easy concept formation problems. Since there are different problems at each difficulty level, problems are nested within levels of difficulty. Before this chapter is concluded, an example of this type will be considered. However, for the most part the chapter will be concerned with the more common *subjects-within-groups-within-treatments* design and several of its variants.

The importance of the material of Chapter 9 goes beyond the actual applications of the hierarchical designs. The presentation of these designs and their analyses should further understanding of the establishment of structural models and the translation of models into analysis of variance tables. Every design that the researcher may encounter cannot possibly be discussed. However, the material of this chapter should help establish certain fundamental but widely applied principles of data analysis, which were first introduced in Chapter 8.

9.2 GROUPS WITHIN TREATMENTS

9.2.1 The analysis of variance model. We begin by conceptualizing a treatment populations, differing systematically only with respect to the level of the treatment variable, A. From each population, subjects are randomly selected in groups of size n; the sampling process ends when g groups of n subjects have been sampled from each population. The resulting experimental layout is presented in Table 9–1. The design involves a total of agn subjects with n subjects in each group, g groups at each level of A, and a levels of A. The notational indices are

$$i = 1, 2, \ldots, n$$
$$j = 1, 2, \ldots, g$$
$$k = 1, 2, \ldots, a$$

Next, an equation is required to relate the representative score, Y_{ijk}, to

TABLE 9–1 *Data matrix for the groups-within-treatments design*

	A_1					A_k					A_a	
G_{11}	$\cdots$	G_{j1}	$\cdots$	G_{g1}	G_{1k}	$\cdots$	G_{jk}	$\cdots$	G_{gk}	G_{1a}	$\cdots$	G_{ja}	$\cdots$	G_{ga}
Y_{111}		Y_{1j1}		Y_{1g1}	Y_{11k}		Y_{1jk}		Y_{1gk}	Y_{11a}		Y_{1ja}		Y_{1ga}
Y_{i11}	$\cdots$	Y_{ij1}		Y_{ig1}	Y_{i1k}		Y_{ijk}		Y_{igk}	Y_{i1a}		Y_{ija}		Y_{iga}
Y_{n11}		Y_{nj1}		Y_{ng1}	Y_{n1k}		Y_{njk}		Y_{ngk}	Y_{n1a}		Y_{nja}		Y_{nga}

the parameters of the population from which the sample is drawn. We begin by ignoring the variable A; the design is viewed as a one-factor design, the "factor" being groups (G) with ag levels and n subjects at each level. In accord with the one-factor model of Chapter 4, this view suggests

(9.1) $$Y_{ijk} = \mu + \gamma_{jk} + \varepsilon_{ijk}$$

where $\gamma_{jk} = \mu_{jk} - \mu$, the overall effect of the jth group at the kth level of A, and $\varepsilon_{ijk} = Y_{ijk} - \mu_{jk}$, the residual error component. Equation (9.1) disregards the possibility of an effect due to A_k. Presumably, group means differ not merely because the groups have different compositions, but also because some groups are at one level of A, others are at a different level. This line of reasoning suggests that part of the γ_{jk} effect is due to α_k, the effect due to the level of A in which the group exists. Accordingly, we subtract the contribution of α_k:

(9.2) $$\gamma_{jk} - \alpha_k = (\mu_{jk} - \mu) - (\mu_k - \mu)$$
$$= \mu_{jk} - \mu_k$$
$$= \gamma_{j/k}$$

where $\gamma_{j/k}$ is the pure effect of the jkth group, uninflated by any contribution due to α_k. We may now substitute in Equation (9.1) for γ_{jk}, obtaining

(9.3) $$Y_{ijk} = \mu + \alpha_k + \gamma_{j/k} + \varepsilon_{ijk}$$

Each score is contributed to by a treatment effect, a group effect, and a residual component reflecting error of measurement and individual differences.

A common error in the analysis of the group designs is the failure to consider group effects in the model. In this case, the analysis proceeds as if the design were a completely randomized one-factor design with gn subjects in each of a treatment groups. This failure to separate out the γ component from ε may result in an inflated F ratio, as will be shown in the next section.

To complete the presentation of the underlying theory, and in order to arrive at the *EMS*, the nature of the effects must be considered. It is assumed that the levels of A have been arbitrarily chosen by the experimenter, and consequently, that α_k is a fixed variable. Then, $\sum_k \alpha_k = 0$ and the variance component is defined as $\theta_A^2 = \sum_k^a \alpha_k^2 / (a - 1)$. The group effect, $\gamma_{j/k}$, is viewed as a random variable, since the groups are clearly a random sample from the population of all possible groups of size n which could be composed. As usual, ε_{ijk} is also a random variable. The $\gamma_{j/k}$ and ε_{ijk} are assumed to be sampled from normally distributed populations with mean zero and variances $\sigma_{G/A}^2$ and σ_e^2, respectively.

9.2.2 The analysis of variance. Table 9–2 contains the *SV*, *df*, *SS*, *EMS*, and *F* for the *groups-within-treatments* design. One source exists for each term on the right-hand side of Equation (9.1). To facilitate the analysis, we have

TABLE 9–2 *Analysis of variance for the groups-within-treatments design*

SV	df	SS	EMS	F
Total	$agn - 1$	$\sum_i^n \sum_j^g \sum_k^a Y_{ijk}^2 - C$		
Between G	$ag - 1$	$\dfrac{\sum_j^g \sum_k^a (\sum_i^n Y_{ijk})^2}{n} - C$		
A	$a - 1$	$\dfrac{\sum_k^a (\sum_j^g \sum_i^n Y_{ijk})^2}{gn} - C$	$\sigma_e^2 + n\sigma_{G/A}^2 + ng\theta_A^2$	$\dfrac{MS_A}{MS_{G/A}}$
G/A	$a(g - 1)$	$SS_{B.G} - SS_A$	$\sigma_e^2 + n\sigma_{G/A}^2$	$\dfrac{MS_{G/A}}{MS_{S/G/A}}$
S/G/A	$ag(n - 1)$	$SS_{tot} - SS_{B.G}$	σ_e^2	

first divided our variability between two sources; ignoring the treatment variable, A, we have between-groups and within-groups variability. This breakdown corresponds to a model of the form of Equation (9.1). Subsequently, the between-groups source is further divided into the G/A and A components, as in the development of the model in Section 9.2.1.

The *df* are easily obtainable once we have noted the sources. For the nested term, G/A, we may make use of the fact that the between-groups term is a composite of the A and G/A terms. Therefore, $df_{B.G} - df_A = df_{G/A}$, or

$$(ag - 1) - (a - 1) = a(g - 1)$$

Alternatively, we note that G/A literally represents the summing, over all treatment levels, of the variability of group means about the mean of their treatment level. At each treatment level we have $g - 1$ *df* to represent the variability of g means about the treatment mean. Pooling over a levels, we again arrive at $a(g - 1)$ *df*.

The computational formulas in the *SS* column can be obtained from the *df* column. For example, the $ag - 1$ *df* for between groups requires ag squared quantities minus the correction term. Thus, we have

$$\sum_j^g \sum_k^a (\quad)^2 - C$$

Including the absent index, i, within the parentheses, we have

$$\sum_j^g \sum_k^a \left(\sum_i^n Y_{ijk} \right)^2 - C$$

Dividing by the number of scores within the parentheses, we have the final result,

$$SS_{B.G} = \frac{\sum_j^g \sum_k^a (\sum_i^n Y_{ijk})^2}{n} - C$$

As a second example, expansion of $df_{G/A}$, yielding $ag - a$, leads to the result,

$$SS_{G/A} = \frac{\sum_j^g \sum_k^a (\sum_i^n Y_{ijk})^2}{n} - \frac{\sum_k^a (\sum_j^g \sum_i^n Y_{ijk})^2}{gn}$$

In this instance, it is simpler to note merely that the G/A term is the difference between the between-groups term and the A term. The result, presented in Table 9–2, is algebraically identical to that given above.

The *EMS* follow the rules of thumb developed in Chapter 8. Considering the A line first, we immediately set down σ_e^2 and the null hypothesis term, θ_A^2. Noting that the subscript G/A includes the letter A and that the essential letter, G, represents a random effects variable, we include $\sigma_{G/A}^2$ in the expectation. The remaining two lines should pose no problems. However, note that in this design σ_e^2 is the sum of variances due to measurement errors and individual differences; that is, it includes $\sigma^2_{S/G/A}$. Only when there are between- and within-subjects terms in the analysis, in which case individual differences do not contribute to all sources, do we list separately the variance component due to individual differences.

On the basis of the *EMS*, it is clear that the treatment mean square should be tested against the mean square for groups within A. A little consideration of the error *df* will indicate that this F test may often be lacking in power. For example, suppose the experiment involves three four-man conference groups at each of three levels of A. The total of 36 subjects would generally be considered reasonable for a study involving three treatment levels, since 33 error *df* would be available if the design were a completely randomized one-factor design. However, the error *df* for the hierarchical design actually employed are only $3(3 - 1)$, or six. This number of error *df* is not likely to make the experimenter feel secure if he fails to reject the null hypothesis. One possible way out of the difficulty is to assume a different model; specifically, the experimenter might assume that the group structures do not contribute to error variability, and consequently, that the $\sigma^2_{G/A}$ component may be deleted wherever it appears in the *EMS* column of Table 9–2. In this case, both $MS_{G/A}$ and $MS_{S/G/A}$ are estimates of σ^2_e and may therefore be combined to provide a new error term. The combining, or *pooling* as it is generally referred to, takes the form $(SS_{G/A} + SS_{S/G/A})/(df_{G/A} + df_{S/G/A})$, an error mean square distributed on 33 *df*. For reasons that will shortly be considered, pooling should not be carried out unless the data strongly indicate that $\sigma^2_{G/A}$ is a negligible quantity; appropriate criteria will be discussed in Section 11.2.2.

Since the decision on pooling must await the collection of data, in designing the experiment one must proceed on the assumption that $MS_{G/A}$ will be the error term. Consequently, the investigator planning an experiment that involves a group of subjects as a unit should carefully evaluate $df_{G/A}$, and if it is too small to provide an adequate test of treatment effects, modify his design accordingly.

What are the consequences if the G/A and $S/G/A$ terms are pooled when $\sigma^2_{G/A}$ contributes to the data variability? To answer this, we first derive the expectation of our new error term, which will be labeled $MS_{S/A}$, and which is computed as

$$MS_{S/A} = \frac{SS_{G/A} + SS_{S/G/A}}{df_{G/A} + df_{S/G/A}}$$

$$= \frac{(df_{G/A})(MS_{G/A}) + (df_{S/G/A})(MS_{S/G/A})}{df_{G/A} + df_{S/G/A}}$$

Taking the expectations of the two sides yields

$$E(MS_{S/A}) = \left(\frac{df_{G/A}}{df_{G/A} + df_{S/G/A}}\right) E(MS_{G/A}) + \left(\frac{df_{S/G/A}}{df_{G/A} + df_{S/G/A}}\right) E(MS_{S/G/A})$$

Thus, $E(MS_{S/A})$ is a weighted average of $E(MS_{G/A})$ and $E(MS_{S/G/A})$, where the weights are proportions of *df* associated with the two terms. Assuming that the two expectations to the right of the "equals" sign are unequal, their average must lie between them. Thus, $E(MS_{S/A})$ will be less than $E(MS_{G/A})$, the larger of the two terms being averaged. But this means that the pooled error term has a smaller expectation than the error term that is generally appropriate for testing the A main effect (see Table 9–2). Thus, if the two terms being pooled do not have identical expectations; that is, if $\sigma_{G/A}^2 \neq 0$, the F test of A against the pooled error term will be positively biased; under $H_0(\theta_A^2 = 0)$, the *EMS* ratio will be greater than 1. Furthermore, the ratio of mean squares will not have an F distribution over replications of the experiment since, under H_0, $E(MS_A) \neq E(MS_{S/A})$.

9.2.3. A numerical example.

TABLE 9–3 *Data for a groups-within-treatments design*

	A_1				A_2		
	G_{11}	G_{21}	G_{31}		G_{12}	G_{22}	G_{32}
	5	7	16		24	9	17
	6	18	5		21	23	26
	18	4	9		12	28	24
	12	11	14		16	19	19
$\sum_i Y_{ij1} =$	41	40	44	$\sum_i Y_{ij2} =$	73	79	86
$\sum_i \sum_j Y_{ij1} = 125$				$\sum_i \sum_j Y_{ij2} = 238$			

$$\sum_i \sum_j \sum_k Y_{ijk} = 363$$

Table 9–3 presents data for a groups-within-treatments design. We first obtain the total sum of squares:

$$SS_{\text{tot}} = \sum_{i=1}^4 \sum_{j=1}^3 \sum_{k=1}^2 Y_{ijk}^2 - C$$

$$= 5^2 + 6^2 + \cdots + 24^2 + 19^2 - \frac{(363)^2}{24}$$

$$= 6{,}671.00 - 5{,}490.38$$

$$= 1{,}180.62$$

The measure of between-groups variability is

$$SS_{\text{B·G}} = \frac{\sum_{j=1}^{3} \sum_{k=1}^{2} (\sum_{i=1}^{4} Y_{ijk})^2}{4} - C$$

$$= \frac{(41)^2 + \cdots + (86)^2}{4} - \frac{(363)^2}{24}$$

$$= 6,045.75 - 5,490.38$$

$$= 555.37$$

Turning to the A source, we have

$$SS_A = \frac{\sum_{k=1}^{2} (\sum_{j=1}^{3} \sum_{i=1}^{4} Y_{ijk})^2}{12} - C$$

$$= \frac{(125)^2 + (238)^2}{12} - \frac{(363)^2}{24}$$

$$= 6,022.42 - 5,490.38$$

$$= 532.04$$

Subtracting SS_A from $SS_{\text{B·G}}$, we obtain the sum of squares for the nested group effect:

$$SS_{G/A} = SS_{\text{B.G}} - SS_A$$
$$= 555.37 - 532.04$$
$$= 23.33$$

The total variability may be partitioned into a between-groups and a within-groups component. Therefore,

$$SS_{S/G/A} = SS_{\text{tot}} - SS_{\text{B.G}}$$
$$= 1,180.62 - 555.37$$
$$= 625.25$$

The results of the analysis are summarized in Table 9–4. The difference

TABLE 9–4 *Analysis of variance for the data of Table 9–3*

SV	df	SS	MS	F
Total	23	1,180.62		
Between G	5	555.37		
A	1	532.04	532.04	92.98*
G/A	4	23.33	5.83	
S/G/A	18	625.25	34.74	
				*p < .01

in performance under treatments A_1 and A_2 is clearly significant. Since the ratio of G/A to $S/G/A$ mean squares is less than 1, these two sources might be pooled to provide a new error term to test the A effect; however, the F test of Table 9–4 is so clearly significant that this appears to be an unnecessary step.

9.3 A WITHIN-GROUPS VARIABLE

We next consider the situation where a variable is introduced within each group. There are again g groups at each of the a levels of the variable A. In addition, each group is divided into b levels of the variable B, with n subjects at each level. For example, there may be several conference groups working under a high stress condition and several others working under low stress. Within each conference group, half of the subjects may be high anxious (as measured by the Taylor Manifest Anxiety Scale), while the other half are low anxious. *Stress* is the A variable and *Anxiety* is the B variable.

This hierarchical design is represented in Table 9–5. The indices of notation are

$$i = 1, 2, \ldots, n$$
$$j = 1, 2, \ldots, g$$
$$k = 1, 2, \ldots, a$$
$$m = 1, 2, \ldots, b$$

The design involves bn subjects within each group, bgn subjects at each level of A, and $abgn$ subjects for the entire experiment.

9.3.1 The analysis of variance model. As in Chapter 8, we first require some understanding of the structure of the design; in particular, we desire some reasonable breakdown of the total variability. It helps to relate the present design to one we have previously considered. In this case, we have something analogous to the first design considered in Chapter 8, in which there was one between-subject variable and one within-subject variable. In the present instance, all variability is between subjects since there are no repeated measurements. However, there is one between-groups variable; there are different groups at each level of A. Furthermore, there is one within-groups variable; all groups are represented at all levels of B. This suggests the SV of Table 9–6; it will be helpful to compare this with Table 8–2. Note how the *Between G* and *Within G* sources parallel the *Between S* and *Within S* sources of the earlier table. The sole discrepancy is the presence of an extra term, $S/GB/A$, due to the fact that we have an added level of nesting in the present design.

TABLE 9-5 Data matrix for a hierarchical design containing a within-groups variable

	A_1				A_k				A_a		
	G_{11}	$\cdots$	G_{g1}	$\cdots$	G_{1k}	G_{jk}	G_{gk}	$\cdots$	G_{1a}	G_{ja}	G_{ga}
B_1	Y_{1111}		Y_{1g11}		Y_{11k1}	Y_{1jk1}	Y_{1gk1}		Y_{11a1}	Y_{1ja1}	Y_{1ga1}
	$\cdots$		$\cdots$		$\cdots$	$\cdots$	$\cdots$		$\cdots$	$\cdots$	$\cdots$
	Y_{i111}		Y_{ig11}		Y_{i1k1}	Y_{ijk1}	Y_{igk1}		Y_{i1a1}	Y_{ija1}	Y_{iga1}
	$\cdots$		$\cdots$		$\cdots$	$\cdots$	$\cdots$		$\cdots$	$\cdots$	$\cdots$
	Y_{n111}		Y_{ng11}		Y_{n1k1}	Y_{njk1}	Y_{ngk1}		Y_{n1a1}	Y_{nja1}	Y_{nga1}
$\cdots$											
B_m	Y_{111m}		Y_{1g1m}		Y_{11km}	Y_{1jkm}	Y_{1gkm}		Y_{11am}	Y_{1jam}	Y_{1gam}
	$\cdots$		$\cdots$		$\cdots$	$\cdots$	$\cdots$		$\cdots$	$\cdots$	$\cdots$
	Y_{i11m}		Y_{ig1m}		Y_{i1km}	Y_{ijkm}	Y_{igkm}		Y_{i1am}	Y_{ijam}	Y_{igam}
	$\cdots$		$\cdots$		$\cdots$	$\cdots$	$\cdots$		$\cdots$	$\cdots$	$\cdots$
	Y_{n11m}		Y_{ng1m}		Y_{n1km}	Y_{njkm}	Y_{ngkm}		Y_{n1am}	Y_{njam}	Y_{ngam}
$\cdots$											
B_b	Y_{111b}		Y_{1g1b}		Y_{11kb}	Y_{1jkb}	Y_{1gkb}		Y_{11ab}	Y_{1jab}	Y_{1gab}
	$\cdots$		$\cdots$		$\cdots$	$\cdots$	$\cdots$		$\cdots$	$\cdots$	$\cdots$
	Y_{i11b}		Y_{ig1b}		Y_{i1kb}	Y_{ijkb}	Y_{igkb}		Y_{i1ab}	Y_{ijab}	Y_{igab}
	$\cdots$		$\cdots$		$\cdots$	$\cdots$	$\cdots$		$\cdots$	$\cdots$	$\cdots$
	Y_{n11b}		Y_{ng1b}		Y_{n1kb}	Y_{njkb}	Y_{ngkb}		Y_{n1ab}	Y_{njab}	Y_{ngab}

TABLE 9-6 *Analysis of variance for a hierarchical design containing a within-groups variable*

SV	df	SS	EMS	F
Total	$abgn - 1$	$\sum_i^n \sum_j^g \sum_k^a \sum_m^b Y_{ijkm}^2 - C$		
Between G	$ag - 1$	$\dfrac{\sum_j^g \sum_k^a (\sum_i^n \sum_m^b Y_{ijkm})^2}{nb} - C$		
A	$a - 1$	$\dfrac{\sum_k^a (\sum_i^n \sum_j^g \sum_m^b Y_{ijkm})^2}{ngb} - C$	$\sigma_e^2 + ng\sigma_{G/A}^2 + nbg\theta_A^2$	$\dfrac{MS_A}{MS_{G/A}}$
G/A	$a(g - 1)$	$SS_{B.G} - SS_A$	$\sigma_e^2 + nb\sigma_{G/A}^2$	$\dfrac{MS_{G/A}}{MS_{S/GB/A}}$
Within G	$ag(bn - 1)$	$SS_{tot} - SS_{B.G}$		
B	$b - 1$	$\dfrac{\sum_m^b (\sum_i^n \sum_j^g \sum_k^a Y_{ijkm})^2}{nga} - C$	$\sigma_e^2 + n\sigma_{GB/A}^2 + nga\theta_B^2$	$\dfrac{MS_B}{MS_{GB/A}}$
AB	$(a - 1)(b - 1)$	$\dfrac{\sum_k^a \sum_m^b (\sum_i^n \sum_j^g Y_{ijkm})^2}{ng} - C - SS_A - SS_B$	$\sigma_e^2 + n\sigma_{GB/A}^2 + ng\theta_{AB}^2$	$\dfrac{MS_{AB}}{MS_{GB/A}}$
GB/A	$a(b - 1)(g - 1)$	$\dfrac{\sum_j^g \sum_k^a \sum_m^b (\sum_i^n Y_{ijkm})^2}{n} - \dfrac{\sum_k^a \sum_m^b (\sum_i^n \sum_j^g Y_{ijkm})^2}{ng} + \dfrac{\sum_j^g \sum_k^a (\sum_i^n \sum_m^b Y_{ijkm})^2}{ngb}$	$\sigma_e^2 + n\sigma_{GB/A}^2$	$\dfrac{MS_{GB/A}}{MS_{S/GB/A}}$
S/GB/A	$abg(n - 1)$	$SS_{W.G} - SS_B - SS_{AB} - SS_{GB/A}$	σ_e^2	

A model consistent with this view of the partitioning of the total variability is

(9.4) $Y_{ijkm} = \mu + \alpha_k + \gamma_{j/k} + \beta_m + (\alpha\beta)_{km} + (\gamma\beta)_{jm/k} + \varepsilon_{ijkm}$

Note that there are no interactions involving subjects, nor is there any AG or ABG effect. This is because subjects cross with none of the three other variables, and G does not cross with A. As a further help in establishing models for designs involving nesting, note that the interaction of a nested effect with another variable will also be nested. For example, the interaction of G/A with B is the nested interaction GB/A.

It is assumed that A and B are fixed effect variables, but that the groups are a random sample from a large population of such groups. Consequently,

$$\sum_k \alpha_k = \sum_m \beta_m = \sum_k \sum_m (\alpha\beta)_{km} = 0$$

The variance components for the fixed effects are

$$\theta_A^2 = \frac{\sum_k \alpha_k^2}{a - 1} \qquad \theta_B^2 = \frac{\sum_m \beta_m^2}{b - 1}$$

$$\theta_{AB}^2 = \frac{\sum_k \sum_m (\alpha\beta)_{km}^2}{(a - 1)(b - 1)}$$

$\gamma_{j/k}$, $(\gamma\beta)_{jm/k}$, and ε_{ijkm} are randomly sampled from normally distributed populations with mean zero and variances $\sigma_{G/A}^2$, $\sigma_{GB/A}^2$, and σ_e^2, respectively.

9.3.2 The analysis of variance. Table 9–6 presents the pertinent aspects of the analysis of variance. The SV have already been discussed. The df column provides an excellent check on our breakdown of sources of variance. The initial division into between- and within-groups sources may be checked by noting that $df_{B.G}$ reflects the variability of ag means about the grand mean; therefore, $ag - 1$ is the required number. The $df_{W.G}$ reflects the variability of all bn scores within the group deviated about the group mean and then pooled over the ag groups; the result is $ag(bn - 1)$. Adding the two df quantities, we obtain $abgn - 1$, which is, of course, the correct total. The df for the A, B, and AB terms require no comment. The $df_{G/A}$ reflect the fact that g group means have been subtracted from the mean for the level of A in which the groups are nested (thus requiring $g - 1$ df), that this process has been repeated at all a levels of A, and that the squared deviations have then been pooled. Similarly, GB/A represents the interaction of groups and B for each level of A [$(g - 1)(b - 1)$ df] pooled over a levels of A. Finally, we note that there are $n - 1$ df for the variability of scores within each level of B within each group; pooling over b levels and ag groups gives the $df_{S/GB/A}$. Adding the various df together, we find that there are neither too few nor too many terms in the analysis; the individual terms sum to the appropriate total.

The *SS* formulas should present no problems. As in the past, $SS_{GB/A}$ may be obtained by expanding the *df*,

$$a(g - 1)(b - 1) = abg - ab - ag + a$$

or by noting that the nested interaction is the difference between two interactions. The first approach is the basis for the formula in Table 9–6. The second approach, yields an algebraically identical result:

$$SS_{GB/A} = SS_{GB} - SS_{AB}$$

$$= \left[\frac{\sum_j^g \sum_k^a \sum_m^b (\sum_i^n Y_{ijkm})^2}{n} - C - SS_{B \cdot G} - SS_B \right] - SS_{AB}$$

A third approach is available if we note that

$$SS_{GB/A} = SS_{GB/A_1} + \cdots + SS_{GB/A_a}$$

i.e., the nested interaction is actually the result of computing the *GB* variability at each level of *A* and then pooling over levels. While this last approach seems closest to the definition of *GB/A*, it is most laborious and least recommended.

Remembering that *A* and *B* are fixed-effect variables and that groups are assumed to have random effects, we can readily verify that the entries in the *EMS* column follow the rules developed in Section 8.2.2. Once the *EMS* have been set down, appropriate *F* tests are immediately available.

9.3.3 A numerical example. Table 9–7 presents data and cell totals for a hierarchical design with one within-groups variable. The total variability is as usual

$$SS_{tot} = \sum_i^n \sum_j^g \sum_k^a \sum_m^b Y_{ijkm}^2 - C$$

$$= (4)^2 + (5)^2 + \cdots + (22)^2 - \frac{(328)^2}{24}$$

$$= 5,502.00 - 4,482.67$$

$$= 1,019.33$$

This quantity is then partitioned into between- and within-groups components:

$$SS_{B \cdot G} = \frac{\sum_j^g \sum_k^a (\sum_i^n \sum_m^b Y_{ijkm})^2}{nb} - C$$

$$= \frac{(32)^2 + \cdots + (81)^2}{4} - \frac{(328)^2}{24}$$

$$= 5,188.50 - 4,482.67$$

$$= 705.83$$

TABLE 9-7 Data matrix for a hierarchical design including a within-groups variable

	A_1		A_2		A_3	
	G_{11}	G_{21}	G_{12}	G_{22}	G_{13}	G_{23}
B_1	4	5	3	11	20	18
	6	9	10	6	23	17
	$\sum_i Y_{iJ11} = 10$	14	$\sum_i Y_{iJ21} = 13$	17	$\sum_i Y_{iJ31} = 43$	35
	$\sum_i \sum_j Y_{iJ11} = 24$		$\sum_i \sum_j Y_{iJ21} = 30$		$\sum_i \sum_j Y_{iJ31} = 78$	$\sum_i \sum_j \sum_k Y_{iJk1} = 132$
B_2	8	10	12	14	19	24
	14	15	15	17	26	22
	$\sum_i Y_{iJ12} = 22$	25	$\sum_i Y_{iJ22} = 27$	31	$\sum_i Y_{iJ32} = 45$	46
	$\sum_i \sum_j Y_{iJ12} = 47$		$\sum_i \sum_j Y_{iJ22} = 58$		$\sum_i \sum_j Y_{iJ32} = 91$	$\sum_i \sum_j \sum_k Y_{ijk2} = 196$
	$\sum_i \sum_j \sum_m Y_{iJ1m} = 71$		$\sum_i \sum_j \sum_m Y_{iJ2m} = 88$		$\sum_i \sum_j \sum_m Y_{iJ3m} = 169$	$\sum_i \sum_j \sum_k \sum_m Y_{iJkm} = 328$

$$\sum_i \sum_m Y_{iJkm} = 32 \qquad \sum_i \sum_j \sum_m Y_{iJkm} = 71$$

Group Totals

	A_1		A_2		A_3	
	G_{11}	G_{21}	G_{12}	G_{22}	G_{13}	G_{23}
	32	39	40	48	88	81
	71		88		169	

$$\sum_i \sum_j \sum_k \sum_m Y_{iJkm} = 328$$

and

$$SS_{W.G} = SS_{tot} - SS_{B.G}$$
$$= 1{,}019.33 - 705.83$$
$$= 313.50$$

The A main effect contributes to the variability among groups. Therefore, we compute

$$SS_A = \frac{\sum_k^a (\sum_i^n \sum_j^g \sum_m^b Y_{ijkm})^2}{ngb} - C$$

$$= \frac{(71)^2 + (88)^2 + (169)^2}{8} - \frac{(328)^2}{24}$$

$$= 5{,}168.92 - 4{,}482.67$$

$$= 686.25$$

The residual variability due to differences among groups is

$$SS_{G/A} = SS_{B.G} - SS_A$$
$$= 705.83 - 686.25$$
$$= 19.53$$

We next analyze the within-groups variability. The effect of B may be swiftly computed by use of the single *df* formula:

$$SS_B = \frac{(\sum_i^n \sum_j^g \sum_k^b Y_{ijk1} - \sum_i^n \sum_j^g \sum_k^b Y_{ijk2})^2}{24}$$

$$= \frac{(132 - 196)^2}{24}$$

$$= 170.67$$

The AB interaction is investigated next:

$$SS_{AB} = \frac{\sum_k^a \sum_m^b (\sum_i^n \sum_j^g Y_{ijkm})^2}{ng} - C - SS_A - SS_B$$

$$= \frac{(24)^2 + \cdots + (91)^2}{4} - \frac{(328)^2}{24} - 686.25 - 170.67$$

$$= 5{,}353.50 - 4{,}482.67 - 686.25 - 170.67$$

$$= 13.91$$

Next, we have

$$SS_{GB/A} = \frac{\sum_j^g \sum_k^a \sum_m^b (\sum_i^n Y_{ijkm})^2}{n} - \frac{\sum_k^a \sum_m^b (\sum_i^n \sum_j^g Y_{ijkm})^2}{ng}$$

$$- \frac{\sum_j^g \sum_k^a (\sum_i^n \sum_m^b Y_{ijkm})^2}{nb} + \frac{\sum_k^a (\sum_i^n \sum_j^g \sum_m^b Y_{ijkm})^2}{ngb}$$

$$= \frac{(10)^2 + (14)^2 + \cdots + (46)^2}{2} - 5{,}353.50 - 5{,}188.50 + 5{,}168.92$$

$$= 10.92$$

Finally, we have the residual variability among subjects:

$$SS_{S/GB/A} = SS_{W.G} - SS_B - SS_{AB} - SS_{GB/A}$$
$$= 313.50 - 170.67 - 13.91 - 10.92$$
$$= 118.00$$

Table 9–8 summarizes the analysis. Even on the small number of *df* provided in the example, the *A* and *B* main effects are highly significant. No other sources are significant.

TABLE 9–8 *Analysis of variance for the data of Table 9–7*

SV	df	SS	MS	F
Total	23	1,019.33		
Between G	5	705.83		
A	2	686.25	343.13	52.71**
G/A	3	19.53	6.51	
Within G	18	313.50		
B	1	170.67	170.67	46.89*
AB	2	13.91	6.96	1.91
GB/A	3	10.92	3.64	.37
S/GB/A	12	118.00	9.83	

$$** \; p < .005$$
$$* \; p < .01$$

9.4 REPEATED MEASUREMENTS IN A HIERARCHICAL DESIGN

The design of Section 9.3 may be extended by requiring several measures from each subject. For example, we again have g groups under high stress (A_1) and g groups under low stress (A_2). Within each group there are n high anxious subjects (B_1) and n low anxious subjects (B_2). Each member of the group is required to solve each of four problems; *problems* is the within-subjects variable which we shall label C in this section. In general, we have g groups, sampled randomly from a large population of such groups, at each of a levels of the independent variable A. Within each group are b arbitrarily chosen levels of the independent variable B; there are n different subjects at each of these levels. Thus, we have bn subjects in each of ag groups for a total of $abgn$ subjects. A measure is obtained from each subject at each of c levels of the variable C. The design is represented in Table 9–9 at one level

TABLE 9–9 *Data matrix for a hierarchical design with repeated measurements*

			C_1	C_2	$\cdots$	C_p	$\cdots$	C_c
		S_{1111}	Y_{11111}	Y_{11112}		Y_{1111p}		Y_{1111c}
		$\vdots$						
	B_1	S_{i111}	Y_{i1111}	Y_{i1112}		Y_{i111p}		Y_{i111c}
		$\vdots$						
		S_{n111}	Y_{n1111}	Y_{n1112}		Y_{n111p}		Y_{n111c}
	$\vdots$							
		S_{111m}	Y_{111m1}	Y_{111m2}		Y_{111mp}		Y_{111mc}
		$\vdots$						
G_{11}	B_m	S_{i11m}	Y_{i11m1}	Y_{i11m2}		Y_{i11mp}		Y_{i11mc}
		$\vdots$						
		S_{n11m}	Y_{n11m1}	Y_{n11m2}		Y_{n11mp}		Y_{n11mc}
	$\vdots$							
		S_{111b}	Y_{111b1}	Y_{111b2}		Y_{111bp}		Y_{111bc}
		$\vdots$						
A_1	B_b	S_{i11b}	Y_{i11b1}	Y_{i11b2}		Y_{i11bp}		Y_{i11bc}
$\vdots$		$\vdots$						
		S_{n11b}	Y_{n11b1}	Y_{n11b2}		Y_{n11bp}		Y_{n11bc}
$\vdots$								
		S_{1j1m}	Y_{1j1m1}	Y_{1j1m2}		Y_{1j1mp}		Y_{1j1mc}
		$\vdots$						
G_{j1}	B_m	S_{ij1m}	Y_{ij1m1}	Y_{ij1m2}		Y_{ij1mp}		Y_{ij1mc}
		$\vdots$						
		S_{nj1m}	Y_{nj1m1}	Y_{nj1m2}		Y_{nj1mp}		Y_{nj1mc}
$\vdots$								
		S_{1g1m}	Y_{1g1m1}	Y_{1g1m2}		Y_{1g1mp}		Y_{1g1mc}
		$\vdots$						
G_{g1}	B_m	S_{ig1m}	Y_{ig1m1}	Y_{ig1m2}		Y_{ig1mp}		Y_{ig1mc}
$\vdots$		$\vdots$						
		S_{ng1m}	Y_{ng1m1}	Y_{ng1m2}		Y_{ng1mp}		Y_{ng1mc}

of A. The indices are

$$i = 1, 2, \ldots, n$$
$$j = 1, 2, \ldots, g$$
$$k = 1, 2, \ldots, a$$
$$m = 1, 2, \ldots, b$$
$$p = 1, 2, \ldots, c$$

9.4.1 The analysis of variance model. As with earlier designs, the simplest approach is to attempt to relate the present design to one with which we are already familiar. Upon reflection, it should become clear that this design is a simple extension of the design analyzed in Table 9–6; the difference is that we now have c measurements per subject where earlier we had one. It follows that the *Between S* variability must be broken down just as the total variability was in Table 9–6. The first part (*Between S* and its component terms) of Table 9–10 is essentially a replica of the whole of Table 9–6.

The total *df* for the present design is $abcgn - 1$ and the *Between S* terms account for $abgn - 1$ of these. We must still account for the remaining $abgn(c - 1)$ *df*. From Table 9–9, it is apparent that there should be a C main effect and, furthermore, that C crosses with all the other variables in the study. Thus, to complete the list of *SV* in Table 9–10, we cross C with each of the basic terms in the *Between S* part of the table. This yields the *Within S* terms of Table 9–10.

Equation (9.5) presents a general model, consistent with the analysis:

(9.5)

$$
\begin{array}{lllll}
Y_{ijkmp} = \mu + & \alpha_k + & \gamma_{j/k} + & \beta_m + & (\alpha\beta)_{km} + (\gamma\beta)_{jm/k} \\
 & A & G/A & B & AB \quad\quad GB/A \\
 & \text{effect} & \text{effect} & \text{effect} & \text{effect} \quad\quad \text{effect}
\end{array}
$$

$$
\begin{array}{lllll}
+ & \eta_{i/jm/k} + & \delta_p + & (\alpha\delta)_{kp} + & (\gamma\delta)_{jp/k} + (\beta\delta)_{mp} \\
 & S/GB/A & C & AC & GC/A \quad\quad BC \\
 & \text{effect} & \text{effect} & \text{effect} & \text{effect} \quad\quad \text{effect}
\end{array}
$$

$$
\begin{array}{llll}
+ & (\alpha\beta\delta)_{kmp} + & (\gamma\beta\delta)_{jmp/k} + & (\eta\delta)_{ip/jm/k} + & \varepsilon_{ijkmp} \\
 & ABC & GBC/A & SC/GB/A & \text{error of} \\
 & \text{effect} & \text{effect} & \text{effect} & \text{measurement}
\end{array}
$$

This is essentially an extension of Equation (9.4) to encompass the main and interaction effects of the within-subject variable, C. In addition, the individual effect ($\eta_{i/jm/k}$) is now listed separately because it only contributes to *some* of the *SV*; in Equation (9.4), such effects were subsumed under the general error component, ε_{ijkm}.

9.4.2 The analysis of variance. The analysis of variance represented in Table 9–10 is less forbidding than it may appear. The key to the analysis lies in Equation (9.5), which dictates the sources of variance. The organization of these sources begins with the subject; in any design an effect must either be between subjects or within subjects. Since there are different subjects in each group, and consequently in each level of *A*, and also in each level of *B*, and since each subject is tested at all levels of *C*, the classification in the present design seems reasonable. The only question might be the status of interaction effects involving between- and within-subjects variables; in previous designs involving such interactions they were shown to be within-subjects effects. If any doubt exists in the present case, a check is available.

The sum of *df* for the sources classified as between subjects is *abgn* − 1, clearly the appropriate number for the total variability among subjects. Within subjects, we should have *c* − 1 *df* for each individual, multiplied by *abgn* to reflect the pooling over subjects. This is the total obtained when the *df* for the within-subjects effects are added together. The reader should follow the two approaches just described to verify that the between- and within-groups classifications are also accurate.

It is extremely helpful to consider not only what the sources are, but also how they are to be organized. The appropriate classification of sources of variance, for example, into between- and within-groups categories, makes the table easier to digest and facilitates calculations of *df* and sums of squares. In addition, fewer mistakes in the selection of error terms will be made if the table is so laid out that each error term lies beneath the sources that it will be used to evaluate.

Once the *SV* have been determined and organized as in Table 9–10, *df* and *SS* columns are readily established. As always, the *df* are determined by a few simple principles, which we have used repeatedly:

(a) The *df* for an unnested main effect are one less than the number of levels of the variable involved. For example, the *A* source represents the variability of *a* means about the grand mean of the data matrix.

(b) The *df* for a nested main effect involve the computation of *df* for each level of nesting and then multiplication by the number of levels. For example, subjects are nested within groups × *B* combinations which, in turn, are nested within levels of *A*. Thus, we have *n* − 1 *df* for each groups × *B* combination, multiplied by the number of such combinations, *abg*. The $df_{S/GB/A}$ are therefore equal to $abg(n - 1)$.

(c) The *df* for interaction effects are merely a product of the *df* for the effects involved. Thus $df_{SC/BG/A}$ equals $abg(n - 1) \times (c - 1)$.

The formulas for sum of squares computations in Table 9–10 are those that are considered most easily generated and executed. Alternatives exist, particularly in the case of nested effects, which may be computed as a difference among interactions rather than by the expanded *df* approach.

The *EMS* are particularly important in the present design, since there are many sources and error terms. It is foolish to attempt to remember what is tested against what in this and similar designs. The best approach is to remember the rules of thumb of Chapter 8, which allow us to quickly

TABLE 9-10 *Analysis of variance for a hierarchical design with repeated measurements*

SV	df	SS	EMS	F
Total	$abgnc - 1$	$\sum_i^n\sum_j^g\sum_k^a\sum_m^b\sum_p^c Y_{ijkmp}^2 - C$		
Between S	$abgn - 1$	$\dfrac{\sum_i^n\sum_j^g\sum_k^a\sum_m^b(\sum_p^c Y_{ijkmp})^2}{c} - C$		
Between G	$ag - 1$	$\dfrac{\sum_j^g\sum_k^a(\sum_i^n\sum_m^b\sum_p^c Y_{ijkmp})^2}{nbc} - C$		
A	$a - 1$	$\dfrac{\sum_k^a(\sum_i^n\sum_j^g\sum_m^b\sum_p^c Y_{ijkmp})^2}{ngbc} - C$	$\sigma_e^2 + nbc\sigma_{G/A}^2 + c\sigma_{S/GB/A}^2$ $+ nbgc\theta_A^2$	$\dfrac{MS_A}{MS_{G/A}}$
G/A	$a(g - 1)$	$SS_{B.G} - SS_A$	$\sigma_e^2 + nbc\sigma_{G/A}^2 + c\sigma_{S/GB/A}^2$	
Within G	$ag(bn - 1)$	$SS_{B.S} - SS_{B.G}$		
B	$b - 1$	$\dfrac{\sum_m^b(\sum_i^n\sum_j^g\sum_k^a\sum_p^c Y_{ijkmp})^2}{ngac} - C$	$\sigma_e^2 + c\sigma_{S/GB/A}^2 + nc\sigma_{GB/A}^2$ $+ nagc\theta_B^2$	$\dfrac{MS_B}{MS_{GB/A}}$
AB	$(a - 1)(b - 1)$	$\dfrac{\sum_k^a\sum_m^b(\sum_i^n\sum_j^g\sum_p^c Y_{ijkmp})^2}{ngc} - C - SS_A - SS_B$	$\sigma_e^2 + c\sigma_{S/GB/A}^2 + nc\sigma_{GB/A}^2$ $+ ngc\theta_{AB}^2$	$\dfrac{MS_{AB}}{MS_{GB/A}}$
GB/A	$a(g - 1)(b - 1)$	$\dfrac{\sum_j^g\sum_k^a\sum_m^b(\sum_i^n\sum_p^c Y_{ijkmp})^2}{nc} - \dfrac{\sum_j^g\sum_k^a(\sum_i^n\sum_m^b\sum_p^c Y_{ijkmp})^2}{nbc}$ $- \dfrac{\sum_k^a\sum_m^b(\sum_i^n\sum_j^g\sum_p^c Y_{ijkmp})^2}{ngc} + \dfrac{\sum_k^a(\sum_i^n\sum_j^g\sum_m^b\sum_p^c Y_{ijkmp})^2}{ngbc}$	$\sigma_e^2 + c\sigma_{S/GB/A}^2 + nc\sigma_{GB/A}^2$	$\dfrac{MS_{GB/A}}{MS_{S/GB/A}}$
S/GB/A	$abg(n - 1)$	$SS_{W.G} - SS_B - SS_{AB} - SS_{GB/A}$	$\sigma_e^2 + c\sigma_{S/GB/A}^2$	

Source	df	SS	E(MS)	F
Within S	$abgn(c-1)$	$SS_{tot} - SS_{B.S}$		
C	$c-1$	$\dfrac{\sum_p^c(\sum_i^n\sum_j^g\sum_k^a\sum_m^b Y_{ijkmp})^2}{ngab} - C$	$\sigma_e^2 + nb\sigma_{GC/A}^2 + \sigma_{SC/GB/A}^2 + ngba\theta_C^2$	$\dfrac{MS_C}{MS_{GC/A}}$
AC	$(a-1)(c-1)$	$\dfrac{\sum_k^a\sum_p^c(\sum_i^n\sum_j^g\sum_m^b Y_{ijkmp})^2}{ngb} - C - SS_A - SS_C$	$\sigma_e^2 + nb\sigma_{GC/A}^2 + \sigma_{SC/GB/A}^2 + ngb\theta_{AC}^2$	$\dfrac{MS_{AC}}{MS_{GC/A}}$
GC/A	$a(g-1)(c-1)$	$\dfrac{\sum_j^g\sum_k^a\sum_p^c(\sum_i^n\sum_m^b Y_{ijkmp})^2}{nb} - \dfrac{\sum_j^g\sum_k^a(\sum_i^n\sum_m^b\sum_p^c Y_{ijkmp})^2}{nbc} - \dfrac{\sum_k^a\sum_p^c(\sum_i^n\sum_j^g\sum_m^b Y_{ijkmp})^2}{nbg} + \dfrac{\sum_k^a(\sum_i^n\sum_j^g\sum_m^b\sum_p^c Y_{ijkmp})^2}{ngbc}$	$\sigma_e^2 + nb\sigma_{GC/A}^2 + \sigma_{SC/GB/A}^2$	$\dfrac{MS_{GC/A}}{MS_{SC/GB/A}}$
BC	$(b-1)(c-1)$	$\dfrac{\sum_m^b\sum_p^c(\sum_i^n\sum_j^g\sum_k^a Y_{ijkmp})^2}{nga} - C - SS_B - SS_C$	$\sigma_e^2 + n\sigma_{GBC/A}^2 + \sigma_{SC/GB/A}^2 + nga\theta_{BC}^2$	$\dfrac{MS_{BC}}{MS_{GBC/A}}$
ABC	$(a-1)(b-1)(c-1)$	$\dfrac{\sum_k^a\sum_m^b\sum_p^c(\sum_i^n\sum_j^g Y_{ijkmp})^2}{ng} - C - SS_{AB} - SS_{AC} - SS_{BC} - SS_A - SS_B - SS_C$	$\sigma_e^2 + n\sigma_{GBC/A}^2 + \sigma_{SC/GB/A}^2 + ng\theta_{ABC}^2$	$\dfrac{MS_{ABC}}{MS_{GBC/A}}$
GBC/A	$a(g-1)(b-1)(c-1)$	$\dfrac{\sum_j^g\sum_k^a\sum_m^b\sum_p^c(\sum_i^n Y_{ijkmp})^2}{n} - SS_{AC} - SS_{BC} + \dfrac{\sum_k^a\sum_m^b\sum_p^c(\sum_i^n\sum_j^g Y_{ijkmp})^2}{ngc} + \dfrac{\sum_j^g\sum_k^a\sum_p^c(\sum_i^n\sum_m^b Y_{ijkmp})^2}{nbc} - \dfrac{\sum_j^g\sum_k^a\sum_m^b(\sum_i^n\sum_p^c Y_{ijkmp})^2}{nc} - \dfrac{\sum_k^a\sum_m^b(\sum_i^n\sum_j^g\sum_p^c Y_{ijkmp})^2}{ng} - \dfrac{\sum_j^g\sum_k^a\sum_p^c(\sum_i^n\sum_m^b Y_{ijkmp})^2}{ngbc} - SS_{GBC/A}$	$\sigma_e^2 + n\sigma_{GBC/A}^2 + \sigma_{SC/GB/A}^2$	$\dfrac{MS_{GBC/A}}{MS_{SC/GB/A}}$
$SC/GB/A$	$abg(n-1)(c-1)$	$SS_{w.s} - SS_C - SS_{AC} - SS_{GC/A} - SS_{BC} - SS_{ABC} - SS_{GBC/A}$	$\sigma_e^2 + \sigma_{SC/GB/A}^2$	

generate *EMS* for any design; the *F* tests follow immediately. The reader should note that regardless of how many levels of nesting exist, the letters designating the source must be contained somewhere among the subscripts if a component is to be considered for inclusion in the *EMS*. However, only those subscripts to the left of the first slash are "essential." The component is included in the *EMS* only if, after ignoring the letters that describe the source, the essential subscripts all represent random effect variables. For example, in assessing the *EMS* for G/A, we consider (in addition to error variance and the null hypothesis term, $\sigma^2_{G/A}$)

$$\sigma^2_{GB/A}, \; \sigma^2_{S/GB/A}, \; \sigma^2_{GC/A}, \; \sigma^2_{GBC/A}, \; \sigma^2_{SC/GB/A}$$

Of these, only $\sigma^2_{S/GB/A}$ contains no fixed-effect letter subscript in the "essential" position after ignoring G and A. As a second example, consider the BC source. We can consider

$$\theta^2_{ABC}, \; \sigma^2_{GBC/A}, \; \sigma^2_{SC/GB/A}$$

Ignoring B and C, we find that $\sigma^2_{GBC/A}$ and $\sigma^2_{SC/GB/A}$ enter into the *EMS*.

9.4.3 A numerical example. Table 9–11 presents data for a hierarchical design with repeated measurements. Squaring each score and summing, we obtain

$$SS_{\text{tot}} = \sum_i^n \sum_j^g \sum_k^a \sum_m^b \sum_p^c Y^2_{ijkmp} - C$$

$$= (4)^2 + (3)^2 + \cdots + (15)^2 - \frac{(4 + 3 + \cdots + 15)^2}{48}$$

$$= 3{,}515.00 - 2{,}806.02$$

$$= 708.98$$

We next obtain a measure of the overall variability among subjects:

$$SS_{\text{B·S}} = \frac{\sum_i^n \sum_j^g \sum_k^a \sum_m^b (\sum_p^c Y_{ijkmp})^2}{c} - C$$

$$= \frac{(4 + 5 + 8)^2 + \cdots + (9 + 11 + 15)^2}{3} - \frac{(4 + 3 + \cdots + 15)^2}{48}$$

$$= 3{,}094.33 - 2{,}806.02$$

$$= 288.31$$

Part of the variability among subjects is due to the fact that some subjects are in one group, some are in a different group. To measure this variability among group means, we have

$$SS_{B.G} = \frac{\sum_j^g \sum_k^a (\sum_i^n \sum_m^b \sum_p^c Y_{ijkmp})^2}{nbc} - C$$

$$= \frac{(4 + 5 + \cdots + 5 + 9)^2 + (4 + 7 + \cdots + 2 + 12)^2}{12} + \cdots$$

$$+ \frac{(3 + 5 + \cdots + 11 + 15)^2}{12} - \frac{(4 + 3 + \cdots + 15)^2}{48}$$

$$= 2{,}953.75 - 2{,}806.02$$

$$= 147.73$$

The overall variability among group means may be attributed to the effects of the treatment variable, A, and to a residual group variability. We have

$$SS_A = \frac{\sum_k^a (\sum_i^n \sum_j^g \sum_m^b \sum_p^c Y_{ijkmp})^2}{ngbc} - C$$

$$= \frac{(4 + 5 + \cdots + 2 + 12)^2 + (7 + 7 + \cdots + 11 + 15)^2}{24}$$

$$- \frac{(4 + 3 + \cdots + 15)^2}{48}$$

$$= 2{,}942.71 - 2{,}806.02$$

$$= 136.69$$

TABLE 9–11 *Data for a hierarchical design with repeated measurements*

			B_1				B_2		
			C_1	C_2	C_3		C_1	C_2	C_3
		S_{1111}	4	5	8	S_{1112}	3	6	10
	G_{11}								
		S_{2111}	3	6	10	S_{2112}	4	5	9
A_1									
		S_{1211}	4	7	8	S_{1212}	1	6	8
	G_{21}								
		S_{2211}	3	6	9	S_{2212}	4	2	12
		S_{1121}	7	7	11	S_{1122}	9	8	16
	G_{12}								
		S_{2121}	4	8	14	S_{2122}	7	10	19
A_2									
		S_{1221}	3	5	9	S_{1222}	10	12	13
	G_{22}								
		S_{2221}	2	7	8	S_{2222}	9	11	15

and

$$SS_{G/A} = SS_{B.G} - SS_A$$
$$= 147.73 - 136.69$$
$$= 11.04$$

Part of the variability among subject means still remains to be accounted for. Since there are different subjects at each level of B, this treatment variable contributes to $SS_{B.S}$. We have

$$SS_B = \frac{\sum_m^b (\sum_i^n \sum_j^g \sum_k^a \sum_p^c Y_{ijkmp})^2}{ngac} - C$$

$$= \frac{(4 + 5 + \cdots + 7 + 8)^2 + (3 + 6 + \cdots + 11 + 15)^2}{24}$$

$$- \frac{(4 + 3 + \cdots + 15)^2}{48}$$

$$= 2{,}860.21 - 2{,}806.02$$

$$= 54.19$$

The variable B crosses with A and groups. We therefore compute

$$SS_{AB} = \frac{\sum_k^a \sum_m^b (\sum_i^n \sum_j^g \sum_p^c Y_{ijkmp})^2}{ngc} - C - SS_A - SS_B$$

$$= \frac{(4 + 5 + \cdots + 6 + 9)^2 + \cdots + (9 + 8 + \cdots + 11 + 15)^2}{12}$$

$$- C - SS_A - SS_B$$

$$= 3{,}064.58 - 2{,}806.02 - 136.69 - 54.19$$

$$= 67.68$$

and

$$SS_{GB/A} = \frac{\sum_j^g \sum_k^a \sum_m^b (\sum_i^n \sum_p^c Y_{ijkmp})^2}{nc} - \frac{\sum_j^g \sum_k^a (\sum_i^n \sum_m^b \sum_p^c Y_{ijkmp})^2}{nbc}$$

$$- \frac{\sum_k^a \sum_m^b (\sum_i^n \sum_j^g \sum_p^c Y_{ijkmp})^2}{ngc} + \frac{\sum_k^a (\sum_i^n \sum_j^g \sum_m^b \sum_p^c Y_{ijkmp})^2}{ngbc}$$

$$= \frac{(4 + 5 + \cdots + 6 + 10)^2}{6} + \cdots + \frac{(10 + 12 + \cdots + 11 + 15)^2}{6}$$

$$- 2{,}953.75 - 3{,}064.58 + 2{,}942.71$$

$$= 3{,}090.17 - 2{,}953.75 - 3{,}064.58 + 2{,}942.71$$

$$= 14.55$$

Note that only $\sum_j^g \sum_k^a \sum_m^b (\sum_i^n \sum_p^c Y_{ijkmp})^2/nc$ is computed. The remaining parts of $SS_{GB/A}$ have been calculated previously in order to obtain other sum of square terms. A check on the above calculation is provided by

$$SS_{GB/A} = \frac{\sum_j^g \sum_k^a \sum_m^b (\sum_i^n \sum_p^c Y_{ijkmp})^2}{nc} - C - SS_{B.G} - SS_B - SS_{AB}$$

$$= 3{,}090.17 - 2{,}806.02 - 147.73 - 54.19 - 67.68$$

$$= 14.55$$

The residual variability among subjects may now be computed:

$$SS_{S/GB/A} = SS_{B.S} - SS_{B.G} - SS_B - SS_{AB} - SS_{GB/A}$$
$$= 288.31 - 147.73 - 54.19 - 67.68 - 14.55$$
$$= 4.16$$

We now turn to the variability within subjects which is due to the presence of repeated measurements. We obtain the overall within-subjects term first:

$$SS_{W.S} = SS_{tot} - SS_{B.S}$$
$$= 708.98 - 288.31$$
$$= 420.67$$

The variability due to treatment C is

$$SS_C = \frac{\sum_p^c (\sum_i^n \sum_j^g \sum_k^a \sum_m^b Y_{ijkmp})^2}{ngab} - C$$

$$= \frac{(4 + 3 + \cdots + 10 + 9)^2 + (5 + 6 + \cdots + 12 + 11)^2 + (8 + 10 + \cdots + 13 + 15)^2}{16}$$
$$- 2{,}806.02$$

$$= 3{,}143.19 - 2{,}806.02$$

$$= 337.17$$

For the AC interaction, we have

$$SS_{AC} = \frac{\sum_k^a \sum_p^c (\sum_i^n \sum_j^g \sum_m^b Y_{ijkmp})^2}{ngb} - C - SS_A - SS_C$$

$$= \frac{(4 + 3 + \cdots + 1 + 4)^2}{8} + \cdots + \frac{(11 + 14 + \cdots + 13 + 15)^2}{8}$$
$$- 2{,}806.02 - 136.69 - 337.17$$

$$= 3{,}281.38 - 2{,}806.02 - 136.69 - 337.17$$

$$= 1.50$$

Both the C and AC terms are tested against GC/A; the sum of squares for this is

$$SS_{GC/A} = \frac{\sum_j^g \sum_k^a \sum_p^c (\sum_i^n \sum_m^b Y_{ijkmp})^2}{nb} - \frac{\sum_j^g \sum_k^a (\sum_i^n \sum_m^b \sum_p^c Y_{ijkmp})^2}{nbc}$$

$$- \frac{\sum_k^a \sum_p^c (\sum_i^n \sum_m^b \sum_j^g Y_{ijkmp})^2}{nbg} + \frac{\sum_k^a (\sum_i^n \sum_j^g \sum_m^b \sum_p^c Y_{ijkmp})^2}{ngbc}$$

$$= \frac{(4 + 3 + 3 + 4)^2 + \cdots + (9 + 8 + 13 + 15)^2}{4} - 2{,}953.75$$

$$- 3{,}281.38 + 2{,}942.71$$

$$= 3{,}311.75 - 2{,}953.75 - 3{,}281.38 + 2{,}942.71$$

$$= 19.33$$

For the BC interaction, we have

$$SS_{BC} = \frac{\sum_m^b \sum_p^c (\sum_i^n \sum_j^g \sum_k^a Y_{ijkmp})^2}{nga} - C - SS_B - SS_C$$

$$= \frac{(4 + 3 + \cdots + 3 + 2)^2}{8} + \cdots + \frac{(10 + 9 + \cdots + 13 + 15)^2}{8}$$

$$- 2{,}806.02 - 54.19 - 337.17$$

$$= 3{,}205.38 - 2{,}806.02 - 54.19 - 337.17$$

$$= 8.00$$

Next, we have

$$SS_{ABC} = \frac{\sum_k^a \sum_m^b \sum_p^c (\sum_i^n \sum_j^g Y_{ijkmp})^2}{ng} - C - SS_{AB} - SS_{AC} - SS_{BC} - SS_A$$

$$- SS_B - SS_C$$

$$= \frac{(4 + 3 + 4 + 3)^2 + \cdots + (16 + 19 + 13 + 15)^2}{4} - 2{,}806.02$$

$$- 67.68 - 1.50 - 8.00 - 136.69 - 54.19 - 337.17$$

$$= 3{,}411.75 - 2{,}806.02 - 67.68 - 1.50 - 8.00 - 136.69$$

$$- 54.19 - 337.17$$

$$= .50$$

The error variability for the two preceding terms is

$$SS_{GBC/A} = \frac{\sum_j^g \sum_k^a \sum_m^b \sum_p^c (\sum_i^n Y_{ijkmp})^2}{n} + \frac{\sum_j^g \sum_k^a (\sum_i^n \sum_m^b \sum_p^c Y_{ijkmp})^2}{nbc}$$

$$+ \frac{\sum_k^a \sum_m^b (\sum_i^n \sum_j^g \sum_p^c Y_{ijkmp})^2}{ngc} + \frac{\sum_k^a \sum_p^c (\sum_i^n \sum_j^g \sum_m^b Y_{ijkmp})^2}{ngb}$$

$$- \frac{\sum_j^g \sum_k^a \sum_m^b (\sum_i^n \sum_p^c Y_{ijkmp})^2}{nc} - \frac{\sum_j^g \sum_k^a \sum_p^c (\sum_i^n \sum_m^b Y_{ijkmp})^2}{nb}$$

$$- \frac{\sum_k^a \sum_m^b \sum_p^c (\sum_i^n \sum_j^g Y_{ijkmp})^2}{ng} - \frac{\sum_k^a (\sum_i^n \sum_j^g \sum_m^b \sum_p^c Y_{ijkmp})^2}{ngbc}$$

$$= \frac{(4+3)^2 + \cdots + (13+15)^2}{1} + 2{,}953.75 + 3{,}064.58$$

$$+ 3{,}281.38 - 3{,}090.17 - 3{,}311.75 - 3{,}411.75 - 2{,}942.71$$

$$= 3{,}464.50 + 2{,}953.75 + 3{,}064.58 + 3{,}281.38 - 3{,}090.17$$
$$- 3{,}311.75 - 3{,}411.75 - 2{,}942.71$$

$$= 7.83$$

An alternative formula, which provides a partial check on $SS_{GBC/A}$, is

$$SS_{GBC/A} = \frac{\sum_j^g \sum_k^a \sum_m^b \sum_p^c (\sum_i^n Y_{ijkmp})^2}{n} - C - SS_A - SS_{G/A} - SS_B - SS_C$$

$$- SS_{GB/A} - SS_{GC/A} - SS_{BC} - SS_{AB} - SS_{AC} - SS_{ABC}$$

$$= 3{,}464.50 - 2{,}806.02 - 136.69 - 11.04 - 54.19 - 337.17$$
$$- 14.55 - 19.33 - 8.00 - 67.68 - 1.50 - .50$$

$$= 7.83$$

The $SS_{SC/GB/A}$ is the residual within-subjects variability. Therefore,

$$SS_{SC/GB/A} = SS_{\text{w.s}} - SS_C - SS_{AC} - SS_{GC/A} - SS_{BC} - SS_{ABC} - SS_{GBC/A}$$
$$= 420.67 - 337.17 - 1.50 - 19.33 - 8.00 - .50 - 7.83$$
$$= 46.34$$

Table 9–12 presents the completed analysis.

9.5 CONCLUDING REMARKS

Our presentation has been limited to designs in which subjects are nested within levels of some variable, G, which is in turn nested within levels of some variable, A. There are numerous variations on the designs that have been considered. There could be several between-groups variables, several

TABLE 9–12 *Analysis of variance for the data of Table 9–11*

SV	df	SS	MS	F
Total	47	708.98	15.08	
Between S	15	288.31	19.22	
Between G	3	147.73	49.24	
A	1	136.69	136.69	24.76*
G/A	2	11.04	5.52	
Within G	12	140.58	11.72	
B	1	54.19	54.19	7.44
AB	1	67.68	67.68	9.30
GB/A	2	14.55	7.28	14.00
S/GB/A	8	4.16	.52	
Within S	32	420.67	13.15	
C	2	337.17	168.59	34.90**
AC	2	1.50	.75	.16
GC/A	4	19.33	4.83	2.46
BC	2	8.00	4.00	2.04
ABC	2	.50	.25	.13
GBC/A	4	7.83	1.96	.68
SC/GB/A	16	46.34	2.90	

$$* \ p \ < \ .05$$
$$** \ p \ < \ .01$$

between-subjects-within-groups variables, and several within-subjects variables. Less frequently, within-subjects variables may be nested within each other. An example was given in Section 9.1; problems were nested within levels of difficulty. In the general case, there might be *an* subjects, randomized among the *a* levels of the treatment variable *A*. We obtain *bc* measures from each subject. However, the *b* levels of *B* at C_1 are not the same as the *b* levels at C_2, and so on; *B* is nested within levels of *C*. In terms of our previous example, *B* is problems and *C* is difficulty level. Noting that there are *S/A*, *A*, *B/C*, *C*, and the possible interactions *AC*, *SB/AC*, *SC/A*, and *AB/C*, we arrive at the structural equation

$$Y_{ijkm} = \mu + \underset{\substack{A \\ \text{effect}}}{\alpha_j} + \underset{\substack{S/A \\ \text{effect}}}{\eta_{i/j}} + \underset{\substack{C \\ \text{effect}}}{\gamma_m} + \underset{\substack{B/C \\ \text{effect}}}{\beta_{k/m}} + \underset{\substack{AC \\ \text{effect}}}{(\alpha\gamma)_{jm}}$$

(9.6)

$$+ \underset{\substack{AB/C \\ \text{effect}}}{(\alpha\beta)_{jk/m}} + \underset{\substack{SC/A \\ \text{effect}}}{(\eta\gamma)_{im/j}} + \underset{\substack{SB/AC \\ \text{effect}}}{(\eta\beta)_{ik/jm}} + \varepsilon_{ijkm}$$

which is sufficient to generate the sources of Table 9–13, p. 256. The *EMS* are based on the assumption that *A*, *B*, and *C* are all fixed variables.

In closing, we note two pitfalls of the designs considered in this chapter. As pointed out earlier, there is the danger that certain F tests will lack power due to too few error *df*. This will most often occur when some nested group effect is the error term. Unless there are grounds for pooling to obtain an error term on more *df*, there is little the experimenter can do after the data are collected. This being the case, it is important to work out the actual analysis of sources and *df* prior to the collection of data and to modify the design in whatever ways seem necessary to obtain powerful tests of effects of interest. The second pitfall is the occasional failure to replicate groups within levels of A, essentially confounding groups and levels of A. If one class is taught by one method and another by a second, is a difference in class means due to the different methods or to differences in the personal interactions within the two classes? We need some measure of variability among classes taught by the same method in order to determine the effect of the treatment. Experimenters often do not realize that the failure to replicate groups within levels is not particularly different from running one subject at each level of A in a simple completely randomized one-factor design.

EXERCISES

9.1 Analyze the following data set.

A_1		A_2		A_3	
G_{11}	G_{21}	G_{12}	G_{22}	G_{13}	G_{23}
5	8	11	32	12	18
21	23	15	18	22	31
14	10	16	26	36	38
12	17	23	25	18	37
8	15	27	26	34	28
16	20	31	17	41	32

9.2 There are three measures on each subject in the groups nested within levels of A. Analyze the data.

			B_1	B_2	B_3				B_1	B_2	B_3
		S_{111}	5	8	12			S_{112}	18	21	28
	G_{11}	S_{211}	7	8	14		G_{12}	S_{212}	17	19	23
		S_{311}	8	10	17			S_{312}	14	18	24
		S_{411}	7	11	16			S_{412}	16	21	27
A_1						A_2					
		S_{121}	10	12	19			S_{122}	15	21	32
	G_{21}	S_{221}	8	10	20		G_{22}	S_{222}	14	18	29
		S_{321}	7	11	17			S_{322}	15	23	28
		S_{421}	8	12	18			S_{422}	17	22	31

TABLE 9–13 Analysis of variance for a design with nesting within subjects

SV	df	SS	EMS	F
Total	$abcn - 1$	$\sum_i^n \sum_j^a \sum_k^b \sum_m^c Y_{ijkm}^2 - C$		
Between S	$an - 1$	$\dfrac{\sum_i^n \sum_j^a (\sum_k^b \sum_m^c Y_{ijkm})^2}{bc} - C$		
A	$a - 1$	$\dfrac{\sum_j^a (\sum_i^n \sum_k^b \sum_m^c Y_{ijkm})^2}{nbc} - C$	$\sigma_e^2 + bc\sigma_{S/A}^2 + nbc\theta_A^2$	$\dfrac{MS_A}{MS_{S/A}}$
S/A	$a(n - 1)$	$SS_{B.S} - SS_A$	$\sigma_e^2 + bc\sigma_{S/A}^2$	
Within S	$an(bc - 1)$	$SS_{tot} - SS_{B.S}$		
C	$c - 1$	$\dfrac{\sum_m^c (\sum_i^n \sum_j^a \sum_k^b Y_{ijkm})^2}{nab} - C$	$\sigma_e^2 + b\sigma_{SC/A}^2 + nab\theta_C^2$	$\dfrac{MS_C}{MS_{SC/A}}$
AC	$(a - 1)(c - 1)$	$\dfrac{\sum_j^a \sum_m^c (\sum_i^n \sum_k^b Y_{ijkm})^2}{nb} - C - SS_A - SS_C$	$\sigma_e^2 + b\sigma_{SC/A}^2 + nb\theta_{AC}^2$	$\dfrac{MS_{AC}}{MS_{SC/A}}$

TABLE 9-13 (continued)

Source	df	SS	E(MS)	F
SC/A	$a(n-1)(c-1)$	$\dfrac{\sum_i^n \sum_j^a \sum_m^c (\sum_k^b Y_{ijkm})^2}{b} - \dfrac{\sum_i^n \sum_j^a (\sum_k^b \sum_m^c Y_{ijkm})^2}{bc}$	$\sigma_e^2 + b\sigma_{SC/A}^2$	
B/C	$c(b-1)$	$\dfrac{\sum_k^b \sum_m^c (\sum_i^n \sum_j^a Y_{ijkm})^2}{na} - \dfrac{\sum_m^c (\sum_i^n \sum_j^a \sum_k^b Y_{ijkm})^2}{nab}$	$\sigma_e^2 + \sigma_{SB/AC}^2 + na\theta_{B/C}^2$	$\dfrac{MS_{B/C}}{MS_{SB/AC}}$
AB/C	$c(a-1)(b-1)$	$\dfrac{\sum_j^a \sum_k^b \sum_m^c (\sum_i^n Y_{ijkm})^2}{n} - \dfrac{\sum_k^b \sum_m^c (\sum_i^n \sum_j^a Y_{ijkm})^2}{na} - \dfrac{\sum_j^a \sum_m^c (\sum_i^n \sum_k^b Y_{ijkm})^2}{nb} + \dfrac{\sum_m^c (\sum_i^n \sum_j^a \sum_k^b Y_{ijkm})^2}{nab}$	$\sigma_e^2 + \sigma_{SB/AC}^2 + n\theta_{AB/C}^2$	$\dfrac{MS_{AB/C}}{MS_{SB/AC}}$
SB/AC	$ac(n-1)(b-1)$	$SS_{\text{w.s}} - SS_C - SS_{AC} - SS_{SC/A} - SS_{B/C} - SS_{AB/C}$	$\sigma_e^2 + \sigma_{SB/AC}^2$	

9.3 Six high schools are chosen at random for an experiment on teaching machines, which are introduced into three of the schools and not into the other three. Two introductory French classes are randomly selected from each school. There are ten students in each class, each of whom is given a mid-term and final examination. These constitute the measures. Give the *SV*, *df*, *EMS* and error terms.

9.4 Thirty-two subjects are randomly assigned to eight 4-man groups, four of which are task-oriented and four of which are ego-oriented. Each group is required to solve three problems under stress and three other problems under no stress. The score for each group is number of trials it takes the group to solve the problems. Give the *SV*, *df*, *EMS*, and error terms.

9.5 A study in economic bargaining involves thirty-six 3-man groups—a seller and two buyers. One of the buyers is a stooge who can side with the other buyer by keeping his own bids low, or with the seller by bidding the price of the commodity up. In $\frac{1}{3}$ of the groups, the stooge aids the seller on 75 percent of the trials; 50 percent in another 12 groups; 25 percent in another 12 groups. An average concession score (distance from initial bid) was obtained from each *S* under two conditions: trials when stooge sided with *S*, trials when stooge was against *S*. These two measures were obtained from each of four trial blocks yielding eight measures/*S*. Give *SV*, *df*, *EMS*.

9.6 An experiment was performed on the effects of socio-economic status and intelligence upon self-evaluation. Three school districts, each of a different social stratum, were chosen for participation in the experiment. Ten schools were selected at random from among the elementary schools in each district (i.e., a total of 30 schools). Ten high and ten low IQ subjects were selected from each school. Each subject was given a self-evaluation scale, the dependent variable. The experimenter performed the following analysis.

SV	df
District (*D*)	2
Intelligence level (*I*)	1
D × *I*	2
Within error	594
Total	599

Can you suggest an alternative analysis (including error terms)? What inference might be added? Changed? Why?

Latin Square Designs

10.1 INTRODUCTION

The basic Latin square design may be represented as

$$
\begin{array}{c}
 \; C_1 \; \; C_2 \; \; C_3 \\
\begin{array}{c} B_1 \\ B_2 \\ B_3 \end{array}
\begin{bmatrix}
A_1 & A_2 & A_3 \\
A_2 & A_3 & A_1 \\
A_3 & A_1 & A_2
\end{bmatrix}
\end{array}
$$

where A, B, and C are variables in an experiment. There are two aspects of this layout that are worth noting: First, this is a Latin square because each level of A appears exactly once in each row (level of B) and in each column (level of C); the above layout is one of several that meet this definition. Secondly, the Latin square is an incomplete design in the sense that it includes only nine of the possible 27 combinations of levels of A, B, and C that would appear in a complete factorial design; more generally, assuming a rows and columns in the square, we have a^2 of the possible a^3 treatment combinations. We will later show that this "incompleteness" has direct and important consequences for the applicability of the design.

The Latin square has several potential advantages over other designs. We have already touched upon one, its ability to investigate several variables with less expenditure of time and subjects than complete factorial designs would involve. This would be an important consideration in designing pilot studies or whenever the supply of available subjects is limited.

Even more important is the efficiency of the Latin square relative to other designs. For example, consider a treatment × blocks design with a blocks and a treatment levels. The blocks might correspond to levels of performance on some measure of ability related to the dependent variable. We could introduce a second concomitant variable, perhaps a measure of motivation, with no additional experimental effort; the rows of our Latin square layout would be blocks based on level of intelligence and the columns would be blocks based on level of motivation. Since two sources of individual differences have been removed, we should have a very small error term and, consequently, a very precise test of the treatment variable, A.

Perhaps the most prevalent use of the Latin square design is in situations where it is desirable to test each subject under all levels of the treatment variable, A. Then, the levels of B correspond to individual subjects or to groups of subjects and the levels of C correspond to positions in time; for example, successive days of the experiment. The repeated measurements design of Chapter 7, in which the order of the A_j was independently randomized for each subject, permitted the removal of error variance due to individual differences. In the Latin square design, in which the orders of the A_j are chosen to meet the Latin square requirement, we are able to remove still another source of error variance, the variability due to temporal effects. Thus, if subjects were tested under each level of A on different days, and if the order was Latin squared, variability due to days could be removed from the error variance. We would then have a more precise test of the A effects as well as an independent evaluation of the variance due to days.

By this time, dazzled by the promised savings in effort and error variance, the reader may wonder why any investigator would contemplate any design other than the Latin square or some extension of it. Of course, there is a catch. As has been the case throughout this text, the potential advantages are accompanied by potential disadvantages. Information about interactions among the row, column, and treatment variables will frequently be impossible to extract. Furthermore, if such interactions exist, they may obscure inferences about the main effects of the treatments of interest. Whether the Latin square can reasonably be applied in a particular situation depends upon whether interactions exist in the population, which variables interact, and which variables have random or fixed effects. In Section 10.3 we will consider the nature of the interaction problem in detail and in subsequent sections we will comment upon specific types of design applications. First, however, we will briefly describe how the actual design layout is selected.

10.2 SELECTING A LATIN SQUARE

Consider the following arrangement of treatments:

$$\begin{bmatrix} A_1 & A_2 & A_3 \\ A_2 & A_3 & A_1 \\ A_3 & A_1 & A_2 \end{bmatrix}$$

The characteristic that defines this arrangement as a Latin square is the occurrence of each treatment exactly once in each row and in each column. The above square is only one of several possible. It is generally referred to as a *standard square*, since the first row and first column are in standard order (i.e., A_1, A_2, A_3). Interchanging rows and columns, we might obtain

$$\begin{bmatrix} A_1 & A_3 & A_2 \\ A_3 & A_2 & A_1 \\ A_2 & A_1 & A_3 \end{bmatrix}$$

This is one of $3!2! - 1$ nonstandard squares obtainable from the standard square previously exhibited. If there are four treatment levels, there are four possible standard squares. Since each of these can, by permutation of rows and columns, give rise to $4!3! - 1$ nonstandard squares, the total number of 4×4 Latin squares is $4(4!3! - 1) + 4$, or 576. As we add levels of A, even more marked increases occur in the possible number of squares.

Sets of standard squares are available in several sources. A very extensive presentation is the set of tables in Fisher and Yates. (1955).

A standard square should be chosen at random from the complete set. Then randomly permute all the rows, all the columns, and, for squares larger than 4×4, the letters. For example, suppose we choose at random the following 6×6 standard square

$$\begin{bmatrix} A_1 & A_2 & A_3 & A_4 & A_5 & A_6 \\ A_2 & A_4 & A_5 & A_3 & A_6 & A_1 \\ A_3 & A_6 & A_1 & A_5 & A_4 & A_2 \\ A_4 & A_1 & A_2 & A_6 & A_3 & A_5 \\ A_5 & A_3 & A_6 & A_1 & A_2 & A_4 \\ A_6 & A_5 & A_4 & A_2 & A_1 & A_3 \end{bmatrix}$$

We note the order of appearance of the numbers 1 through 6 in a table of random numbers; we might obtain $\langle 1, 3, 2, 4, 6, 5 \rangle$. Then the rows are permuted so that the first row is still first, the third row is now second, the second row is now third, and so on. The result of the permutation of rows is

$$\begin{bmatrix} A_1 & A_2 & A_3 & A_4 & A_5 & A_6 \\ A_3 & A_6 & A_1 & A_5 & A_4 & A_2 \\ A_2 & A_4 & A_5 & A_3 & A_6 & A_1 \\ A_4 & A_1 & A_2 & A_6 & A_3 & A_5 \\ A_6 & A_5 & A_4 & A_2 & A_1 & A_3 \\ A_5 & A_3 & A_6 & A_1 & A_2 & A_4 \end{bmatrix}$$

We next permute all columns; we might have the random sequence $\langle 4, 1, 6, 2, 5, 3 \rangle$. Then the square becomes

$$\begin{bmatrix} A_4 & A_1 & A_6 & A_2 & A_5 & A_3 \\ A_5 & A_3 & A_2 & A_6 & A_4 & A_1 \\ A_3 & A_2 & A_1 & A_4 & A_6 & A_5 \\ A_6 & A_4 & A_5 & A_1 & A_3 & A_2 \\ A_2 & A_6 & A_3 & A_5 & A_1 & A_4 \\ A_1 & A_5 & A_4 & A_3 & A_2 & A_6 \end{bmatrix}$$

The random sequence dictating the permuting of letters might be $\langle 2, 4, 5, 1, 6, 3 \rangle$. The A_1's in the above square are replaced by A_2's, the A_2's by A_4's, the A_3's by A_5's, and so on. Then we have

$$
\begin{bmatrix}
A_1 & A_2 & A_3 & A_4 & A_6 & A_5 \\
A_6 & A_5 & A_4 & A_3 & A_1 & A_2 \\
A_5 & A_4 & A_2 & A_1 & A_3 & A_6 \\
A_3 & A_1 & A_6 & A_2 & A_5 & A_4 \\
A_4 & A_3 & A_5 & A_6 & A_2 & A_1 \\
A_2 & A_6 & A_1 & A_5 & A_4 & A_3
\end{bmatrix}
$$

This square may be viewed as a random selection from the population of 6×6 squares.

10.3 THE NONADDITIVITY PROBLEM

Suppose that an additive model provides a valid representation of the population of possible observations. Then population interaction effects are negligible and each score is viewed as the sum of μ, an error component, and main effects associated with the row, column, and treatment (A) variables. In the course of this chapter, we will note situations in which such a model might be appropriate. For the present, we need only remark that, in such circumstances, efficient tests of main effects, readily interpretable, will be available.

Unfortunately, the best of all possible worlds does not always prevail. When treatment effects are not additive—that is, when interaction effects are present—the interpretation of F tests may be difficult. Even under nonadditivity, this is not always so, but it is best to be aware of the possible problems. We begin by considering the simplest Latin square, a 2×2 design.

Suppose that we had the following set of cell totals:

$$
\begin{array}{cc}
 & \quad C_1 \quad\;\; C_2 \\
B_1 & \begin{bmatrix} (A_1) & (A_2) \\ 5 & 8 \\ (A_2) & (A_1) \\ 6 & 11 \end{bmatrix} \\
B_2 &
\end{array}
$$

Using the single df approach of Section 5.10, $SS_A = [(5 + 11) - (6 + 8)]^2/4 = 1$. Now consider SS_{BC}. The calculations are identical to those for SS_A. The finding that the SS_A is equal to the SS_{BC} is typical of the 2×2 Latin square. In fact, it can be shown that the sum of squares for the main effect of any one variable will always be equal to the sum of squares for the interaction effect of the other two variables. Thus, the B main effect is confounded with the AC interaction ($SS_B = SS_{AC}$) and the C effect cannot be distinguished from the AB interaction ($SS_C = SS_{AB}$). As a consequence of this confounding of main and interaction effects in the 2×2 design, the F test of a main effect can be interpreted only if it is assumed that the population variance due to the interaction of the other two variables is zero. For example, the SS_A can be interpreted to reflect variability due to differences in the effects of A_1 and A_2 only if we assume that σ^2_{BC} is negligible.

Confounding exists in larger squares, but it is not complete. To illustrate the nature of the confounding, we will compare the relation of main and interaction effects in a 3^3 factorial design with those in a 3×3 Latin square. We first consider the factorial arrangement, noting the example presented in Table 10–1. If $\frac{2}{9}$ is added to every score at C_2 and subtracted from every

TABLE 10–1 *Data and subtotals from a* 3^3 *factorial design*

	C_1			C_2			C_3			
	A_1	A_2	A_3	A_1	A_2	A_3	A_1	A_2	A_3	
B_1	4	1	6	2	4	1	1	4	2	
B_2	5	2	3	1	7	5	3	8	6	
B_3	6	1	4	3	3	4	1	5	4	
	15	4	13	6	14	10	5	17	12	Grand
	C_1 Total = 32			C_2 Total = 30			C_3 Total = 34			Total = 96

Subtotals for AB Cells

	A_1	A_2	A_3	B Totals
B_1	7	9	9	25
B_2	9	17	14	40
B_3	10	9	12	31
A Totals	26	35	35	96

score at C_3, the C main effect has been removed (i.e., SS_C is now zero), as may be seen in Table 10–2. However, note that the AB interaction effects are clearly unchanged. This is because C and AB effects are independent quantities in the complete factorial design.

We next turn to the Latin square approach. Table 10–3 presents a Latin square that involves three levels of each of three variables. If $\frac{4}{9}$ is added to all the C_1 scores, $\frac{1}{9}$ is added to all the C_2 scores, and $\frac{5}{9}$ is subtracted from all the C_3 scores, the C main effects are removed from the data matrix. The BC interaction effects have not been changed; the distance between any two rows in any column is exactly what it was before the adjustment. This

TABLE 10–2 *Data of Table 10–1 after adjustment for C effects*

	C_1			C_2			C_3			
	A_1	A_2	A_3	A_1	A_2	A_3	A_1	A_2	A_3	
B_1	4	1	6	$2\frac{2}{9}$	$4\frac{2}{9}$	$1\frac{2}{9}$	$\frac{7}{9}$	$3\frac{7}{9}$	$1\frac{7}{9}$	
B_2	5	2	3	$1\frac{2}{9}$	$7\frac{2}{9}$	$5\frac{2}{9}$	$2\frac{7}{9}$	$7\frac{7}{9}$	$5\frac{7}{9}$	
B_3	6	1	4	$3\frac{2}{9}$	$3\frac{2}{9}$	$4\frac{2}{9}$	$\frac{7}{9}$	$4\frac{7}{9}$	$3\frac{7}{9}$	
	15	4	13	$6\frac{2}{3}$	$14\frac{2}{3}$	$10\frac{2}{3}$	$4\frac{1}{3}$	$16\frac{1}{3}$	$11\frac{1}{3}$	Grand
	C_1 Total = 32			C_2 Total = 32			C_3 Total = 32			Total = 96

Subtotals for *AB* Cells

	A_1	A_2	A_3	*B* Totals
B_1	7	9	9	25
B_2	9	17	14	40
B_3	10	9	12	31
A Totals	26	35	35	96

TABLE 10–3 *Data from a Latin square design*

	C_1	C_2	C_3	*B* Totals
B_1	(A_1) 4	(A_2) 4	(A_3) 2	10
B_2	(A_3) 3	(A_1) 1	(A_2) 8	12
B_3	(A_2) 1	(A_3) 4	(A_1) 1	6
A Totals	8	9	11	28

TABLE 10–4 *Data of Table 10–3 after adjustment for C effects*

	C_1	C_2	C_3	*B* Totals
B_1	$4\frac{4}{9}$	$4\frac{1}{9}$	$1\frac{4}{9}$	10
B_2	$3\frac{4}{9}$	$1\frac{1}{9}$	$7\frac{4}{9}$	12
B_3	$1\frac{4}{9}$	$4\frac{1}{9}$	$\frac{4}{9}$	6
A Totals	$9\frac{1}{3}$	$9\frac{1}{3}$	$9\frac{1}{3}$	28

is apparent in Table 10–4. The reader should verify that the adjustment has also not affected the variability due to the *AC* interaction effects. However, the *AB* interaction variability is changed. Retabling the data, prior to the adjustment for C effects we had

$$
\begin{array}{c}
 & B_1 & B_2 & B_3 \\
A_1 & \begin{array}{c}(C_1)\\4\end{array} & \begin{array}{c}(C_2)\\1\end{array} & \begin{array}{c}(C_3)\\1\end{array} \\
A_2 & \begin{array}{c}(C_2)\\4\end{array} & \begin{array}{c}(C_3)\\8\end{array} & \begin{array}{c}(C_1)\\1\end{array} \\
A_3 & \begin{array}{c}(C_3)\\2\end{array} & \begin{array}{c}(C_1)\\3\end{array} & \begin{array}{c}(C_2)\\4\end{array}
\end{array}
$$

We now have

$$
\begin{array}{c}
 & B_1 & B_2 & B_3 \\
A_1 & \begin{array}{c}(C_1)\\4\tfrac{4}{9}\end{array} & \begin{array}{c}(C_2)\\1\tfrac{1}{9}\end{array} & \begin{array}{c}(C_3)\\\tfrac{4}{9}\end{array} \\
A_2 & \begin{array}{c}(C_2)\\4\tfrac{1}{9}\end{array} & \begin{array}{c}(C_3)\\7\tfrac{4}{9}\end{array} & \begin{array}{c}(C_1)\\1\tfrac{4}{9}\end{array} \\
A_3 & \begin{array}{c}(C_3)\\1\tfrac{4}{9}\end{array} & \begin{array}{c}(C_1)\\3\tfrac{4}{9}\end{array} & \begin{array}{c}(C_2)\\4\tfrac{1}{9}\end{array}
\end{array}
$$

The sums of squares for the AB interaction are not the same for the two data sets. The situation is not quite the same as in the 2×2 Latin square. In that case, the SS_{AB} would be reduced to zero if variability due to C were removed from the matrix. In the larger squares, confounding is not complete. However, neither are main and interaction effects completely independent as in the factorial design considered earlier. In the factorial example, removal of variability due to C left the variability due to the AB interaction completely unaffected.

The implication of the preceding discussion is that the expected mean square for a main effect in a Latin square design will have variance of interaction effects involving the other two variables as one of its components, unless such interaction effects are absent in the population. To understand why this happens, again consider the layout of Table 10.3. Assuming one score in each of the nine cells, and assuming a very general nonadditive model, we have

$$
\textbf{(10.1)} \qquad Y_{jkm} = \mu + \alpha_j + \beta_k + \gamma_m + (\alpha\beta)_{jk} + (\alpha\gamma)_{jm} + (\beta\gamma)_{km}
$$
$$
+ (\alpha\beta\gamma)_{jkm} + \varepsilon_{jkm}
$$

where α, β, and γ correspond to A, B and C respectively. The average score at B_1 is given by $\overline{Y}_{.1.} = (Y_{111} + Y_{212} + Y_{313})/3$. Employing Equation (10.1), we have

$$
\textbf{(10.2)} \qquad \overline{Y}_{.1.} = \tfrac{1}{3}[(\mu + \alpha_1 + \beta_1 + \gamma_1 + (\alpha\beta)_{11} + (\alpha\gamma)_{11} + (\beta\gamma)_{11}
$$
$$
+ (\alpha\beta\gamma)_{111} + \varepsilon_{111}) + (\mu + \alpha_2 + \beta_1 + \gamma_2 + (\alpha\beta)_{21}
$$
$$
+ (\alpha\gamma)_{22} + (\beta\gamma)_{12} + (\alpha\beta\gamma)_{212} + \varepsilon_{212}) + (\mu + \alpha_3 + \beta_1
$$
$$
+ \gamma_3 + (\alpha\beta)_{31} + (\alpha\gamma)_{33} + (\beta\gamma)_{13} + (\alpha\beta\gamma)_{313} + \varepsilon_{313})]
$$

Assuming that α, β, and γ are fixed effects,

$$\alpha_1 + \alpha_2 + \alpha_3 = 0$$

$$\gamma_1 + \gamma_2 + \gamma_3 = 0$$

$$(\alpha\beta)_{11} + (\alpha\beta)_{21} + (\alpha\beta)_{31} = 0$$

$$(\beta\gamma)_{11} + (\beta\gamma)_{12} + (\beta\gamma)_{13} = 0$$

These four results can be proven algebraically once we assume fixed effects. However, in general

$$(\alpha\gamma)_{11} + (\alpha\gamma)_{22} + (\alpha\gamma)_{33} \neq 0$$

$$(\alpha\beta\gamma)_{111} + (\alpha\beta\gamma)_{212} + (\alpha\beta\gamma)_{313} \neq 0.$$

The sum of interaction effects can be shown to be zero when we sum over all levels at one index *and hold all other indices constant*. The $(\alpha\beta)_{jk}$ and $(\alpha\beta\gamma)_{jkm}$ effects are varying with respect to two indices as we sum them. The result of these developments is

(10.3) $\overline{Y}_{.1.} = \mu + \beta_1 + \frac{1}{3}[(\alpha\gamma)_{11} + (\alpha\gamma)_{22} + (\alpha\gamma)_{33}]$

$$+ \tfrac{1}{3}[(\alpha\beta\gamma)_{111} + (\alpha\beta\gamma)_{212} + (\alpha\beta\gamma)_{313}]$$

$$+ \tfrac{1}{3}[\varepsilon_{111} + \varepsilon_{212} + \varepsilon_{313}]$$

The means at B_2 and B_3 are also contributed to by an average of three $\alpha\gamma$ components as well as by an average of three $\alpha\beta\gamma$ components. Thus, the SS_B involves not only the variability of the β_k but also the variability among $(\frac{1}{3})[(\alpha\gamma_{11}) + (\alpha\gamma)_{22} + (\alpha\gamma)_{33}]$, $(\frac{1}{3})[(\alpha\gamma)_{31} + (\alpha\gamma)_{12} + (\alpha\gamma)_{23}]$, and $(\frac{1}{3})[(\alpha\gamma)_{21} + (\alpha\gamma)_{32} + (\alpha\gamma)_{13}]$ as well as the variability among $(\frac{1}{3})[(\alpha\beta\gamma)_{111} + (\alpha\beta\gamma)_{212} + (\alpha\beta\gamma)_{313}]$, $(\frac{1}{3})[(\alpha\beta\gamma)_{321} + (\alpha\beta\gamma)_{122} + (\alpha\beta\gamma)_{223}]$, and $(\frac{1}{3})[(\alpha\beta\gamma)_{231} + (\alpha\beta\gamma)_{332} + (\alpha\beta\gamma)_{133}]$. Related conclusions hold for SS_A and SS_C.

The confounding of main and interaction effects is a potential source of trouble in the use of Latin square designs. However, the existence of this problem does not preclude the use of the design. Transformations may be found which eliminate interaction effects, permitting a clear interpretation of tests of main effects. Even in the presence of interaction effects, reasonable interpretations of the F ratio will often be available.

10.4 BETWEEN-SUBJECTS DESIGNS

10.4.1 Basic calculations. Table 10–5 presents *SV*, *df*, and *SS* for an $a \times a$ Latin square with n subjects in each cell. Note that, despite the abundance of terms in Equation (10.1), we distinguish only five sources of variance. In particular, we cannot separately extract independent terms representing each

TABLE 10–5 *Analysis of variance for a single Latin square, n subjects in a cell*

SV	df	SS
Total	$a^2n - 1$	$\sum_i \sum_j \sum_k \sum_m Y_{ijkm}^2 - C$
A	$a - 1$	$\dfrac{\sum_j (\sum_i \sum_k \sum_m Y_{ijkm})^2}{an} - C$
B	$a - 1$	$\dfrac{\sum_k (\sum_i \sum_j \sum_m Y_{ijkm})^2}{an} - C$
C	$a - 1$	$\dfrac{\sum_m (\sum_i \sum_j \sum_k Y_{ijkm})^2}{an} - C$
B. cells res.	$(a - 1)(a - 2)$	$\dfrac{\sum_j \sum_k \sum_m (\sum_i Y_{ijkm})^2}{n} - C - SS_A - SS_B - SS_C$
S/cells		$SS_{\text{tot}} - SS_A - SS_B - SS_C - SS_{B.\text{ cells res.}}$

of the possible interactions. This is so because there are only $(a^2 - 1)$ *df* to account for the between-cell variability. We lose $3(a - 1)$ of these when we compute the sums of squares for the three main effects. The residual, $(a - 1)(a - 2)$, is less than $(a - 1)(a - 1)$, the *df* required to account for even a single interaction term.

Note that the $SS_{B.\text{cells res.}}$ may be viewed as the sum of squares for the interaction of any two Latin square variables further adjusted for the third variable. That is,

$$SS_{B.\text{cells res.}} = SS_{AB} - SS_C$$

$$= SS_{AC} - SS_B$$

$$= SS_{BC} - SS_A$$

where the component sums of squares are computed in the usual manner.

Table 10–6 presents a data matrix for a 4×4 square; the entries are

TABLE 10–6 *Data for a single Latin square design*

	C_1	C_2	C_3	C_4	$\sum_i \sum_j \sum_m Y_{ijkm}$
B_1	(A_3) 12	(A_1) 24	(A_2) 9	(A_4) 48	93
B_2	(A_1) 36	(A_3) 18	(A_4) 57	(A_2) 21	132
B_3	(A_2) 27	(A_4) 69	(A_1) 33	(A_3) 15	144
B_4	(A_4) 51	(A_2) 24	(A_3) 27	(A_1) 45	147
$\sum_i \sum_j Y_{ijk} = 126$		135	126	129	$\sum_i \sum_j \sum_k Y_{ijk} = 516$

A Subtotals

A_1	A_2	A_3	A_4
138	81	72	225

cell totals based on nine subjects each. We first calculate

$$C = \frac{(516)^2}{144}$$

$$= 1{,}849$$

The *A*, *B*, and *C* terms are obtained as in factorial designs:

$$SS_A = \frac{(138)^2 + (81)^2 + (72)^2 + (225)^2}{36} - C$$

$$= 412.5$$

$$SS_B = \frac{(93)^2 + (132)^2 + (144)^2 + (147)^2}{36} - C$$

$$= 51.5$$

$$SS_C = \frac{(126)^2 + (135)^2 + (126)^2 + (129)^2}{36} - C$$

$$= 1.5$$

The residual variability among the cell means is obtained as

$$SS_{B.\ cells\ res.} = \frac{(12)^2 + (24)^2 + \cdots + (27)^2 + (45)^2}{9} - C - SS_A$$

$$- SS_B - SS_C$$

$$= 35.5$$

and the within-cell variability is the difference between the SS_{tot} and the four terms above.

10.4.2 Expected mean squares and *F* ratios. Wilk and Kempthorne (1957) have considered the expected mean squares under the general nonadditive model represented by Equation (10.1). Their results are not simple and cannot be completely generated by the "rules of thumb" of Section 8.2.2. Nevertheless, they merit consideration because they provide the basis for decisions about the use of the Latin square design and interpretation of results obtained from the application of the design.

Fixed effects. Suppose that all three variables—*A*, *B*, and *C*—were fixed-effects variables. The expected mean squares would, as usual, contain an error variance component (σ_e^2), and a null hypothesis term (θ_A^2 or θ_B^2 or θ_C^2). In addition, if all interactions were nonzero in the population sampled, they would also contribute to the expected mean squares. We would have

SV	EMS
A	$\sigma_e^2 + na\theta_A^2 + n\theta_{BC}^2 + \dfrac{n(a-2)}{a}\theta_{ABC}^2$
B	$\sigma_e^2 + na\theta_B^2 + n\theta_{AC}^2 + \dfrac{n(a-2)}{a}\theta_{ABC}^2$
C	$\sigma_e^2 + na\theta_C^2 + n\theta_{AB}^2 + \dfrac{n(a-2)}{a}\theta_{ABC}^2$
B. cells res.	$\sigma_e^2 + n\theta_{AB}^2 + n\theta_{AC}^2 + n\theta_{BC}^2 + \dfrac{n(a-3)}{a}\theta_{ABC}^2$
S/ABC	σ_e^2

The interaction component in each main effect source is explained in Section 10.3; in particular, by the discussion following Equation (10.1). As we showed there, part of the sum of squares for any main effect is attributable to a portion of the variance due to the first-order interaction of the remaining two variables and part to a portion of the second-order variance attributable to interaction among all three variables. The remaining portion of all the first-order interaction variances, as well as some portion of the second-order interaction variance, are all lumped together in the residual term. The rather peculiar looking coefficients in the second-order interaction component is a result of the incompleteness of the design and falls out of the Wilk-Kempthorne derivations.

Consider the dilemma these expectations pose for the investigator. The within-cell (S/ABC) term, if used as a measure of error variance, yields positive bias; under the null hypothesis, the ratio of expected mean squares is greater than 1. The result will generally be too many Type I errors. The use of the between-cell residual as an error term poses the opposite problem—negative bias. The ratio of expected mean squares will be less than 1 when H_0 is true. Treatment effects will have to be very large if they are to be detected as significant.

With luck, a transformation may be found that will remove nonadditivity. Rather than trust to such luck, the design should be used primarily in cases where, a priori, intuition and pilot data indicate that unbiased F tests of the treatments of interest can be compiled. This does not mean that complete nonadditivity is required. Let us consider a relevant example.

Suppose that we decide to increase the efficiency of a test of some treatment, A, by arbitrarily dividing subjects into four blocks on the basis of level of ability. In addition, we employ four experimenters who can run subjects concurrently, thereby reducing the number of weeks needed to complete the experiment. If we have four levels of A, a complete factorial design would involve 64 cells and take an unreasonable amount of time to complete. However, if we designate blocks as B and experimenters as C and use a 4×4 Latin square, we have a design that is potentially very efficient with respect to both error variance and running time. The F test of A is unbiased and efficient if the BC and ABC interactions are negligible, which is not an unreasonable assumption. Then $E(MS_A) = \sigma_e^2 + na\theta_A^2$ and the within-cell error term is completely adequate. If, in addition, we can assume that there is little variability due to experimenters, a partial test of the treatments × blocks (AB) interaction is available, for then we have $E(MS_C) = \sigma_e^2 + n\theta_{AB}^2$. A more powerful test is available if we also can assume that experimenters and treatments do not interact, for then $E(MS_{B.\,\text{cells res.}}) = \sigma_e^2 + n\theta_{AB}^2$ and $df_{B.\,\text{cells res.}}$ is greater than df_C.

Suppose that we wish to investigate two treatment variables, A and C. We might desire to introduce experimenters as a third variable to reduce the time for each experimenter, or we might introduce blocks as a third variable to increase efficiency, as in Chapter 6. Or it may be that the experiment involves some third variable that is not of intrinsic interest but is a necessary part of a well-planned experiment, perhaps position of the correct alternative in a multiple-choice discrimination task. In any event, if this variable, B, does not interact with either A or C or both, then S/ABC yields an unbiased error term for testing A and C effects. Furthermore, a test of B. cells res. against the within-cell term provides a test of some portion of the AC interaction variance, again assuming that all other interactions are negligible.

In the preceding examples, we assumed strong a priori evidence that certain interaction terms were negligible. Occasionally, practical considerations will necessitate the use of the Latin square design when there is doubt about the status of critical interaction components. If the F test of B. cells res. is nonsignificant, significant tests of treatments can reasonably be assumed to reflect main and not interaction effects. If there is evidence of nonadditivity in the preliminary test of the residual term, and if it is feared that the interaction present is one that is confounded with the treatment of interest, transformation to additivity should be sought. The criterion of the adequacy of the transformation will be the magnitude of the F test of B. cells res.; the smaller the F, the better the transformation.

Random effects. Frequently, the levels of one or more of our three variables may be assumed to have been randomly sampled from some population of levels. We wish to generalize to the population of levels, which necessitates treating the variable, or variables, in question as random-effects variables. For example, first-grade schoolchildren are divided into classes on the basis of a readiness-to-read test. Designate this variable, level of readiness, as C. A might be methods of instruction in reading. Several different schools are involved in the study and these comprise the levels of the random-effects variable, B. Under a general nonadditive model that assumes the possibility of all interactions contributing variance, the expected mean squares presented in the preceding section must be extended. For each treatment source of variance, we again have an error component, a null hypothesis component, and first- and second-order interaction components due to confounding. In addition, the "rules of thumb" of Chapter 8 are applied. Thus, in our example (B random), an AB component must be added to the expected mean square for A and a BC component to the expected mean square for C; both also contribute to B. cells res. We would have

SV	EMS
A	$\sigma_e^2 + na\theta_A^2 + n\sigma_{BC}^2 + n\sigma_{AB}^2 + \dfrac{n(a-2)}{a}\sigma_{ABC}^2$
B (random)	$\sigma_e^2 + na\sigma_B^2 + n\theta_{AC}^2 + \dfrac{n(a-1)}{a}\sigma_{ABC}^2$
C	$\sigma_e^2 + na\theta_e^2 + n\sigma_{AB}^2 + n\sigma_{BC}^2 + \dfrac{n(a-2)}{a}\sigma_{ABC}^2$
B. cells res.	$\sigma_e^2 + n\sigma_{AB}^2 + n\theta_{AC}^2 + n\sigma_{BC}^2 + \dfrac{n(a-2)}{a}\sigma_{ABC}^2$
S/ABC	σ_e^2

If we can reasonably assume that θ^2_{AC} is the only interaction term that will contribute variance, σ^2_{AB}, σ^2_{AC} and σ^2_{ABC} drop out and the F tests of A and C against S/ABC are unbiased. Furthermore, under this assumption, B. cells res. may be viewed as a measure of the AC interaction and tested against S/ABC. Finally, B against B. cells res. provides an unbiased F test of the random-effects variable, though one considerably less powerful than the others since the error term is on fewer df and its variance will be larger than that of the S/ABC term if $\theta^2_{AC} > 0$.

Unbiased F tests of main effects can also be obtained if $\theta^2_{AC} = 0$ and any of the other interaction components are sizable. Under this assumption, the ratio of expected mean squares for A or C against B. cells res. will be 1 when H_0 is true. However, for reasons given in the preceding paragraph, this test will tend to be inefficient and lacking in power. A transformation to additivity would be preferred under these conditions.

Despite the possible problems, this is a very nice example of the usefulness of the Latin square. If there were a methods of instruction and a levels of reading readiness, we would require a^2 classes for each school in order to have a complete factorial design; this may be very impractical if not impossible. The Latin square gives us a way out with much less expenditure of effort. Even nonadditivity does not rule out the use of the approach; as we indicated above, the bias of F tests depends upon the particular combinations of interactions that are present. In the worst of cases, with all interactions suspect, transformations of the data may be sought. This loses information about the AC interaction, which may well be of interest, but provides a test of the important instructional (A) variable. Even if nonadditivity cannot be obtained, if instructions have a large enough effect to be of practical interest, they should prove significant when tested against B. cells res. despite the likely negative bias of the test.

10.4.3 A modified Latin square design. Suppose that we wish to compare two *methods of psychotherapy* (P); furthermore, we wish to know whether the number of contact *hours per week* (H) is important, particularly whether differences in the effects of the therapies depend upon hours of contact ($P \times H$ interaction). Subjects are chosen from the populations at four outpatient *clinics* (C) and are placed in one of four categories depending upon the severity of *symptoms* (S). Assessment of improvement by a panel of clinicians provides the dependent variable. We might lay out the design as follows:

$$
\begin{array}{c}
\begin{array}{cccc}
C_1 & C_2 & C_3 & C_4
\end{array} \\
\begin{array}{c}
S_1 \\
S_2 \\
S_3 \\
S_4
\end{array}
\left[
\begin{array}{cccc}
P_1H_2 & P_1H_1 & P_2H_1 & P_2H_2 \\
P_1H_1 & P_1H_2 & P_2H_2 & P_2H_1 \\
P_2H_2 & P_2H_1 & P_1H_2 & P_1H_1 \\
P_2H_1 & P_2H_2 & P_1H_1 & P_1H_2
\end{array}
\right]
\end{array}
$$

We have a 4×4 Latin square in which the subject populations at each of four clinics have been grouped according to level of S and then assigned to one of four PH combinations. It is a Latin square because each of the four treatment combinations appears exactly once in each clinic and at each level of S. As usual with such designs, it is incomplete; the complete design would require 64 cells and each clinic would require enough patients to provide 16 combinations of S, P, and H.

Turn now to the general case in which there are ad levels of B, C, and treatment combinations with n subjects within each of the a^2d^2 cells. If the Latin square variables do not interact (that is, B and C do not interact with each other or with A, D, or AD), the appropriate model is

(10.4) $$Y_{ijkmp} = \mu + \alpha_j + \beta_k + \gamma_m + \delta_p + (\alpha\delta)_{jp} + \varepsilon_{ijkmp}$$

where β_k designates the effect of B_k, γ_m designates the effect of C, and α_j and δ_p designate the effects of A_j and D_p respectively. The analysis of variance is summarized in Table 10–7. The key point in the partitioning of the total variance is that the df and sum of squares for treatment combinations can be further analyzed into several components of interest.

The use of this design permits us to study the main and interaction effects of two treatment variables. In fact, the design could be extended to more variables. For example, there might be two levels of each of three treatment variables; then each row or column would contain eight treatment combinations. The seven df for treatment combinations would be partitioned into seven components, representing three main effects, three first-order interactions and one second-order interaction. Nor is the approach limited to two levels of the treatment variables. We could have a 12×12 square with four levels of A and three levels of D; the 11 df for treatment combinations would yield an A term on three df, a D effect on two df, and an AD term on six df.

The appropriateness of the model represented by Equation (10.4) may be partly verified by testing the $B.$ cells res. against S/cells. A nonsignificant result supports the assumption that B, C, and *treatment combinations* contribute additively to the data; $MS_{S/\text{cells}}$ would then be the error term. Of course, nonsignificance in the preliminary test, $MS_{B.\,\text{cells res.}}/MS_{S/\text{cells}}$, does not guarantee the absence of interactions among B, C, and *treatment combinations*. We may have made a Type II error; furthermore, the residual contains only a portion of interaction variance, the remainder being confounded with the B, C, and *treatment combination* effects. Nevertheless, together with prior evidence, the test result provides a reasonable ground for proceeding.

If the preliminary test should prove significant, the considerations raised in Section 10.4.2 become relevant. We run the risk that variance due to the BC interaction contributes to the A, D, and AD mean squares, and that the interaction of A or D with B (or C) contributes to the C (or B) mean square. The interpretation of F tests will depend upon what we are

TABLE 10–7 *Latin squaring treatment combinations*

SV	df	SS	EMS*
Total	$a^2d^2n - 1$	$\sum_i\sum_j\sum_k\sum_m\sum_p Y_{ijkmp}^2 - C$	
B	$ad - 1$	$\dfrac{\sum_k(\sum_i\sum_j\sum_m\sum_p Y_{ijkmp})^2}{adn} - C$	$\sigma_e^2 + adn\theta_B^2$
C	$ad - 1$	$\dfrac{\sum_m(\sum_i\sum_j\sum_k\sum_p Y_{ijkmp})^2}{adn} - C$	$\sigma_e^2 + adn\theta_C^2$
Treat. Comb.	$ad - 1$	$\dfrac{\sum_j\sum_p(\sum_i\sum_k\sum_m Y_{ijkmp})^2}{adn} - C$	
A	$a - 1$	$\dfrac{\sum_j(\sum_i\sum_k\sum_m\sum_p Y_{ijkmp})^2}{ad^2n} - C$	$\sigma_e^2 + ad^2n\theta_A^2$
D	$d - 1$	$\dfrac{\sum_p(\sum_i\sum_j\sum_k\sum_m Y_{ijkmp})^2}{a^2dn} - C$	$\sigma_e^2 + ad^2n\theta_D^2$
AD	$(a - 1)(d - 1)$	$SS_{\text{Treat.Comb.}} - SS_A - SS_0$	$\sigma_e^2 + adn\theta_{AD}^2$
B. cells res.	$(ad - 1)(ad - 2)$	$SS_{BC} - SS_C - SS_{\text{Treat.Comb.}}$	σ_e^2
S/cells	$a^2d^2(n - 1)$	$SS_{\text{tot}} - SS_B - SS_C - SS_{\text{Treat.Comb.}} - SS_{B.\text{cells res.}}$	σ_e^2

*Assuming Equation (10.4). See discussion in text.

able to assume about these interaction components and which variables are treated as random or fixed in their effects. For example, if all variables can be regarded as fixed and we can reasonably assume that BC interaction effects are negligible, then regardless of the status of other interaction effects, A, D, and AD may be tested without bias against $S/$cells. Other contingencies have been discussed in Section 10.4.2.

Numerical example. We will employ the data of Table 10–8 to illustrate the calculations. The entries are cell totals based on four subjects each. As usual, we first calculate the correction term:

$$C = \frac{(350)^2}{64}$$

$$= 1,914.063$$

Then

$$SS_B = \frac{(88)^2 + \cdots + (82)^2}{16} - C$$

$$= 3.187$$

$$SS_C = \frac{(82)^2 + \cdots + (108)^2}{16} - C$$

$$= 37.187$$

We next turn to the *ad* treatment combinations:

$$SS_{T.C.} = \frac{(52)^2 + \cdots + (42)^2}{16} - C$$

$$= 417.687$$

This variability can be divided into three components: A, D, and AD. Taking advantage of the fact that each of these component terms is distributed on 1 *df*, we may use the short-cut formula of Section 5.10:

$$SS_A = \frac{(134 + 122 - 52 - 42)^2}{64}$$

$$= 410.063$$

$$SS_D = \frac{(52 + 134 - 122 - 42)^2}{64}$$

$$= 7.563$$

and

$$SS_{AD} = \frac{(52 + 122 - 134 - 42)^2}{64}$$

or

$$SS_{AD} = SS_{T.C.} - SS_A - SS_B$$

$$= .063$$

TABLE 10–8 Data for a Latin square of treatment combinations

	C_1	C_2	C_3	C_4	$\sum_i \sum_j \sum_m \sum_p Y_{ijkmp}$
B_1	(A_1D_1) 10	(A_2D_1) 34	(A_2D_2) 28	(A_1D_2) 16	88
B_2	(A_2D_1) 32	(A_1D_2) 8	(A_1D_1) 12	(A_2D_2) 40	92
B_3	(A_2D_2) 28	(A_1D_1) 16	(A_1D_2) 6	(A_2D_1) 38	88
B_4	(A_1D_2) 12	(A_2D_2) 26	(A_2D_1) 30	(A_1D_1) 14	82
$\sum_i \sum_j \sum_k \sum_p Y_{ijkmp} = 82$		84	76	108	$\sum_i \sum_j \sum_k \sum_m \sum_p Y_{ijkmp} = 350$

AD Subtotals

A_1D_1	A_2D_1	A_1D_2	A_2D_2
52	134	42	122

Next, we obtain

$$SS_{B.\text{ cells res.}} = \frac{10^2 + 17^2 + \cdots + 15^2 + 7^2}{4} - C - SS_B - SS_C - SS_{T.C.}$$

$$= 18.876$$

We have not bothered to provide the individual scores but the SS_{tot} is calculated in the usual way and

$$SS_{S/\text{cells}} = SS_{\text{tot}} - SS_B - SS_C - SS_{T.C.} - SS_{B.\text{ cells res.}}$$

10.4.4 Using several Latin squares. Suppose that in our example of the study of *methods of psychotherapy* (P) and *hours of therapy* (H) each week we desired to investigate the interaction of P with *severity of symptoms* (S). In the more general notation of the preceding section, we desire information on the interaction of D with B and/or C as well as with A; the previous design provided a direct test only of AD. One way to approach the problem is to use the design of Section 10.4.2 as a building block. B and C are again rows and columns and again only A is Latin squared. There are a levels of A, B, and C. We randomly sample d squares from the population of $a \times a$ squares, where d may take on any value. Then we have one Latin square under treatment D_1, another at D_2, and so on. The layout might be

$$
D_1 \quad
\begin{array}{c}
\\ B_1 \\ B_2 \\ B_3 \\ B_4
\end{array}
\begin{array}{cccc}
C_1 & C_2 & C_3 & C_4 \\
\left[\begin{array}{cccc} A_4 & A_2 & A_1 & A_3 \\ A_1 & A_3 & A_2 & A_4 \\ A_3 & A_1 & A_4 & A_2 \\ A_2 & A_4 & A_3 & A_1 \end{array}\right]
\end{array}
$$

$$
D_2 \quad
\begin{array}{c}
\\ B_1 \\ B_2 \\ B_3 \\ B_4
\end{array}
\begin{array}{cccc}
C_1 & C_2 & C_3 & C_4 \\
\left[\begin{array}{cccc} A_2 & A_3 & A_1 & A_4 \\ A_3 & A_4 & A_2 & A_1 \\ A_1 & A_2 & A_4 & A_3 \\ A_4 & A_1 & A_3 & A_2 \end{array}\right]
\end{array}
$$

There are n subjects in each of the $a^2 d$ cells.

Assuming that the effects of the dimensions of the square (i.e., A, B, and C) are additive, the appropriate model would be:

(10.5) $\quad Y_{ijkmp} = \mu + \alpha_j + \beta_k + \gamma_m + \delta_p + (\alpha\delta)_{jp} + (\beta\delta)_{kp}$

$$+ (\gamma\delta)_{mp} + \varepsilon_{ijkmp}$$

Table 10–9 summarizes the analysis of variance. The *pooled residual* term represents summing over the levels of D of the between-cell residual for each square. In practice, we find the sum of squares among the $a^2 d$ cell means and then adjust for the main and interaction effects previously computed.

TABLE 10–9 Analysis of variance for several squares

SV	df	SS	EMS*
Total	$a^2dn - 1$	$\sum_i \sum_j \sum_k \sum_m \sum_p Y_{ijkmp}^2 - C$	
B	$a - 1$	$\dfrac{\sum_k (\sum_i \sum_j \sum_m \sum_p Y_{ijkmp})^2}{adn} - C$	$\sigma_e^2 + adn\theta_B^2$
C	$a - 1$	$\dfrac{\sum_m (\sum_i \sum_j \sum_k \sum_p Y_{ijkmp})^2}{adn} - C$	$\sigma_e^2 + adn\theta_C^2$
A	$a - 1$	$\dfrac{\sum_j (\sum_i \sum_k \sum_m \sum_p Y_{ijkmp})^2}{adn} - C$	$\sigma_e^2 + adn\theta_A^2$
D	$d - 1$	$\dfrac{\sum_p (\sum_i \sum_j \sum_k \sum_m Y_{ijkmp})^2}{a^2n} - C$	$\sigma_e^2 + a^2n\theta_D^2$
BD	$(a - 1)(d - 1)$	$\dfrac{\sum_k \sum_p (\sum_i \sum_j \sum_m Y_{ijkmp})^2}{an} - C - SS_B - SS_D$	$\sigma_e^2 + an\theta_{BC}^2$
CD	$(a - 1)(d - 1)$	$\dfrac{\sum_m \sum_p (\sum_i \sum_j \sum_k Y_{ijkmp})^2}{an} - C - SS_C - SS_D$	$\sigma_e^2 + an\theta_{CD}^2$
AD	$(a - 1)(d - 1)$	$\dfrac{\sum_j \sum_p (\sum_i \sum_k \sum_m Y_{ijkmp})^2}{an} - C - SS_A - SS_D$	$\sigma_e^2 - an\theta_{AD}^2$
Pooled residual	$d(a - 1)(a - 2)$	$\dfrac{\sum_j \sum_k \sum_m \sum_p (\sum_i Y_{ijkmp})^2}{n} - C - SS_A - SS_B - SS_C - SS_D - SS_{AD} - SS_{BD} - SS_{CD}$	σ_e^2
S/cells	$a^2d(n - 1)$	$SS_{tot} - SS_A - SS_B - SS_C - SS_D - DD_{AD} - SS_{BD} - SS_{CD} - SS_{pooled\ res.}$	σ_e^2

*Assuming Equation (10.5).

As in the past, we can test the additivity assumptions; in this case, we compute $MS_{\text{pooled res.}}/MS_{S/\text{cells}}$. Nonsignificance would tend to support the model represented by Equation (10.5) and the expected mean squares of Table 10–9. Significance raises the usual problems. There is little to be added to our previous comments (in particular, see Section 10.4.2) except to note that not only will the interaction of two Latin square variables contribute to the main effect of the third, but that also the interaction of two Latin square variables with D will be confounded with the interaction of the third variable with D; for example, if BCD effects are not negligible in the population, their variance will contribute to the AD mean square.

One advantage of the present design is that the use of many squares tends to minimize the danger due to interaction among Latin square variables. To see why this is so, again consider Section 10.3, particularly Equation (10.3) and the discussion following it. With a single $a \times a$ Latin square, the means at the different levels of B involve a distinct, nonoverlapping, sets of values of $(\alpha\gamma)_{jm}$. When we add more squares, the sets of $(\alpha\gamma)_{jm}$ values contributing to each $\overline{Y}_{.k.}$ tend to overlap; for example, $(\alpha\gamma)_{23}$ might contribute to both $\overline{Y}_{.1.}$ and $\overline{Y}_{.2.}$. This reduces the relative contribution of σ^2_{AC} to the MS_B.

Despite the advantage just cited, it is occasionally profitable to replicate the *same* square at each level of D. Returning to our example of the investigation of several types of therapy, suppose we had strong prior grounds for believing that *clinics* would not interact with the other variables but that severity of symptoms might interact with P, the Latin square variable, and H, the between-squares variable; furthermore, the interaction is of some interest. In general terms, we assume that interactions with C are negligible but we are interested in possible interactions of A and D with B. If we use the same square at all levels of D, there is one change in the analysis of Table 10–9; the pooled residual can be partitioned into two terms which, under the current model, we designate AB' and ABD':

$$SS_{\text{pooled res.}} = SS_{AB'} + SS_{ABD'}$$
$$d(a - 1)(a - 2) = (a - 1)(a - 2) + (a - 1)(a - 2)(d - 1)$$

The prime sign indicates that only part of the variance is contained in the term in question. The remainder of the AB variability contributes to SS_C and the ABD variability is partly confounded with CD. $SS_{AB'}$ is computed as $SS_{B.\text{cells res.}}$ was previously, where the cell in question contains dn scores.

10.4.5 The balanced Latin square design. Suppose that we are interested in the main and interaction effects of three treatment variables: A, B, and C. We are willing and able to run a complete factorial design involving a levels of all variables; that is, we propose a^3 cells with n subjects in each. However, there is a fourth variable, D, which it is advisable or necessary to include in the study although its effects are not of interest. For example, we might like to control for intelligence level by introducing a blocks variable, or we might wish to use several experimenters in order to shorten the duration of the experiment. Still another possibility is that we must use several schools or clinics in order to obtain sufficient subjects. Suppose that A and B are instructional variables in some classroom learning experiment, C is grade level, and D represents schools. To carry out the experiment as a complete four-factor design, we require a^3 cells for each school, presumably a^2 classes at each grade within a school. Generally, this will be impossible. We can compromise by having all combinations of A, B, and C but only $1/a$ of the total number of $ABCD$ cells. Assuming that $a = 3$, one possible layout is:

$$
\begin{array}{ccc}
 & C_1 \; C_2 \; C_3 \\
B_1 & \begin{bmatrix} A_2 & A_1 & A_3 \end{bmatrix} \\
D_1 \;\; B_2 & \begin{bmatrix} A_1 & A_3 & A_2 \end{bmatrix} \\
B_3 & \begin{bmatrix} A_3 & A_2 & A_1 \end{bmatrix}
\end{array}
\qquad
\begin{array}{ccc}
 & C_1 \; C_2 \; C_3 \\
B_1 & \begin{bmatrix} A_1 & A_3 & A_2 \end{bmatrix} \\
D_2 \;\; B_2 & \begin{bmatrix} A_3 & A_2 & A_1 \end{bmatrix} \\
B_3 & \begin{bmatrix} A_2 & A_1 & A_3 \end{bmatrix}
\end{array}
\qquad
\begin{array}{ccc}
 & C_1 \; C_2 \; C_3 \\
B_1 & \begin{bmatrix} A_3 & A_2 & A_1 \end{bmatrix} \\
D_3 \;\; B_2 & \begin{bmatrix} A_2 & A_1 & A_3 \end{bmatrix} \\
B_3 & \begin{bmatrix} A_1 & A_3 & A_2 \end{bmatrix}
\end{array}
$$

We achieve the balanced layout by first selecting an $a \times a$ square at random (see Section 10.2). Next we rotate the columns so that the mth column of the original square becomes the $m - 1$st column of the new square, the original first column becoming the ath column of the new square. We repeat this rotation once again to generate still another square and continue the process until we have a squares. The result is that at any level of one Latin square variable, all combinations of the other two Latin square variables are present. Then, when we compute the $\overline{Y}_{.k.}$, for example, each one is an average over all $(\alpha\gamma)_{jm}$ effects, and B is not confounded with AC.

There are still some potential sources of confounding of which we should be aware. The means at the different levels of D will be based on different $(\alpha\beta\gamma)_{jkm}$ sets, thus D and ABC will be partially confounded. Consequently, the design should not be used if the main effect of D is of interest. Interactions of D with any of the Latin square variables could create problems in interpreting some of the F tests. For example, the B main effect could be viewed as a contrast among a sets of $(\alpha\gamma\delta)_{jmp}$ components; thus, if $\sigma^2_{ACD} > 0$, ACD will be partially confounded with B. Other interactions with D, if sizable, will contribute to other main and interaction terms involving Latin square variables.

Assuming that interactions involving D are negligible, the appropriate model is

$$
\begin{aligned}
\textbf{(10.6)} \quad Y_{ijkmp} = {}& \mu + \alpha_j + \beta_k + \gamma_m + \delta_p + (\alpha\beta)_{jk} + (\alpha\gamma)_{jm} \\
& + (\beta\gamma)_{km} + (\alpha\beta\gamma)_{jkm} + \varepsilon_{ijkmp}
\end{aligned}
$$

An analysis consistent with the model is presented in Table 10–10. Note that the between-cell residual may be taken as a measure of the ABC interaction, and that it is in fact that portion which remains after we have adjusted for the component of ABC confounded with D.

TABLE 10-10 *Analysis of balanced Latin square design*

SV	df	SS	EMS*
Total	$a^3n - 1$	$\sum_i \sum_j \sum_k \sum_m \sum_p Y_{ijkmp}^2 - C$	
A	$a - 1$	$\dfrac{\sum_j(\sum_i \sum_k \sum_m \sum_p Y_{ijkmp})^2}{a^2 n} - C$	$\sigma_e^2 + a^2 n\theta_A^2$
B	$a - 1$	$\dfrac{\sum_k(\sum_i \sum_j \sum_m \sum_p Y_{ijkmp})^2}{a^2 n} - C$	$\sigma_e^2 + a^2 n\theta_B^2$
C	$a - 1$	$\dfrac{\sum_m(\sum_i \sum_j \sum_k \sum_p Y_{ijkmp})^2}{a^2 n} - C$	$\sigma_e^2 + a^2 n\theta_C^2$
AB	$(a - 1)^2$	$\dfrac{\sum_j \sum_k(\sum_i \sum_m \sum_p Y_{ijkmp})^2}{an} - C - SS_A - SS_B$	$\sigma_e^2 + an\theta_{AB}^2$
AC	$(a - 1)^2$	$\dfrac{\sum_j \sum_m(\sum_i \sum_k \sum_p Y_{ijkmp})^2}{an} - C - SS_A - SS_C$	$\sigma_e^2 + an\theta_{AC}^2$
BC	$(a - 1)^2$	$\dfrac{\sum_k \sum_m(\sum_i \sum_j \sum_p Y_{ijkmp})^2}{an} - C - SS_B - SS_C$	$\sigma_e^2 + an\theta_{BC}^2$
D	$a - 1$	$\dfrac{\sum_p(\sum_i \sum_j \sum_k \sum_m Y_{ijkmp})^2}{a^2 n} - C$	$\sigma_e^2 + a^2 n\theta_D^2$
B. cell res.	$(a - 1)^3 - (a - 1)$	$\dfrac{\sum_j \sum_k \sum_m \sum_p(\sum_i Y_{ijkmp})^2}{n} - C - SS_A - SS_B - SS_C - SS_{AB}$ $-SS_{AC} - SS_{BC} - SS_D$	σ_e^2
S/cells	$a^3(n - 1)$	$SS_{tot} - SS_A - SS_B - SS_C - SS_{AB} - SS_{AC} - SS_{BC}$ $-SS_D - SS_{B.\text{cells res.}}$	σ_e^2

*Assuming Equation (10.6).

10.5 REPEATED MEASUREMENTS DESIGNS

10.5.1 Preliminary remarks. The Latin square is used by psychologists primarily as a repeated measurements design. In many instances, subjects are in small supply and it is essential that all subjects undergo all levels of the independent variable. In such a situation, the Latin square is potentially more efficient than the subjects × treatment design. They both permit us to extract variance due to individual differences as a separate source. However, the Latin square permits still greater efficiency by also permitting us to account separately for variance due to temporal effects—for example, fatigue, practice, boredom.

The general layout has not changed from the preceding section. The difference is that each row, previously B_k, now represents one or more subjects who undergo the treatment levels, the A_j, in a specified sequence. Each column, C_m, represents a position within the sequence, a stage of practice. We view the rows as a random-effects variable and C and A as fixed-effects variables. It is important to distinguish between sequence and ordinal position effects. Consider five rats tested in a runway on five successive days, each day under a different drug. Each rat undergoes a different sequence of drugs, the set of five sequences forming a Latin square. Now, regardless of the sequence, we expect reduced running times over days due to the rats' increasing familiarity with the experiment situation. If additivity holds, the C source of variance will reflect these practice effects. To the extent that they are present, this design is more efficient than the subjects × treatments design, for such temporal effects are included in the error term in the latter design.

Presumably, the row means will also differ since they represent different subjects. There are other reasons why they might differ. Suppose that one rat receives a severe depressant on the first day and never wholly recovers; his average running time over the five days will be considerably higher than that of a rat who receives the severe depressant on the last day of the experiment. Such effects are frequently referred to as "carryover" effects; when present, they distort our estimates of the treatment effects because the treatment effect will be either inflated or deflated by the carryover from the preceding treatment or treatments. In many experiments, such effects will be minimal. For example, consider the typical signal detection experiment in which the subject is required to report the presence or absence of a tone against a noise background; I suspect that the signal-to-noise intensity ratio could be varied over trial blocks with little fear of carryover effects. In many experiments, a recovery period between presentations of the treatments will suffice; assuming that no drug is lethal or does severe physical damage, our experiment with the rats might be carried out, free of carryover effects, with sufficiently long time periods between tests under different drugs. Of course, there will be situations where carryover effects persist for longer time periods than can practically be incorporated into the experiment. Studies of learning, where the treatment variable is type of material to be learned, may well fall into this category. There may even be situations where the carryover effects are themselves of interest. Procedures have been developed for estimating carryover effects and adjusting treatment effects for their presence (Cochran and Cox, 1957, pp. 135-140); while of considerable interest, such procedures go beyond the scope of the present text and will not be included.

In this section, we will assume that the designs are applied when carryover effects are negligible. Even so, there are still other sources of difficulty in interpreting the results of the analysis. We still have the bias problems discussed in Section 10.4.2. Because the Latin square is an incomplete design, interactions and main effects are partially confounded. Furthermore, interactions involving the random-effects variable, rows, will contribute to main effects—e.g., $\sigma^2_{\text{rows} \times A}$, if nonzero, is a component of $E(MS_A)$. There are several possibilities depending upon the constellation of nonzero interactions. We will refer to the F test as *unbiased* if, assuming H_0 to be true, the numerator and denominator expected mean squares are equal; *negatively biased* (power is lost) if the denominator expectation is larger; and *positively biased* (α is inflated) if the numerator expectation is larger. Bear in mind also that, even in cases we designate as unbiased, there may be positive bias due to heterogeneity of covariance (see the discussion in Section 7.2.2). In these instances, the conservative adjustment for *df* might be employed.

The designs we will consider are for the most part the designs of Section 10.4, the difference being that subjects are now nested in rows rather than in cells. This will result in only slight changes in the breakdown of the sources of variance; most of the computations will proceed as before. In some ways, the analyses will be similar to those of Chapter 8; when several subjects undergo a particular sequence of treatments, we may view rows as a between-subjects variable and A and C as within-subjects variables. In presenting the models, we will not bother to reiterate the definitions and assumptions used throughout the last few chapters. Notation such as η_i, α_j, and $(\eta\alpha)_{ij}$ are interpreted as usual and the usual distributional assumptions still hold for random-effects variables.

10.5.2 A single Latin square. There will be situations in which so few subjects will be available, or in which running time per subject is so great, that we must content ourselves with a subjects, one for each sequence of a treatments. We first assume an additive model in which all interactions are assumed to be negligible:

$$(10.7) \qquad Y_{ijk} = \mu + \eta_i + \alpha_j + \gamma_k + \varepsilon_{ijk}$$

The analysis of variance is summarized in Table 10–11. Under the rather restrictive assumptions represented by Equation (10.7), unbiased F tests are available.

TABLE 10–11 *Analysis of variance for a single Latin square*

SV	df	SS	EMS	F
Total	$a^2 - 1$	$\sum_i \sum_j \sum_k Y_{ijk}^2 - C$		
S	$a - 1$	$\dfrac{\sum_i (\sum_j \sum_k Y_{ijk})^2}{a} - C$	$\sigma_e^2 + n\sigma_S^2$	$\dfrac{MS_S}{MS_{res}}$
Columns	$a - 1$	$\dfrac{\sum_k (\sum_i \sum_j Y_{ijk})^2}{a} - C$	$\sigma_e^2 + n\theta_C^2$	$\dfrac{MS_C}{MS_{res}}$
A	$a - 1$	$\dfrac{\sum_j (\sum_i \sum_k Y_{ijk})^2}{a} - C$	$\sigma_e^2 + n\theta_A^2$	$\dfrac{MS_A}{MS_{res}}$
Residual	$(a-1)(a-2)$	$SS_{tot} - SS_S - SS_C - SS_A$	σ_e^2	

Suppose that interactions involving subjects are present but that treatments are not differentially affected by practice; that is, we still assume $\theta_{AC}^2 = 0$. Now, due to the structural confounding in the Latin square, σ_{SC}^2 contributes to $E(MS_A)$; so also does σ_{SA}^2, under the rules developed in Chapters 7 and 8. The F test of treatments will nevertheless be unbiased in the sense that numerator and denominator have the same expectations *under* H_0. That is,

$$E(MS_A) = E(MS_{B.\,cells\,res.}) = \sigma_e^2 + \left(\frac{a-2}{a}\right)\sigma_{SAC}^2 + \sigma_{SA}^2 + \sigma_{SB}^2$$

Of course, the presence of interaction components in the error term will reduce efficiency, and nonadditivity introduces the possibility of positive bias due to heterogeneity of covariance.

The situation is considerably worse if $\theta_{AC}^2 > 0$, regardless of the status of the row interaction components. The AC interaction variance contributes to the error variability but does not contribute to the A mean square. The result is a negatively biased F ratio and a consequent loss of power.

Tukey's single df test. If the single Latin square must be used, a test for nonadditivity should be employed. Treatments and position in time are only too likely to interact, negatively biasing the F test. Tukey (1955) has provided a modification of the single df test described in Section 7.2.6. The following steps are involved:

(i) Compute

$$\bar{Y}_{...} = \text{mean of all } a^2 \text{ scores}$$
$$\bar{Y}_{i..} = \text{mean of row } i$$
$$\bar{Y}_{.j.} = \text{mean for } A_j$$
$$\bar{Y}_{..k} = \text{mean for column } k$$
$$d_{ijk} = (\bar{Y}_{i..} - \bar{Y}_{...}) + (\bar{Y}_{.j.} - \bar{Y}_{...}) + (\bar{Y}_{..k} - \bar{Y}_{...})$$

(ii) $SS_{\text{nonadd}} = \left\{ \sum_{i,k} [(Y_{ijk} - \overline{Y}...) - d_{ijk}] d_{ijk}^2 \right\}^2 / SS_{\text{B. cell res.}}$

(iii) $$F = \frac{SS_{\text{nonadd}}}{(SS_{\text{B. cells res.}} - SS_{\text{nonadd}})/[(a - 1)(a - 2) - 1]}$$

   $df = 1, (a - 1)(a - 2) - 1$

The investigator would employ the transformation that most reduces the magnitude of the F ratio computed in step (iii).

Relative Efficiency. If the additive model of Equation (10.7) is a valid representation of the data, the Latin square will generally be more efficient than the subjects × treatments design of Chapter 7. Even under non-additivity, the Latin square may prove more efficient, since some of the same interaction variability will also contribute to the subjects × treatments error term. We previously provided an intuitive argument for the Latin square's greater efficiency; we will now derive an expression relating the two error terms, assuming nonadditivity in both designs.

Assume that there are a subjects and a treatment levels in both designs. Presumably, the variability *within* subjects is, in the long run, the same for both designs. That is

$$E_{ST}(SS_{SA}) + E_{ST}(SS_A) = E_{LS}(SS_{\text{B. cells res.}}) + E_{LS}(SS_A) + E_{LS}(SS_C)$$

The subscripts "*ST*" and "*LS*" designate the two designs. Substituting components of variance, we have

$$(a - 1)^2 \sigma_{e_{ST}}^2 + (a - 1)(\sigma_{e_{ST}}^2 + a\theta_A^2) = (a - 1)(a - 2)\sigma_{e_{LS}}^2$$
$$+ (a - 1)(\sigma_{e_{LS}}^2 + a\theta_A^2)$$
$$+ (a - 1)(\sigma_{e_{LS}}^2 + a\theta_C^2)$$

Then, canceling the θ_A^2 terms and rearranging the remaining terms:

$$\sigma_{e_{ST}}^2 = \sigma_{e_{LS}}^2 + \left(\frac{1}{a}\right)(\sigma_{e_{LS}}^2 + a\theta_C^2 - \sigma_{e_{LS}}^2)$$

or

(10.8) $MS_{ST} = MS_{\text{B. cells res.}} + (1/a)(MS_C - MS_{\text{B. cells res.}})$

We can expect a smaller error term in the Latin square design if the variability due to C (position in time) exceeds its error term. However, the Latin square error term has fewer df than the subjects × treatments error term. As we noted in Section 6.2 [see Equation (6.7)], Fisher has suggested the following measure of Relative Efficiency of Design 1 to Design 2:

$$\text{R.E.} = \left[\frac{MS_{\text{error 2}}}{MS_{\text{error 1}}}\right]\left[\frac{df_1 + 1}{df_1 + 3}\right]\left[\frac{df_2 + 3}{df_2 + 1}\right]$$

Note the adjustment for error *df*. In the present case, assuming that we have used the Latin square design and wish to decide whether it will be profitable to use in future studies, we estimate the subjects × treatments error magnitude by Equation (10.8) and compute

$$(10.9) \quad \text{R.E.} = \left[\frac{MS_{B.\text{ cells res.}} + (1/a)(MS_C - MS_{B.\text{ cells res.}})}{MS_{B.\text{ cells res.}}} \right]$$

$$\times \left[\frac{(a-1)(a-2)+1}{(a-1)(a-2)+3} \right] \left[\frac{(a-1)^2+3}{(a-1)^2+1} \right]$$

Missing Scores. Occasionally, a subject may miss a session and one cell in the square will be empty. We can estimate this missing value, X_{ijk}, using the method developed in Chapter 7. Following the additive model,

$$E(X_{ijk}) = \mu + \eta_i + \alpha_j + \gamma_k$$

and, therefore,

$$X = \left(\frac{T_{...} + X}{a^2} \right) + \left(\frac{T_{i..} + X}{a} - \frac{T_{...} + X}{a^2} \right) + \left(\frac{T_{.j.} + X}{a} - \frac{T_{...} + X}{a^2} \right)$$

$$+ \left(\frac{T_{..k} + X}{a} - \frac{T_{...} + X}{a^2} \right)$$

Simplifying, and solving

$$X = \frac{a(T_i + T_j + T_k) - 2T_{...}}{(a-1)(a-2)}$$

where

$$T_{...} = \text{obtained grand total}$$

$$T_{i..} = \text{obtained total for subject } i$$

$$T_{.j.} = \text{obtained total for } A_j$$

$$T_{..k} = \text{obtained total for } C_k.$$

If several scores are missing, the iterative procedure of Chapter 7 may be employed.

10.5.3 Replicating a Latin square. *The model.* It is assumed that a single $a \times a$ Latin square has been randomly sampled from the population of squares. A total of *an* subjects are randomly distributed among the *a* rows (sequences) of the square, with *n* subjects in each row; there are *a* scores for each subject. A representative score would be Y_{ijkm}, where

i indexes the subject within the row ($i = 1, 2, \ldots, n$),
j indexes the level of the treatment variable, $A(j = 1, 2, \ldots, a)$,
k indexes the column in the square ($k = 1, 2, \ldots, a$),
m indexes the row within the square ($m = 1, 2, \ldots, a$).

As an example of the design, a might equal 4 and n might equal 2. Then, we might represent the design by

$$
\begin{array}{c}
\\
S_{11}, S_{21} \\
S_{12}, S_{22} \\
S_{13}, S_{23} \\
S_{14}, S_{24}
\end{array}
\begin{array}{cccc}
C_1 & C_2 & C_3 & C_4 \\
\begin{bmatrix} A_2 & A_4 & A_3 & A_1 \\ A_1 & A_3 & A_2 & A_4 \\ A_4 & A_2 & A_1 & A_3 \\ A_3 & A_1 & A_4 & A_2 \end{bmatrix}
\end{array}
$$

There are many models which might relate Y_{ijkm} and the population parameters. If carryover effects are negligible, we may ignore sequence and sequence interaction effects. If, in addition, subjects do not interact with A or C, the appropriate model is

(10.10) $Y_{ijkm} = \mu + \eta_{i/m} + \alpha_j + \gamma_k + (\alpha\gamma)_{jk} + \varepsilon_{ijkm}$

As usual, $\eta_{i/m}$ and ε_{ijkm} are assumed to be normally distributed random variables; the other effects are fixed.

The analysis of variance. Table 10–12 summarizes the analysis of variance for the replicated square design, assuming the model of Equation (10.10).

The total variability is analyzed in a manner similar to that of Chapters 8 and 9. Since there are a scores for each of an subjects, there are a total of $a^2n - 1$ df. Part of this variability is due to variability among subjects; this accounts for $an - 1$ df. The remaining $an(a - 1)$ df represent the within-subjects variability. Subjects may differ because they are in different rows in the square. In addition, there will be variability among subjects within rows due to individual differences. Thus the between-subjects variability is partitioned into two components: R and S/R. The interpretation of the test of R against S/R will be considered shortly when we deal with *EMS*.

The partitioning of the within-subjects variability follows that of the single Latin square of Section 10.5.2, except that because of the replication within rows, there is an additional source, the (within-cells) *residual*, $a(n - 1)(a - 1)$ df.

The *EMS* follow directly from Equation (10.10). The source labeled R may be viewed as the *between-subjects component of the AC interaction* since the test of R against S/R implies

$$
\frac{E(MS_R)}{E(MS_{S/R})} = \frac{\sigma_e^2 + a\sigma_{S/R}^2 + n\theta_{AC}^2}{\sigma_e^2 + a\sigma_{S/R}^2}
$$

and tests

$$
H_0 : \theta_{AC}^2 = 0
$$

Still assuming the validity of Equation (10.10), and therefore the *EMS* of Table 10–12, we have also unbiased and quite precise tests of A, C, and B. cells res. The last term is intepreted as a test of the within-subjects component

TABLE 10-12 Analysis of variance for the replicated square design

SV	df	SS	EMS	F
Total	$a^2n - 1$	$\sum_i \sum_j \sum_k \sum_m Y_{ijkm}^2 - C$		
Between S	$an - 1$	$\dfrac{\sum_i \sum_m (\sum_j \sum_k Y_{ijkm})^2}{a} - C$		
R	$a - 1$	$\dfrac{\sum_m (\sum_i \sum_j \sum_k Y_{ijkm})^2}{an} - C$	$\sigma_e^2 + a\sigma_{S/R}^2 + n\theta_{AC}^2$	$\dfrac{MS_R}{MS_{S/R}}$
S/R	$a(n - 1)$	$SS_{B.S} - SS_R$	$\sigma_e^2 + a\sigma_{S/R}^2$	
Within S	$an(a - 1)$	$SS_{tot} - SS_{B.S}$		
C	$a - 1$	$\dfrac{\sum_k (\sum_i \sum_j \sum_m Y_{ijkm})^2}{an} - C$	$\sigma_e^2 + na\theta_e^2$	$\dfrac{MS_C}{MS_{res}}$
A	$a - 1$	$\dfrac{\sum_j (\sum_i \sum_k \sum_m Y_{ijkm})^2}{an} - C$	$\sigma_e^2 + na\theta_A^2$	$\dfrac{MS_A}{MS_{res}}$
B. cells res.	$(a - 1)(a - 2)$	$\dfrac{\sum_j \sum_k \sum_m (\sum_i Y_{ijkm})^2}{n} - C - SS_R - SS_A - SS_C$	$\sigma_e^2 + n\theta_{AC}^2$	$\dfrac{MS_{B.cells\,res.}}{MS_{res}}$
Residual	$a(n - 1)(a - 1)$	$SS_{w.s} - SS_C - SS_A - SS_{B.cells\,res.}$	σ_e^2	

of AC, since

$$\frac{E(MS_{B.\ \text{cells res.}})}{E(MS_{W.\ \text{cells res.}})} = \frac{\sigma_e^2 + n\theta_{AC}^2}{\sigma_e^2}$$

If strong prior evidence against the presence of other interactions is lacking, it is somewhat more difficult to draw inferences. If the two preliminary tests just described yield nonsignificant results, we will assume that interaction variance is negligible, and that tests of A and C against the W. cells res. will be unbiased. In fact, if the preliminary test yields a small enough result, the two residual terms might be pooled to provide more *df* for a more powerful test; the criterion for pooling will be discussed in the next chapter. If the preliminary test of the B. cells res. is significant, we may interpret it as reflecting AC variability alone [the model of Equation (10.10)] and tests of A and C will still be unbiased. This assumption is dangerous unless we have strong prior grounds for making it; if we proceed to test A and C against W. cells res. when R interactions are present, the F tests will be positively biased because such sequence interactions will contribute to the numerator mean square but not to the error term. If sequence interactions are suspected, attempts should be made to find a transformation of the data that results in additivity; the criteria would be the preliminary F tests described above. Failing this, it is preferable to test A and C against B. cells res.; the tests will be inefficient (few *df*, large error variance) but unbiased if $\theta_{AC}^2 = 0$ or negatively biased if $\theta_{AC}^2 > 0$.

Numerical example. Table 10–13 presents an illustration of the replicated square design. The analysis of variance follows that of the simple Latin square design with only a few exceptions. We calculate

$$C = \frac{(373)^2}{18}$$

$$= 7{,}729.389$$

and

$$SS_{\text{tot}} = (9)^2 + (12)^2 + \cdots + (33)^2 + (26)^2 - C$$

$$= 887.611$$

The variability between subjects is given by

$$SS_{B.S} = \frac{(59)^2 + \cdots + (76)^2 + (59)^2}{3} - C$$

$$= 82.944$$

This is partitioned into

$$SS_R = \frac{(118)^2 + (120)^2 + (135)^2}{6} - C$$

$$= 28.778$$

and

$$SS_{S/R} = SS_{B.S} - SS_R$$

$$= 54.166$$

We next obtain a measure of the within-subjects variability:

$$SS_{W.S} = SS_{tot} - SS_{B.S}$$

$$= 804.667$$

TABLE 10–13 *Data for a replicated square design*

		A_2	A_3	A_1	$\sum_j \sum_k Y_{ijkm}$
R_1	S_{11}	9	23	27	59
	S_{21}	12	25	22	59
	$\sum_i Y_{ijk1} = 21$		48	49	$\sum_i \sum_j \sum_k Y_{ijk1} = 118$

		A_1	A_2	A_3	
R_2	S_{12}	15	20	22	57
	S_{22}	8	26	29	63
	$\sum_i Y_{ijk2} = 23$		46	51	$\sum_i \sum_j \sum_k Y_{ijk2} = 120$

		A_3	A_1	A_2	
R_3	S_{13}	16	27	33	76
	S_{23}	12	21	26	59
	$\sum_i Y_{ijk3} = 28$		48	59	$\sum_i \sum_j \sum_k Y_{ijk3} = 135$

	$\sum_i \sum_m Y_{ijkm} = 72$		142	159	$\sum_i \sum_j \sum_k \sum_m Y_{ijkm} = 373$

A Subtotals

A_1	A_2	A_3
120	126	127

This can be further partitioned into several components:

$$SS_A = \frac{(120)^2 + (126)^2 + (127)^2}{6} - C$$

$$= 4.778$$

$$SS_C = \frac{(72)^2 + (142)^2 + (159)^2}{6} - C$$

$$= 708.778$$

$$SS_{B.\text{ cells res.}} = \frac{(21)^2 + (48)^2 + \cdots + (48)^2 + (59)^2}{2} - C - SS_R$$
$$- SS_A - SS_C$$

$$= 8.777$$

and

$$SS_{\text{res}} = SS_{\text{W.S}} - SS_A - SS_C - SS_{B.\text{ cells res.}}$$

$$= 82.335$$

The *df* for *B.* cells res. are $(a - 1)(a - 2)$, or 2, and those for the residual are $a(a - 1)(n - 1)$, or 6. Therefore,

$$\frac{MS_{B.\text{ cells res.}}}{MS_{\text{res.}}} = \frac{4.389}{13.723}$$

Since the *F* is clearly not significant, it seems reasonable to pool the *B.* cells res. and residual terms to obtain increased power. The result is

$$MS_{\text{error}} = \frac{82.335 + 8.777}{6 + 2}$$

$$= 11.389$$

The analysis is summarized in Table 10–14.

TABLE 10–14 *Analysis of variance for the data of Table 10–13*

SV	df	SS	MS	F
Total	17	887.611		
Between S	5	82.944		
R	2	28.778	14.389	.797
S/R	3	54.166	18.055	
Within S	12	804.667		
A	2	4.778	2.389	.210
C	2	708.778	354.389	31.117*
Error	8	91.112	11.389	
				*$p < .001$

10.5.4 Using several Latin squares. In Section 10.4.4, we noted that the use of several Latin squares reduces the possibility of confounding main with interaction effects. For this reason, it is frequently preferable to randomly sample n squares and assign one subject to each of the an sequences rather than to replicate one square n times. The latter approach, discussed in the preceding section, would be most useful when the AC interaction is of major interest and evidence available before the experiment supports the validity of Equation (10.10).

Assuming an additive model, we have the SV, df, SS, and EMS of Table 10–15. The calculations are quite simple. We will therefore omit a numerical example. The one point to note is that if AC interaction effects are not negligible, they will contribute to the pooled residual mean square but not to MS_A. The result will be negative bias and a consequent loss of power. Subject interactions contribute to both the treatment and error mean squares and will cause no problem.

TABLE 10–15 *Analysis of several squares with repeated measurements*

SV	df	SS	EMS
Total	$a^2 n - 1$	$\sum_i \sum_j \sum_k \sum_m Y_{ijkm}^2 - C$	
S	$an - 1$	$\dfrac{\sum_i (\sum_j \sum_k \sum_m Y_{ijkm})^2}{a} - C$	$\sigma_e^2 + a\sigma_S^2$
A	$a - 1$	$\dfrac{\sum_j (\sum_i \sum_k \sum_m Y_{ijkm})^2}{na} - C$	$\sigma_e^2 + na\theta_A^2$
C	$a - 1$	$\dfrac{\sum_k (\sum_i \sum_j \sum_m Y_{ijkm})^2}{na} - C$	$\sigma_e^2 + na\theta_C^2$
Residual	$(an - 2)(a - 1)$	$SS_{\text{tot}} - SS_S - SS_A - SS_C$	σ_e^2

10.5.5 Investigating additional independent variables. *Latin squaring treatment combinations.* Suppose that we are interested in the main and interaction effects of two treatment variables, A and D. There are several variations of the Latin square design that merit consideration. One approach is to test each subject under all ad combinations; the layout is similar to that of Section 10.4.3. This approach is particularly useful when we have few subjects but they are available for sufficient time to test them under all conditions. We could use a single $ad \times ad$ Latin square, or we could replicate the square n times or select n such squares each replicated only once. The analyses follow those of the preceding three sections; the only difference is that, where we had an A source of variance before, we now have a *treatment combinations* source of variance on $(ad - 1)$ *df*. This variability is then further partitioned in an A, a D, and an AD source.

Replicating one square. The replicated square design, described in Section 10.5.3, can be extended to permit the investigation of additional treatment variables. As before, there could be exactly a sequences. However, n of the subjects in a particular sequence would be at D_1, n more at D_2, and so on, giving a total of adn subjects, dn in each of a sequences. As an example of the application of this design, consider the measurement of reaction time under each of four levels of stress (A). Three experimental groups of eight subjects each are tested under all stress levels; the groups differ with respect to level of anxiety (D). In this case, $a = 4$, $d = 3$, and $n = 2$. In general,

i indexes the subjects within a row at a level of D,
j indexes the level of A,
m indexes the level of D,
k indexes the level of C (position in time),
p indexes the row in a level of D.

The layout of the design in our example might be

$$
\begin{array}{c c c}
 & & \begin{array}{cccc} C_1 & C_2 & C_3 & C_4 \end{array} \\
D_1 & \begin{array}{c} S_{111}, S_{211} \\ S_{112}, S_{212} \\ S_{113}, S_{213} \\ S_{114}, S_{214} \end{array} & \begin{bmatrix} A_4 & A_2 & A_1 & A_3 \\ A_2 & A_3 & A_4 & A_1 \\ A_1 & A_4 & A_3 & A_2 \\ A_3 & A_1 & A_2 & A_4 \end{bmatrix} \\
D_2 & \begin{array}{c} S_{121}, S_{221} \\ S_{122}, S_{222} \\ S_{123}, S_{223} \\ S_{124}, S_{224} \end{array} & \begin{bmatrix} A_4 & A_2 & A_1 & A_3 \\ A_2 & A_3 & A_4 & A_1 \\ A_1 & A_4 & A_3 & A_2 \\ A_3 & A_1 & A_2 & A_4 \end{bmatrix} \\
D_3 & \begin{array}{c} S_{131}, S_{231} \\ S_{132}, S_{232} \\ S_{133}, S_{233} \\ S_{134}, S_{234} \end{array} & \begin{bmatrix} A_4 & A_2 & A_1 & A_3 \\ A_2 & A_3 & A_4 & A_1 \\ A_1 & A_4 & A_3 & A_2 \\ A_3 & A_1 & A_2 & A_4 \end{bmatrix}
\end{array}
$$

The design under discussion permits the evaluation of main effects and interactions of A and D using fewer subjects than a completely randomized design, fewer measurements per subject than a simple repeated measurements design, and a potentially more efficient test of A and AD than the mixed design of Chapter 8.

The analysis of variance. Table 10–16 presents the *SV*, *df*, *SS*, *EMS* and *F* ratios. The between-subjects variability is partitioned in the same way as in Section 8.3, where we dealt with two between-subjects variables. The within-subjects effects are generated if one remembers that there is confounding between any main effect associated with the square (i.e., R, C, A)

TABLE 10-16 *Analysis of variance for the replicated square design with a between-squares factor*

SV	df	SS	EMS	F
Total	$a^2dn - 1$	$\sum_i\sum_j\sum_k\sum_m\sum_p Y_{ijkmp}^2 - C$		
Between S	$adn - 1$	$\dfrac{\sum_i\sum_m\sum_p\left(\sum_j\sum_k Y_{ijkmp}\right)^2}{a} - C$		
R	$a - 1$	$\dfrac{\sum_p\left(\sum_i\sum_j\sum_k\sum_m Y_{ijkmp}\right)^2}{adn} - C$	$\sigma_e^2 + a\sigma_{S/RD}^2 + adn\theta_{AC}^2$	$\dfrac{MS_R}{MS_{S/RD}}$
D	$d - 1$	$\dfrac{\sum_m\left(\sum_i\sum_j\sum_k\sum_p Y_{ijkmp}\right)^2}{a^2n} - C$	$\sigma_e^2 + a\sigma_{S/RD}^2 + a^2n\theta_D^2$	$\dfrac{MS_D}{MS_{S/RD}}$
RD	$(a - 1)(d - 1)$	$\dfrac{\sum_m\sum_p\left(\sum_i\sum_j\sum_k Y_{ijkmp}\right)^2}{an} - C - SS_R - SS_D$	$\sigma_e^2 + a\sigma_{S/RD}^2 + an\theta_{ADC}^2$	$\dfrac{MS_{RD}}{MS_{S/RD}}$
S/RD	$ad(n - 1)$	$SS_{B.S} - SS_R - SS_D - SS_{RD}$	$\sigma_e^2 + a\sigma_{S/RD}^2$	
Within S	$adn(a - 1)$	$SS_{tot} - SS_{B.S}$		

TABLE 10–16 (continued)

	df	SS	E(MS)	F
A	$a - 1$	$\dfrac{\sum_j(\sum_i\sum_k\sum_m\sum_p Y_{ijkmp})^2}{adn} - C$	$\sigma_e^2 + adn\theta_A^2$	$\dfrac{MS_A}{MS_{\text{res}}}$
C	$a - 1$	$\dfrac{\sum_k(\sum_i\sum_j\sum_m\sum_p Y_{ijkmp})^2}{adn} - C$	$\sigma_e^2 + adn\theta_C^2$	$\dfrac{MS_C}{MS_{\text{res}}}$
AD	$(a-1)(d-1)$	$\dfrac{\sum_j\sum_m(\sum_i\sum_k\sum_p Y_{ijkmp})^2}{an} - C - SS_A - SS_D$	$\sigma_e^2 + an\theta_{AD}^2$	$\dfrac{MS_{AD}}{MS_{\text{res}}}$
DC	$(d-1)(a-1)$	$\dfrac{\sum_k\sum_m(\sum_i\sum_j\sum_p Y_{ijkmp})^2}{an} - C - SS_D - SS_C$	$\sigma_e^2 + an\theta_{DC}^2$	$\dfrac{MS_{DC}}{MS_{\text{res}}}$
B. cells res.	$(a-1)(a-2)$	$\dfrac{\sum_j\sum_k\sum_p(\sum_i\sum_m Y_{ijkmp})^2}{dn} - C - SS_R - SS_D - SS_{DC} - SS_C$	$\sigma_e^2 + dn\theta_{AC}^2$	$\dfrac{MS_{\text{B.cells res.}}}{MS_{\text{res}}}$
B. cells res × D	$(a-1)(a-2)(d-1)$	$\dfrac{\sum_j\sum_k\sum_m\sum_p(\sum_i Y_{ijkmp})^2}{n} - C - SS_A - SS_C$ $- SS_{RD} - SS_A - SS_C - SS_{AD} - SS_{DC} - SS_{\text{B.cells res.}}$	$\sigma_e^2 + n\theta_{ADC}^2$	$\dfrac{MS_{\text{B.cells res.}\times D}}{MS_{\text{res}}}$
Residual	$ad(a-1)(n-1)$	$SS_{w.s} - SS_A - SS_C - SS_{AD} - SS_{DC}$ $- SS_{\text{B.cells res.}} - SS_{\text{B.cells res.}\times D}$	σ_e^2	

and the interaction of the other two variables. However, note that D can interact with each of the main effects associated with the square and with B. cells res. as well.

The model that has been chosen is a simple extension of Equation (10.10). Carryover effects are assumed to be negligible; therefore, we have the equation:

(10.11) $\quad Y_{ijkmp} = \mu + \eta_{i/kp} + \alpha_j + \delta_k + \gamma_m + (\alpha\delta)_{jk} + (\alpha\gamma)_{jm}$

$$+ (\delta\gamma)_{km} + (\alpha\delta\gamma)_{jkm} + \varepsilon_{ijkmp}$$

This equation generates the *EMS* for Table 10–16 which, in turn, indicate the appropriate error terms. Furthermore, it should be apparent that if Equation (10.11) is valid, R may be interpreted as a between-subjects measure of AC, RD as a between-subjects measure of ADC, B. cells res. as a within-subjects measure of AC, and B. cells res. $\times D$ as a within-subjects measure of ADC. As in Section 10.5, the F tests of A and C effects against the residual will be positively biased if the sequence of treatment presentations interacts with A or C. If there is reason to suspect such carryover effects, the following preliminary tests may be carried out:

$$F = \frac{MS_R}{MS_{S/RD}} \qquad\qquad F = \frac{MS_{B.\text{ cells res.}}}{MS_{\text{res.}}}$$

$$F = \frac{MS_{RD}}{MS_{S/RD}} \qquad\qquad F = \frac{MS_{B.\text{ cells res.} \times D}}{MS_{\text{res.}}}$$

If any of these are significant, a transformation to additivity should be sought.

Note that if sequence interactions are suspect, our recommended procedure provides no way of testing for AC or ADC effects. There is no test that permits determination of which interaction component is present when, for example, B. cells res. is significant. Such a result only reveals that *some* interaction component is present. A significant B. cells res. can only be interpreted as AC (or B. cells res. $\times D$ as ADC) if there is a strong a priori reason for assuming Equation (10.11) to be valid. Such an assumption might be founded on previous experimentation and on knowledge of the independent and dependent variables; i.e., knowledge of the experimental situation.

A numerical example. Table 10–17 presents data from a 3×3 Latin square replicated four times, twice at each of two levels of B. Subtotals that are required in the calculation of sums of squares have also been computed. The correction term is

$$C = \frac{(416)^2}{36}$$

$$= 4,807.111$$

TABLE 10–17 Data for a replicated Latin square design with a between-squares factor

D_1

	A_1	A_3	A_2	$\sum Y_{ij1mp}$
S_{111}	12	4	10	26
S_{211}	14	6	8	28
$\sum_i Y_{ij1m1} = 26$	10	18		$\sum_j\sum_i Y_{ij1m1} = 54$

	A_2	A_1	A_3	
S_{112}	6	9	3	18
S_{212}	11	17	5	33
$\sum_i Y_{ij1m2} = 17$	26	8		$\sum_j\sum_i Y_{ij1m2} = 51$

	A_3	A_2	A_1	
S_{113}	5	12	18	35
S_{213}	7	6	10	23
$\sum_i Y_{ij1m3} = 12$	18	28		$\sum_j\sum_i Y_{ij1m3} = 58$

$\sum_i\sum_m Y_{ij1mp} = 55$	54	54		$\sum_j\sum_i\sum_m Y_{ij1mp} = 163$

D_2

	A_1	A_3	A_2	$\sum Y_{ij2mp}$
S_{121}	19	11	20	50
S_{221}	21	11	20	52
$\sum_i Y_{ij2m1} = 40$	22	40		$\sum_j\sum_i Y_{ij2m1} = 102$

	A_2	A_1	A_3	
S_{122}	16	18	8	42
S_{222}	14	12	3	29
$\sum_i Y_{ij2m2} = 30$	30	11		$\sum_j\sum_i Y_{ij2m2} = 71$

	A_3	A_2	A_1	
S_{123}	6	17	14	37
S_{223}	9	18	16	43
$\sum_i Y_{ij2m3} = 15$	35	30		$\sum_j\sum_i Y_{ij2m3} = 80$

$\sum_i\sum_m Y_{ij2mp} = 85$	87	81		$\sum_j\sum_i\sum_m Y_{ij2mp} = 253$

A Subtotals

	A_1	A_2	A_3	
D_1	80	53	30	163
D_2	100	105	48	253
	180	158	78	416

C Subtotals

	C_1	C_2	C_3	
D_1	55	54	54	163
D_2	85	87	81	253
	140	141	135	416

The total variability is

$$SS_{\text{tot}} = (12)^2 + (4)^2 + \cdots + (18)^2 + (16)^2 - C$$
$$= 5{,}814.000 - 4{,}807.111$$
$$= 1{,}006.889$$

We next obtain the between-subjects variability:

$$SS_{\text{B.S}} = \frac{(26)^2 + (28)^2 + \cdots + (37)^2 + (43)^2}{3} - C$$
$$= 5{,}218.000 - 4{,}807.111$$
$$= 410.889$$

This is now further partitioned:

$$SS_R = \frac{(54 + 102)^2 + (51 + 71)^2 + (58 + 80)^2}{12} - C$$
$$= 4{,}855.333 - 4{,}807.111$$
$$= 48.222$$

$$SS_D = \frac{(163 - 253)^2}{36}$$
$$= 225.000$$

$$SS_{RD} = \frac{(54)^2 + \cdots + (80)^2}{6} - C - SS_R - SS_D$$
$$= 5{,}121.000 - 4{,}807.111 - 48.222 - 225.000$$
$$= 40.667$$

and

$$SS_{S/RD} = SS_{\text{B.S}} - SS_R - SS_D - SS_{RD}$$
$$= 370.222$$

The within-subjects variability is computed next:

$$SS_{\text{W.S}} = SS_{\text{tot}} - SS_{\text{B.S}}$$
$$= 596.000$$

Partitioning this, we obtain

$$SS_C = \frac{(140)^2 + (141)^2 + (135)^2}{12} - C$$
$$= 4{,}808.833 - 4{,}807.111$$
$$= 1.722$$

$$SS_A = \frac{(180)^2 + (158)^2 + (78)^2}{12} - C$$

$$= 5{,}287.333 - 4{,}807.111$$

$$= 480.222$$

$$SS_{DC} = \frac{(55)^2 + (54)^2 + \cdots + (81)^2}{6} - C - SS_D - SS_C$$

$$= 5{,}035.333 - 4{,}807.111 - 225.000 - 1.722$$

$$= 2.500$$

$$SS_{AD} = \frac{(80)^2 + (53)^2 + \cdots + (48)^2}{6} - C - SS_A - SS_D$$

$$= 5{,}573.000 - 4{,}807.111 - 480.222 - 225.000$$

$$= 60.667$$

$$SS_{B.\text{ cells res.}} = \frac{(26 + 40)^2 + (10 + 22)^2 + \cdots + (28 + 30)^2}{4}$$

$$- C - SS_R - SS_C - SS_A$$

$$= 5{,}338.000 - 4{,}807.111 - 48.222 - 1.722 - 480.222$$

$$= .723$$

$$SS_{B.\text{ cells res.} \times D} = \frac{(26)^2 + (10)^2 + \cdots + (35)^2 + (30)^2}{2} - C - SS_D$$

$$- SS_{B.\text{ cells res.}} - SS_{RD} - SS_R - SS_{DC} - SS_C - SS_{AD} - SS_A$$

$$= 5{,}668.000 - 4{,}807.111 - 225.000 - .723 - 40.667$$

$$- 48.222 - 2.500 - 1.722 - 60.667 - 480.222$$

$$= 1.166$$

and

$$SS_{\text{res}} = SS_{W.S} - SS_C - SS_A - SS_{DC} - SS_{AD} - SS_{B.\text{ cells res.}} - SS_{B.\text{ cells res.} \times D}$$

$$= 48.000$$

Table 10–18 summarizes the results of the analysis.

Using several squares. The design of Section 10.5.4 can also be extended to handle an additional independent variable. Suppose that we randomly sample *dn* squares. Squares are randomly assigned to levels of *D* so that there are *n* squares at each D_k. A total of *adn* subjects are then randomly distributed among the *adn* sequences of the selected squares, one subject in each sequence. If we assume that the dimensions of the square—subjects,

TABLE 10–18 *Analysis of variance for the data of Table 10–17*

SV	df	SS	MS	F
Total	35	1,006.889		
Between S	11	410.889		
R	2	48.222	24.111	1.490
D	1	225.000	225.000	13.906*
RD	2	40.667	20.334	1.257
S/RD	6	97.080	16.180	
Within S	24	596.000		
C	2	1.722	.861	.215
A	2	480.222	240.111	60.028**
DC	2	2.500	1.250	.313
AB	2	60.667	30.334	7.584*
B. cells res.	2	.723	.362	.091
B. cells res. × D	2	1.166	.583	.146
Residual	12	48.000	4.000	

$$**p < .001$$
$$*p < .01$$

positions in time, A—do not interact, the following model is appropriate:

(10.12) $Y_{ijkm} = \mu + \eta_{i/k} + \alpha_j + \gamma_m + \delta_k + (\alpha\delta)_{jk} + (\delta\gamma)_{km} + \varepsilon_{ijkm}$

Table 10–19 presents the analysis of variance. AC and ACD interaction effects, if present, will contribute to the residual error term and not to A or C, causing negative bias in those tests as well as in tests of AD and ACD. However, the advantage of the design lies in the likelihood that confounding will be minimized due to the use of several squares.

10.6 CONCLUDING REMARKS

The assets of the Latin square design are clear—the need for fewer subjects; the reduction of systematic treatment biases through counterbalancing; and the reduction of error variance through the removal of variability due to two factors, the rows and columns of the design. The dangers are less immediately obvious but we should all find them familiar by now. Fixed-effects interactions involving the treatment variable contribute to the between-cells residual term but not to the treatment source; an F ratio employing these two sources under such nonadditivity is negatively biased. When *random* interaction effects are present, they contribute to the treatments source but not to the within-cell error term; an F ratio of these two terms under these conditions is positively biased. When repeated measurements are involved,

TABLE 10-19 *Analysis with several squares and a between-squares factor*

SV	df	SS	EMS
Total	$a^2dn - 1$	$\sum_i \sum_j \sum_k \sum_m \sum_p Y_{ijkmp}^2 - C$	
Between S	$adn - 1$	$\dfrac{\sum_i \sum_k \sum_p (\sum_j \sum_m Y_{ijkmp})^2}{a} - C$	
D	$d - 1$	$\dfrac{\sum_p(\sum_i \sum_j \sum_k \sum_m Y_{ijkmp})^2}{a^2n} - C$	$\sigma_e^2 + a\sigma_{S/D}^2 + a^2n\theta_D^2$
S/D	$d(an - 1)$	$SS_{\text{B.S}} - SS_D$	$\sigma_e^2 + a\sigma_{S/D}^2$
Within S	$adn(a - 1)$	$SS_{\text{tot}} - SS_{\text{B.S}}$	
A	$a - 1$	$\dfrac{\sum_j (\sum_i \sum_k \sum_m \sum_p Y_{ijkmp})^2}{adn} - C$	$\sigma_e^2 + adn\theta_A^2$
C	$a - 1$	$\dfrac{\sum_m(\sum_i \sum_j \sum_k \sum_p Y_{ijkmp})^2}{adn} - C$	$\sigma_e^2 + adn\theta_C^2$
AD	$(a - 1)(d - 1)$	$\dfrac{\sum_j \sum_p(\sum_i \sum_k \sum_m Y_{ijkmp})^2}{an} - C - SS_A - SS_D$	$\sigma_e^2 + an\theta_{AD}^2$
BC	$(a - 1)(d - 1)$	$\dfrac{\sum_m \sum_p(\sum_i \sum_j \sum_k Y_{ijkmp})^2}{an} - C - SS_D - SS_C$	$\sigma_e^2 + an\theta_{CD}^2$
Residual	$b(an - 2)(a - 1)$	$SS_{\text{w.s}} - SS_A - SS_C - SS_{AD} - SS_{DC}$	σ_e^2

there is the danger of carryover effects; the effect of a treatment may depend upon which of several treatments preceded it. In addition, heterogeneity of covariance may cause positive bias.

Despite the problems that we have noted, the advantage of the Latin square design, or some variation of it, will frequently outweigh its disadvantages. Carryover effects will be minimal in many experiments, in others they may be reduced by introducing sufficiently long intervals between treatment levels, in still others we may statistically account for them (Cochran and Cox, 1957, pp. 135-140). If we have prior knowledge of the types of interactions we may encounter in our data, we may still achieve unbiased F tests. If an AC interaction is likely but no others, we would be wise to replicate a single square and to use the within-cell term as our measure of error. In the converse case—for example, the random effects interactions are likely—the between-cell residual makes most sense. Where we suspect nonadditivity but do not know the specific sources, transformations that reduce nonadditivity may be sought. In addition, it should be noted that confounding of interaction with main effects becomes less of a factor when squares are larger and more of them are used.

Finally, it is important to realize that there are situations in which the supply of subjects is severely restricted; the experiment may be unpleasant or dangerous and volunteers consequently few, or the experiment may require a particular type of subject, perhaps a particular clinical case who is difficult to obtain. In such instances, the choice of design is quickly reduced to a choice among repeated measurements designs. The subjects × treatments design, with an independent randomization for each subject, has many of the same problems that plague the Latin square—heterogeneity of covariance or carryover effects. In view of this, the Latin square, with its potentially greater efficiency, may be preferred.

EXERCISES

10.1   Each of 36 subjects was tested on four signal detection problems, which varied with respect to the location of the target. Subjects differed with respect to the sequence of presentations of the problems (four sequences were used) and with respect to the instructions read to them (three sets of instructions). Present the appropriate analysis of variance table.

10.2   A clinical investigator is concerned with the effects of alcohol upon reaction time (r.t.) to perceptual inputs. He runs three experimental groups of 20 Ss each on four tasks, each task on a different day. The groups differ with respect to alcohol intake. The tasks are (a) Simple r.t. to an auditory stimulus, (b) simple r.t. to a visual stimulus, (c) choice r.t. to auditory stimuli, and (d) choice r.t. to visual stimuli. The tasks are presented in 60 sequences comprising 15 Latin squares; median r.t.'s are obtained for each S on each task. Present the analysis of variance.

10.3 Five highly practiced Ss were each required to learn different lists on five different days; the lists varied in terms of the type of items—e.g., words, digits, nonsense syllables. Since prior practice in the task should minimize learning-to-learn and Proactive Inhibition should be slight with the different materials used, the sequence of lists was Latin-squared. Each list had seven items; the dependent measure was errors per position in the list. Present the SV and df.

10.4 Assume that 3-man groups are required to solve four problems, varying in difficulty. The purpose is to investigate the number of communications from each individual as a function of problem difficulty. The presentation is Latin-squared and four groups are run through each sequence. Present SV and df.

10.5 We are interested in gambling behavior under variations in initial stake, (three levels), payoffs (three levels), and probability of winning (three levels). Suggest several alternative designs and discuss their relative merits. Assume that 81 subjects are available.

10.6 Consider a $2 \times 2 \times 2$ experiment. Suppose we have four experimenters, each of whom runs two of the eight cells. In the notation of Chapter 5,

E_1	E_2	E_3	E_4
(1), bc	a, abc	b, c	ab, ac

Several effects are confounded among experimenters. Which effects are confounded how?

REFERENCES

Cochran, W. C. and Cox, G. M. *Experimental Designs*, 2nd ed. New York: Wiley, 1957.

Fisher, R. A. and Yates, F. *Statistical Tables for Biological, Agricultural and Medical Research*. Edinburgh: Oliver & Boyd, 1955, pp. 80–82.

Tukey, J. W., "Test for nonadditivity in the Latin square," *Biometrics*, 11: 111–113, (1965).

Wilk, M. B. and Kempthorne, O., "Nonadditivities in a Latin square," *Journal of the American Statistical Association*, 52: 218–236, (1957).

SUPPLEMENTARY READINGS

Readers who have trouble with the notation and derivations of the Wilk–Kempthorne (1957) article might profit from the following treatment based on their paper:

Gaito, J., "The single Latin square design in psychological research," *Psychometrika*, 23: 369–378 (1958).

A slightly different model for the replicated square design than that presented in this chapter, but one that leads to similar conclusions about the effects of nonadditivity, is discussed in:

Gourlay, N., "*F* test bias for experimental designs of the Latin square type," *Psychometrika*, 20: 237–287 (1955).

Gaito has also discussed the relationship between the designs of Chapters 7 and 10:

Gaito, J., "Repeated measurement. Designs and counterbalancing," *Psychological Bulletin*, 58: 46–54 (1961).

Several papers have introduced ways to balance the Latin square with respect to carryover effects—among these are (see also the pages cited in Cochran and Cox, 1957):

Alimena, B. A., "A method of determining unbiased distribution in Latin square," *Psychometrika*, 27: 315–318 (1962).

Atkinson, G. F., "Designs for sequences of treatments with carryover effects," *Biometrics*, 22: 292–309 (1966).

Benjamin, L. S., "A special Latin square for the use of each subject as his own control," *Psychometrika*, 30: 499–513 (1966).

Bradley, J. V., "Complete counterbalancing of immediate sequential effects in a Latin square design," *Journal of the American Statistical Association*, 53: 525–528 (1958).

Williams, R. N., "Experimental designs for serially correlated observations," *Biometrika*, 39: 151–167 (1952).

Numerous additional incomplete designs that involve confounding but hold the potential for efficient analyses may be found in Cochran and Cox (1957).

Expected Mean Squares

11.1 INTRODUCTION

Throughout this text, the *expected mean square* (*EMS*) has been a key concept. The *EMS* has been used to tackle such problems as the choice of error terms, the relative efficiency of designs, and the direction and extent of bias in *F* tests. These applications of the concept of the *EMS* are far from exhaustive; the purpose of this chapter is to extend the reader's knowledge of the uses of the *EMS*. We begin by examining certain problems in null hypothesis testing that have thus far not been considered. Subsequently, point and interval estimation of the components that comprise our expectations will be discussed. Finally, it will be noted how the *EMS* provides information about the reliability of measurement.

11.2 SIGNIFICANCE TESTS

11.2.1 Pooling in the analysis of variance. In Chapter 9 it was noted that the appropriate error term in the hierarchical designs is often distributed on a rather small number of *df*; consequently, the *F* test lacked power. The problem is not restricted to any one set of designs, and it seems reasonable to consider a possible remedy. As an example, we will use the design of Section 9.3.

There are *g* groups at each of *a* levels of *A*. Each group consists of *bn* subjects randomly assigned to the *b* levels of the variable *B*. Turning back to Table 9-6, we note that *A* is tested against *G/A*, and *B* and *AB* are tested against the *GB/A* term. There is a conspicuous waste here; the $MS_{S/GB/A}$, which is distributed on more *df* than either of our appropriate error terms, serves only to test the relatively uninteresting group effects. However, suppose that group effects were negligible. Specifically, we assume

(11.1) $$Y_{ijkm} = \mu + \alpha_k + \beta_m + (\alpha\beta)_{km} + \varepsilon_{ijkm}$$

305

which substitutes for Equation (9.4). Note that no effects involving groups are assumed to contribute to the subject's score. As a result, we have the new set of *EMS* of Table 11–1. Table 11–1 suggests that G/A, GB/A, and $S/GB/A$ might be combined in some way, since all three are estimates of the same population variance, σ_e^2. We pool these terms as follows:

$$\textbf{(11.2)} \quad MS_{S/AB} = \frac{SS_{G/A} + SS_{GB/A} + SS_{S/GB/A}}{a(g-1) + a(g-1)(b-1) + abg(n-1)}$$

$$= \frac{SS_{G/A} + SS_{GB/A} + SS_{S/GB/A}}{ab(ng-1)}$$

TABLE 11–1 *EMS for a hierarchical design assuming no group effects*

SV	EMS
A	$\sigma_e^2 + ngb\theta_A^2$
G/A	σ_e^2
B	$\sigma_e^2 + nga\theta_B^2$
AB	$\sigma_e^2 + ng\theta_{AB}^2$
GB/A	σ_e^2
S/GB/A	σ_e^2

Note that pooling is defined as adding sums of squares and dividing by the sums of *df* for the terms involved. The pooled term—*S/AB*—is actually an average of the three mean squares that estimate σ_e^2, which should be clearer if we rewrite Equation (11.2) as

$$\textbf{(11.2}') \quad MS_{S/AB} = \frac{a(g-1)}{ab(ng-1)} MS_{G/A} + \frac{a(g-1)(b-1)}{ab(ng-1)} MS_{GB/A}$$

$$+ \frac{abg(n-1)}{ab(ng-1)} MS_{S/GB/A}$$

This is of the general form of Equation (3.1), which provides one definition of a mean.

Table 11–2 contains the analysis of variance that results from pooling. The advantage that potentially accrues from pooling is immediately evident. In our example, after pooling we have considerably more error *df* than prior to pooling and, consequently, the power of the *F* test has been increased, *if assumptions that we have made are correct*. In addition to increased power, there is also increased simplicity. The analysis of variance table is less cluttered and therefore easier to interpret and to discuss in a report of the experiment. The pertinent sources of variance, those that make a major contribution to the data matrix, are the only ones that appear in the table.

TABLE 11-2 *Analysis of variance after pooling of sources in Table 11-1*

SV	df	EMS
A	$a - 1$	$\sigma_e^2 + ngb\theta_A^2$
B	$b - 1$	$\sigma_e^2 + nga\theta_B^2$
AB	$(a - 1)(b - 1)$	$\sigma_e^2 + ng\theta_{AB}^2$
S/AB	$ab(gn - 1)$	σ_e^2

What if the assumptions upon which the pooling is based are incorrect? Positive or negative bias could result, depending upon the magnitude of the pooled mean square in relation to the appropriate error term. For example, if $\sigma_{G/A}^2 = \sigma_{GB/A}^2 > 0$, the F test of A would be positively biased because we would expect the average value of the three mean squares involved in the pooling to be less than the value of the appropriate error term, $MS_{G/A}$. On the other hand, if $\sigma_{GB/A}^2$ were very much larger than $\sigma_{G/A}^2$ which, in turn, was only slightly greater than zero, the expectation of the pooled mean square would be greater than that of $MS_{G/A}$, and we would have negative bias in testing A effects. At the same time, the test of B effects would be positively biased because the expectation of the pooled mean square would be less than that of $MS_{GB/A}$, the appropriate error term for B.

The situation is simpler in completely randomized designs, where all variables are assumed to have fixed effects. We might pool the highest-order interaction with the within-cell error term in the belief that the interaction variance was negligible. If this assumption is incorrect, the expectation of the pooled mean square would clearly be larger than the expectation of the within-cell mean square, thus power would be lost.

Statisticians are not agreed upon the conditions under which pooling should be carried out. In the most extensive works available, Bozivich, Bancroft, and Hartley (1956) and Srivistava and Bozivich (1961) have investigated power and α under various violations of the model assumed in pooling, using computer populations in an approach similar to those of Section 4.2.3. Their work has been limited to hierarchical designs in which the components assumed to be zero are variances of random effects. The results are somewhat complicated; however, a rough rule of thumb would be to pool whenever there are a priori grounds for pooling and also a preliminary F test fails to be significant at the .25 level. The prior belief in the pooling model might be based on knowledge of the experimental situation (e.g., in a particular social situation, it may be unlikely that groups really do contribute to the variability in the data beyond the contribution due to individual differences), or upon an analysis of the results of preliminary tests in related experiments. The preliminary test is a ratio of terms to be pooled; in our example, we would test $MS_{G/A}$ against $MS_{S/GB/A}$ and $MS_{GB/A}$

against $MS_{S/GB/A}$. If either F is nonsignificant at the .25 level, its numerator and denominator may be pooled.

The consequences of pooling inappropriately when fixed-effects components are to be neglected are less clear. I suspect that the same .25 level for the preliminary test will prove sufficiently conservative in most instances. Again, there should be prior evidence as well. In general, pooling, when it is not warranted by the true state of affairs in the population, can result in false inferences and, when there is doubt about the appropriateness of the pooling procedure, should not be employed. Wherever possible, sufficient subjects should be run to assure reasonable power without pooling.

11.2.2 Quasi-F ratios. Consider an experiment in which n subjects (S) are required to detect the presence of each of t targets (T) under each of w levels of white noise (W). The levels of noise are arbitrarily selected. The locations of the targets are randomly selected from a continuum of locations available in a 4-ft. square field. The appropriate SV, df, and EMS are presented in Table 11–3. One important implication of the EMS column is that there is no appropriate error term for the W source of variance. By an appropriate error term, we again mean that the expectation contains all the components of the numerator EMS except the null hypothesis component.

TABLE 11–3 *Analysis of variance for one fixed and two random effect variables*

SV	df	EMS
W	$w - 1$	$\sigma_e^2 + n\sigma_{WT}^2 + t\sigma_{WS}^2 + \sigma_{WTS}^2 + tn\theta_W^2$
T	$t - 1$	$\sigma_e^2 + w\sigma_{TS}^2 + nw\sigma_T^2$
S	$n - 1$	$\sigma_e^2 + w\sigma_{TS}^2 + wt\sigma_S^2$
WT	$(w - 1)(t - 1)$	$\sigma_e^2 + \sigma_{WTS}^2 + n\sigma_{WT}^2$
WS	$(w - 1)(n - 1)$	$\sigma_e^2 + \sigma_{WTS}^2 + t\sigma_{WS}^2$
TS	$(t - 1)(n - 1)$	$\sigma_e^2 + w\sigma_{TS}^2$
WTS	$(w - 1)(t - 1)(n - 1)$	$\sigma_e^2 + \sigma_{WTS}^2$

The error term (for W) problem could be approached by assuming that σ_{WS}^2 or σ_{WT}^2 is zero, but unless this is truly so the F tests will be biased. It is also possible that after some trial and error, a transformation might be found that would make the assumption true, but this is not certain. We require additional ammunition for our attack on the W effect (which, incidentally, is likely to be the effect of most interest to the experimenter). A careful examination of the EMS suggests the appropriate method. We note that

(11.3)
$$
\begin{aligned}
E(MS_{WT}) + E(MS_{WS}) - E(MS_{WTS}) &= (\sigma_e^2 + \sigma_{WTS}^2 + n\sigma_{WT}^2) \\
&\quad + (\sigma_e^2 + \sigma_{WTS}^2 + t\sigma_{WS}^2) \\
&\quad - (\sigma_e^2 + \sigma_{WTS}^2) \\
&= \sigma_e^2 + t\sigma_{WS}^2 + n\sigma_{WT}^2 + \sigma_{WTS}^2
\end{aligned}
$$

and

$$
\textbf{(11.4)} \quad \frac{E(MS_W)}{E(MS_{WT}) + E(MS_{WS}) - E(MS_{WTS})}
$$

$$
= \frac{\sigma_e^2 + t\sigma_{WS}^2 + n\sigma_{WT}^2 + \sigma_{WTS}^2 + tn\theta_W^2}{\sigma_e^2 + t\sigma_{WS}^2 + n\sigma_{WT}^2 + \sigma_{WTS}^2}
$$

which has the virtue of equaling 1 if the null hypothesis ($\theta_W^2 = 0$) is true. Thus,

$$
\frac{MS_W}{MS_{WT} + MS_{WS} - MS_{WTS}}
$$

has the look of an *F* ratio. Is it an *F* ratio? Satterthwaite (1946) has shown that the linear combination of mean squares is distributed approximately as a χ^2 divided by its *df*, and since it can be shown that MS_W also has the χ^2 distribution and that numerator and denominator are independent, we have a statistic which is approximately distributed as *F*. The only remaining question is, What are the appropriate *df*? To answer, we define the *combination of mean squares*, $CMS = MS_A \pm MS_B \pm MS_C \pm \cdots$. Then,

$$
\textbf{(11.5)} \quad df_{CMS} = \frac{(CMS)^2}{\dfrac{(MS_A)^2}{df_A} + \dfrac{(MS_B)^2}{df_B} + \dfrac{(MS_C)^2}{df_C} + \cdots}
$$

rounded to the nearest integer.

In our target detection example, the *CMS* for testing the effects of *W* is $MS_{WT} + MS_{WS} - MS_{WTS}$. According to Equation (11.3), this quantity is distributed on

$$
\frac{(MS_{WT} + MS_{WS} - MS_{WTS})^2}{\dfrac{(MS_{WT})^2}{(w - 1)(t - 1)} + \dfrac{(MS_{WS})^2}{(w - 1)(n - 1)} + \dfrac{(MS_{WTS})^2}{(w - 1)(t - 1)(n - 1)}}
$$

df.

What is the basis for the *CMS* selected to test the *W* effects? Simply enough, we require an *EMS* which contains σ_{WS}^2 and one which contains σ_{WT}^2; furthermore, we attempt to avoid expectations which introduce components not in the numerator expectations. Finally, we subtract whichever expectation is needed to adjust the coefficients of our components. Since we have $2\sigma_e^2$ after adding the *WS* and *WT* terms, we must remove a σ_e^2; similarly, a σ_{WTS}^2 must be removed. The same approach can be applied to a variety of designs.

Note the difference between pooling and combining mean squares. When we pool, we compile a weighted average of several mean squares that are all assumed to have the same expectation. When we combine mean squares, we add and subtract terms that may have different expectations.

11.3 ESTIMATING COMPONENTS OF VARIANCE

Thus far the discussion of the analysis of variance has been concerned solely with tests of significance, with the question, Do the different treatment levels (or treatment combinations) have different effects? However, as noted in Chapter 2, there is an equally important question: How greatly do the effects of different treatment levels (or treatment combinations) differ? Indeed, it may be argued that the first question is the less important; that given sufficient data, the variance of any set of treatment effects is significant. Furthermore, since the psychologist invariably deals with data that are quantitative in nature, it is reasonable to assume that he will profit by examining the magnitude of the quantity. Finally, it is only through the estimation of population effects that the relative influences of several variables can be assessed. In view of the desirability of examining the absolute and relative magnitudes of effects, we now turn to the estimation of components of variance which comprise our expected mean squares. We will first deal with point estimation, then turn to the development of measures of the relative contribution of variables to the total variance, and finally to the development of confidence intervals for the variances of random effects.

11.3.1 Point estimation. Consider Table 11–4, which contains the analysis of data from a hierarchical experiment in which $a = 4$, $g = 5$, $b = 2$, $c = 5$, $n = 6$. Suppose that we are interested in estimating θ_A^2, the variance of the

TABLE 11–4 *Analysis of variance for a hierarchical design*

SV	df	MS	EMS	F
A	3	20.6	$\sigma_e^2 + 60\sigma_{G/A}^2 + 5\sigma_{S/GB/A}^2 + 300\theta_A^2$	4.6*
G/A	16	4.5	$\sigma_e^2 + 60\sigma_{G/A}^2 + 5\sigma_{S/GB/A}^2$	
B	1	15.2	$\sigma_e^2 + 5\sigma_{S/GB/A}^2 + 30\sigma_{GB/A}^2 + 600\theta_B^2$	2.9
AB	3	12.4	$\sigma_e^2 + 5\sigma_{S/GB/A}^2 + 30\sigma_{GB/A}^2 + 150\theta_{AB}^2$	2.4
GB/A	16	5.2	$\sigma_e^2 + 5\sigma_{S/GB/A}^2 + 30\sigma_{GB/A}^2$	
S/GB/A	200	3.0	$\sigma_e^2 + 5\sigma_{S/GB/A}^2$	
C	4	35.2	$\sigma_e^2 + 12\sigma_{GC/A}^2 + \sigma_{SC/GB/A}^2 + 240\theta_C^2$	3.6*
AC	12	7.6	$\sigma_e^2 + 12\sigma_{GC/A}^2 + \sigma_{SC/GB/A}^2 + 60\theta_{AC}^2$	.8
GC/A	64	9.8	$\sigma_e^2 + 12\sigma_{GC/A}^2 + \sigma_{SC/GB/A}^2$	
BC	4	4.1	$\sigma_e^2 + 6\sigma_{GBC/A}^2 + \sigma_{SC/GB/A}^2 + 120\theta_{BC}^2$	1.2
ABC	12	5.2	$\sigma_e^2 + 6\sigma_{GBC/A}^2 + \sigma_{SC/GB/A}^2 + 30\theta_{ABC}^2$	1.5
GBC/A	64	3.5	$\sigma_e^2 + 6\sigma_{GBC/A}^2 + \sigma_{SC/GB/A}^2$	
SC/GB/A	800	4.6	$\sigma_e^2 + \sigma_{SC/GB/A}^2$	

$*p < .025$

means of the four treatment populations defined by the levels of A. Referring to Table 11–4, we note that

$$E(MS_A) - E(MS_{G/A}) = (\sigma_e^2 + 60\sigma_{G/A}^2 + 5\sigma_{S/GB/A}^2 + 300\theta_A^2)$$
$$- (\sigma_e^2 + 60\sigma_{G/A}^2 + 5\sigma_{S/GB/A}^2)$$
$$= 300\theta_A^2$$

Then,

$$\frac{E(MS_A - MS_{G/A})}{300} = \theta_A^2$$

or, substituting numerical values from Table 11–7,

$$\hat{\theta}_A^2 = \frac{20.6 - 4.5}{300}$$

$$= .054$$

Note the use of $\hat{\ }$ to denote "estimate of."

Some of our estimates will turn out to be negative. For example, we find that

$$\hat{\theta}_{AC}^2 = \frac{MS_{AC} - MS_{GC/A}}{60}$$

$$= -.037$$

an unreasonable result, since variances cannot be negative. We conclude that our best estimate of θ_{AC}^2 is zero; the negative result is a chance happening.

What has been gained from the estimation process? Even the partial analysis just performed sheds some light beyond that provided by the F ratios. The significant A effect is less impressive now. $\hat{\theta}_A^2$ is quite small, and its significance seems to testify more to the amount of data collected than to any inherent differences in the effects of the four treatment levels. In contrast, note that $\hat{\theta}_C^2$ equals .106, and is two times as large as θ_A^2. This sort of summary of effects is not provided by the F ratio, which does not tell us whether the effect is large or small but only (with some probability) that it is or is not zero. Nor can F ratios be compared to provide knowledge of the relative effectiveness of variance sources unless the *df* are identical for the two ratios. Moreover, estimates of the magnitude of the effect are prerequisite to the establishment of quantitative behavioral laws.

11.3.2 Relative magnitude of effects. A measure that has intuitive appeal is the proportion of the total variance associated with an independent variable. We will use ω^2 to designate this relative magnitude of the effect of a variable. The computation of ω^2 depends upon the design, the assumed structural equation that relates data to population parameters, and whether effects are random or fixed. We will not compute ω^2 for all possible designs, but we will illustrate the principle involved in several different designs.

Completely randomized, one fixed factor. Assuming Equation (4.1), we have

$$Y_{ij} - \mu = \alpha_j + \varepsilon_{ij}$$

Squaring both sides and taking expectations over the set of treatment populations, we have

$$E(Y_{ij} - \mu)^2 = \frac{\sum \alpha_j^2}{a} + E(\varepsilon_{ij})^2$$

Note that the cross-product term vanishes due to the presumed independence of α and ε. The last equation may be rewritten as

(11.6)
$$\sigma_Y^2 = \left(\frac{a-1}{a}\right)\theta_A^2 + \sigma_e^2$$

$$= \delta_A^2 + \sigma_e^2$$

Note that whereas $\theta_A^2[(\sum \alpha_j^2)/(a-1)]$ is not truly the variance of the α_j, $\delta_A^2[(\sum \alpha_j^2)/a]$ is.

We define

(11.7)
$$\omega_A^2 = \frac{\delta_A^2}{\sigma_Y^2}$$

the proportion of the total population variance attributable to the effects of A.

Given the *EMS* for this design, point estimates are readily obtained:

$$\hat{\delta}_A^2 = \left(\frac{a-1}{a}\right)\hat{\theta}_A^2$$

(11.8)
$$= \left(\frac{a-1}{a}\right)\left(\frac{MS_A - MS_{S/A}}{n}\right)$$

$$\hat{\sigma}_e = MS_{S/A}$$

Substituting Equation (11.8) into Equation (11.7), and replacing population parameters by estimates*:

(11.9)
$$\hat{\omega}_A^2 = \frac{\left(\dfrac{a-1}{a}\right)\left(\dfrac{1}{n}\right)(MS_A - MS_{S/A})}{\left(\dfrac{a-1}{a}\right)\left(\dfrac{1}{n}\right)(MS_A - MS_{S/A}) + MS_{S/A}}$$

* If we consider ω_A^2 to be $E(\omega^2)$, Equation (11.9) provides an approximation since $E(\delta_A^2)/E(\sigma_e^2) \neq E(\omega^2)$; that is, the ratio of expected values is not generally equal to the expected value of a ratio. However, the approximation is reasonably accurate and considerably simpler than the correct expression.

Multiplying numerator and denominator by *an* and dividing by $MS_{S/A}$ yields

(11.10)
$$\hat{\omega}_A^2 = \frac{(a-1)(F_A-1)}{(a-1)(F_A-1)+na}$$

where

$$F_A = \frac{MS_A}{MS_{S/A}}$$

Completely randomized, two fixed factors. In this case, the structural equation (Equation (5.1)) leads to

(11.11)
$$\sigma_y^2 = \delta_A^2 + \delta_B^2 + \delta_{AB}^2 + \sigma_e^2$$

where

$$\delta_A^2 = \left(\frac{a-1}{a}\right)\theta_A^2 \qquad \delta_B^2 = \left(\frac{b-1}{b}\right)\theta_B^2 \qquad \delta_{AB}^2 = \frac{(a-1)(b-1)}{ab}\theta_{AB}^2$$

Then,

$$\hat{\delta}_A^2 = \left(\frac{a-1}{a}\right)\left(\frac{1}{bn}\right)(MS_A - MS_{S/AB})$$

(11.12)
$$\hat{\delta}_B^2 = \left(\frac{b-1}{b}\right)\left(\frac{1}{an}\right)(MS_B - MS_{S/AB})$$

$$\hat{\delta}_{AB}^2 = \frac{(a-1)(b-1)}{ab}\left(\frac{1}{n}\right)(MS_{AB} - MS_{S/AB})$$

We can now obtain

$$\hat{\omega}_A^2 = \frac{\hat{\delta}_A^2}{\delta_y^2}$$

$$= \left[\left(\frac{a-1}{a}\right)\left(\frac{1}{bn}\right)(MS_A - MS_{S/AB})\right]$$

$$\times \left[\frac{1}{\left(\frac{a-1}{a}\right)\left(\frac{1}{bn}\right)(MS_A - MS_{S/AB}) + \left(\frac{b-1}{b}\right)\left(\frac{1}{an}\right)(MS_B - MS_{S/AB})}\right.$$

$$\left. \phantom{\frac{1}{X}} + \left(\frac{a-1}{a}\right)\left(\frac{b-1}{b}\right)\left(\frac{1}{n}\right)(MS_{AB} - MS_{S/AB}) + MS_{S/AB}\right]$$

Multiplying numerator and denominator by *abn* and dividing by $MS_{S/AB}$ yields

(11.13)
$$\hat{\omega}_A^2 = \frac{(a-1)(F_A-1)}{(a-1)(F_A-1)+(b-1)(F_B-1)+(a-1)(b-1)(F_{AB}-1)+abn}$$

The equations for ω_B^2 and $\hat{\omega}_{AB}^2$ follow directly. They have the same denominator as $\hat{\omega}_A^2$; the numerators are $(b - 1)(F_B - 1)$ and $(a - 1)(b - 1) \times (F_{AB} - 1)$, respectively.

Subject × treatments. If we assume additivity, $\sigma_{SA}^2 = 0$, we have

(11.14) $$\sigma_y^2 = \delta_A^2 + \sigma_S^2 + \sigma_e^2$$

We find the point estimate of δ_A^2 to be

$$\hat{\delta}_A^2 = \left(\frac{a - 1}{a}\right)\hat{\theta}_A^2 = \left(\frac{a - 1}{a}\right)\left(\frac{1}{n}\right)(MS_A - MS_{SA})$$

and for σ_S^2 we have

$$\hat{\sigma}_S^2 = \left(\frac{1}{a}\right)(MS_S - MS_{SA})$$

and

$$\hat{\sigma}_e^2 = MS_{SA}$$

We have

$$\hat{\omega}_A^2 = \frac{\hat{\delta}_A^2}{\hat{\sigma}_y^2}$$

and substituting into this on the basis of Equation (11.14) and the preceding point estimates gives

(11.15) $$\hat{\omega}_A^2 = \frac{(a - 1)(F_A - 1)}{(a - 1)(F_A - 1) + nF_S + n(a - 1)}$$

Note that the introduction of subjects, a random-effects variable, results in an expression whose form differs slightly from that of previous designs.

The relative contribution of individual difference is also readily calculated:

(11.16) $$\hat{\omega}_S^2 = \frac{nF_S}{(a - 1)(F_A - 1) + nF_S + n(a - 1)}$$

Equations (11.15) and (11.16) are appropriate only if an additive model is valid. If subjects and treatments interact, we cannot find an appropriate estimate of σ_S^2 and are therefore unable to derive expressions for $\hat{\omega}_A^2$ and $\hat{\omega}_S^2$. Nonadditivity does not rule out the possibility of obtaining estimates of ω^2 in all designs. In each new case, the investigator must determine whether he can obtain an estimate of the variance of the effect of interest and whether σ_y^2 or, equivalently, its component variances, can be estimated. The estimates are then plugged into the equation defining $\hat{\omega}^2$.

Generally, if estimates of ω^2 are attainable, they are not difficult to derive. Certainly, the information gained is usually worth the effort. The significance level of an F statistic reflects numerator and denominator df; given sufficient df, any source of variance will prove significant. Direct estimates of the variance of effects and of its contribution to the total population variance will frequently give a truer picture of the importance of a variable, providing a frame of reference within which to interpret significant and nonsignificant results. They also provide a way of judging the relative importance of several variables even when these variables are tested on different *df* and/or against different error terms. Our measure of relative effect, $\hat{\omega}^2$, also permits comparisons across experiments, providing a securer base than relative magnitudes of *F* ratios for comparing the findings of several investigators. Finally, ω^2 has a special application in psychometric research. We will turn now to this topic.

11.3.3 Reliability and analysis of variance. Suppose that we have several judges rate several individuals with respect to some personality trait. Perhaps we require these ratings because we wish to correlate the personality measure with some performance measure, or perhaps we wish to select subjects having various degrees of the trait for a subsequent experiment. If the judges are in close agreement with respect to their ratings of each subject, we feel reasonably confident that we have a good estimate of where each subject stands with respect to the trait being measured. If the judges vary greatly in their ratings of each subject, we are less confident in our assessment of each individual with regard to the trait in question. True differences among individuals may be obscured by the variability in our measuring instrument, the set of judges.

The same problem of the degree of internal consistency in a measuring instrument arises in many other circumstances. We may be concerned with developing a test that discriminates among the arithmetic abilities of individuals; we want the items to correlate well with each other. We wish to choose a set of pictures that represent different points along a scale of attractiveness; they will be used later in an experiment on aesthetic judgement of different personality types. Here, we are again interested in the degree of consistency among the judges rating the pictures.

The problem, a common one, is the assessment of the consistency of measurements (items, ratings) of individual subjects or stimuli in the hope of establishing an internally consistent measuring instrument and, thus, clearly discriminating among the individuals or objects to be measured. One way to do this is to compute correlation coefficients for all pairs of items and examine the average intercorrelation. This is a laborious procedure. The analysis of variance can provide equivalent information in the form of a reliability (internal consistency) coefficient, and with much less effort.

We assume *n* subjects (or, in general, objects to be measured) drawn at random from an infinite population of individuals and a set of *a* items (or judges or other measuring instruments) drawn at random from an infinite population of items. Then the score for subject *i* on item *j* may be represented as

(11.17) $$Y_{ij} = \mu_i + \varepsilon_{ij}$$

We may think of μ_i as the "true" score for the individual, his expected score over the population of items (judges, etc.) from which we have sampled. ε_{ij} is the error of measurement associated with the ijth observation. The error of measurement is due to chance variability in performance; the subject guesses correctly on one item and incorrectly on another; he is more alert or better motivated at one moment than at another. Thus, one subject scores higher on one member of a pair of items while another subject scores higher on the other member of the pair; the variability is not a reflection of true differences in ability of the subjects nor of true differences in the difficulty of the items. Indeed, the model presented does not envisage any systematic differences among items; there is no α_j component associated with item j, raising or lowering all scores on that item by a constant amount. Thus, this model is invalid if, for example, some items are intrinsically more difficult than others or if some judges consistently rate higher than others.

We may subtract μ, the grand mean of the subject-item population, from both sides of Equation (11.17). Defining, as usual, $\eta_i = \mu_i - \mu$, we have

(11.17′)
$$(Y_{ij} - \mu) = \eta_i + \varepsilon_{ij}$$

We assume that the ε_{ij} are independently and normally distributed with variance σ_e^2 and that the η_i are independently and normally distributed with variance σ_S^2. Squaring both sides of Equation (11.17′) and taking expectations over the population of subjects and items, we have

(11.18)
$$\sigma_y^2 = \sigma_S^2 + \sigma_e^2$$

A reasonable definition of reliability is that it is the proportion of total variance attributable to the individuals or objects being measured. That is,

(11.19)
$$R_{11} = \frac{\sigma_S^2}{\sigma_y^2}$$

where R_{11} indicates the reliability coefficient. Note that R_{11} is merely a special instance of $\hat{\omega}^2$. Consequently, we derive a computational formula for it in the same way. Our model suggests the partitioning of variance displayed in Table 11–5.

TABLE 11–5 *Partitioning of variance for a measure of* R_{11}

SV	df	EMS
S	$n - 1$	$\sigma_e^2 + a\sigma_S^2$
Within S	$n(a - 1)$	σ_e^2

Our variance estimates are

(11.20)
$$\hat{\sigma}_S^2 = \frac{MS_{B.S} - MS_{W.S}}{a}$$

and

(11.21)
$$\hat{\sigma}_e^2 = MS_{W.S}$$

On the basis of Equation (11.19), it follows that

(11.22)
$$R_{11} = \frac{\hat{\sigma}_S^2}{\hat{\sigma}_S^2 + \hat{\sigma}_e^2}$$

and substituting from Equations (11.20) and (11.21), we have

(11.23)
$$R_{11} = \frac{(MS_{B.S} - MS_{W.S})/a}{(MS_{B.S} - MS_{W.S})/a + MS_{W.S}}$$

$$= \frac{F - 1}{F - 1 + a}$$

The reliability coefficient defined by Equation (11.23) may be thought of as a measure of the degree to which the a items are measuring the same trait. More specifically, R_{11} can be shown to be an average of the $[a(a - 1)]/2$ intercorrelations which can be computed among items when the average within-items variance is used as the denominator for all correlations.

Suppose that we wish to assess how well another set of a items, drawn from the same item population, will correlate with the present set. How reliable is the subject's average score, $\overline{Y}_i$?

Returning to Equation (11.17), we sum over j and divide by a to obtain

(11.24)
$$\overline{Y}_{i.} = \mu_i + \bar{\varepsilon}_{i.}$$

Since μ_i is assumed to be independent of ε_{ij}, it can be shown that

(11.25)
$$\sigma_{\overline{Y}_{i.}}^2 = \sigma_S^2 + \sigma_{\bar{\varepsilon}_{i.}}^2$$

The quantity $\sigma_{\bar{\varepsilon}_{i.}}^2$ is the expectation over an infinite number of replications of the experiment of $(\sum_j \varepsilon_{ij}/a)^2$. Since the ε_{ij} are independently distributed, $E(\varepsilon_{ij}\varepsilon_{ij'}) = 0$. Consequently,

$$E\left(\frac{\sum_j \varepsilon_{ij}}{a}\right)^2 = E\left(\frac{\sum_j \varepsilon_{ij}^2}{a^2}\right) = \frac{1}{a^2}\sum_j E(\varepsilon_{ij}^2) = \frac{1}{a^2}\sum_j \sigma_{\varepsilon_{ij}}^2$$

Since we assume homogeneity of variance (i.e., $\sigma_{\varepsilon_{ij}}^2 = \sigma_{\varepsilon_{ij'}}^2 = \sigma_e^2$),

$$E\left(\frac{\sum_j \varepsilon_{ij}}{a}\right)^2 = \frac{\sigma_e^2}{a}$$

Then, Equation (11.25) becomes

(11.26)
$$\sigma_{\overline{Y}_{i.}}^2 = \sigma_S^2 + \frac{\sigma_e^2}{a}$$

Then the reliability of the mean, R_{aa}, is

(11.27)
$$R_{aa} = \frac{\hat{\sigma}_S^2}{\hat{\sigma}_S^2 + \hat{\sigma}_e^2/a}$$

Substituting from the *EMS* column of Table 11–9, we have

(11.28)
$$R_{aa} = \frac{(MS_{B.S} - MS_{W.S})/a}{(MS_{B.S} - MS_{W.S})/a + MS_{W.S}/a}$$

$$= 1 - \frac{1}{F}.$$

We may interpret R_{aa} as a measure of the correlation between the mean scores for two sets of a randomly sampled items administered at different times to the same sample of subjects. If R_{aa} were high, we could expect that a second sample of a items would provide a similar inference about the relative performance of our subjects.

Equation (11.17) implies that any differences among items are chance happenings. We have pooled the A and SA sources in the usual repeated measurements design (see Chapter 7) to give a single estimate of error. If there is reason to believe that systematic differences among items exist, the model represented by Equation (7.1) is more appropriate. This might be the case if the items had been intentionally selected at different levels of difficulty. As a second example, subjects might be rated by a raters, each of whom may be suspected to have a somewhat different set of standards. In any event, Equation (7.1), which is restated here, is now the basis for our reliability measure:

$$Y_{ij} = \mu + \eta_i + \alpha_j + \varepsilon_{ij}$$

The appropriate analysis of variance table is now Table 7–2; the estimate of error is $\hat{\sigma}_e^2 = MS_{SA}$. Equation (11.16) provides the appropriate formula for R_{11}.

The approach sketched in this section may be extended to a variety of problems. We might actually administer parallel forms of a test at different times; that is, two sets of items randomly drawn from the same item population. If our concern is whether or not the two forms are consistently measuring the same attributes, we can assess the proportion of total variance associated with the *forms* main effect. Alternatively, we may be interested in test-retest reliability: Does a measuring instrument give consistent results over time? Again, the analysis of variance provides an answer. In fact, a well designed psychometric study could permit us to separate the relative contributions of error variance due to inconsistency among items from inconsistency due to variability over time or forms of a test. We require a statement of the model. Then the general approach of Section 11.3.2 is applied.

11.3.4 Interval estimation. As we pointed out in Chapter 3, we require not only an estimate of the population variance, but some index of its reliability as well. Confidence intervals provide such indices. In this section, we will consider confidence intervals for population variances of *random* effects and for the reliability coefficient, R_{11}. We assume that the effects are normally distributed. The assumption is critical here, unlike the case in which our sole concern is testing H_0.

We will begin by obtaining confidence intervals for σ^2, the variance of a single population from which a sample of n scores has been obtained; the variance of the sample is denoted by S^2. If the scores are normally distributed, it can be shown that $(n-1)S^2/\sigma^2$ is distributed as χ^2 on $n-1$ df. That is, if samples of size n are repeatedly taken from a population with variance σ^2 and if the frequency with which various values of $(n-1)S^2/\sigma^2$ appear is noted, the frequency distribution would approximate the χ^2 on $n-1$ df. Therefore, we may assert that

$$(11.29) \qquad P\left[\chi^2_{1-\alpha,n-1} \leq \frac{(n-1)S^2}{\sigma^2} \leq \chi^2_{\alpha,n-1}\right] = 1 - 2\alpha$$

where χ^2_α is that value of χ^2 on $n-1$ df exceeded by 100α percent of the population of χ^2 values. For example, suppose that there is a sample of size 5 and a 90-percent confidence interval is desired. Then, $1 - 2\alpha = .90$ and $\alpha = .05$. We turn to Table A–4 in the Appendix and find that the χ^2 on 4 df required for significance at the .05 level is $\chi^2_{.05,4} = 9.49$; the χ^2 exceeded by 95 percent of the population is $\chi^2_{.95,4} = .711$. Then, substituting in Equation (11.29),

$$(11.30) \qquad P\left(.711 \leq \frac{4S^2}{\sigma^2} \leq 9.49\right) = .90$$

At this point, we have a confidence interval containing $(n-1)S^2/\sigma^2$; what we want is a confidence interval containing σ^2. A little algebraic manipulation is required. We note that if $X < Y$, then $1/Y < 1/X$. Applying this knowledge to Equation (11.30), we have

$$(11.31) \qquad P\left(\frac{1}{11.1} \leq \frac{\sigma^2}{4S^2} \leq \frac{1}{.711}\right) = .90$$

Multiplying through by $4S^2$ provides the final result:

$$(11.32) \qquad P\left(\frac{4S^2}{11.1} \leq \sigma^2 \leq \frac{4S^2}{.711}\right) = .90$$

and, more generally,

$$(11.33) \qquad P\left[\frac{(n-1)S^2}{\chi^2_{\alpha,n-1}} \leq \sigma^2 \leq \frac{(n-1)S^2}{\chi^2_{1-\alpha,n-1}}\right] = 1 - 2\alpha$$

Equation (11.33) can be generalized to provide confidence intervals for any *EMS* involving random-effect variables. For example, consider Table 11–5 again. The $1 - 2\alpha$ confidence limits for $\sigma_e^2 + a\sigma_S^2$

(11.34)
$$L_l = \frac{(n - 1)MS_S}{\chi_{\alpha,n-1}^2} \text{ (lower limit)}$$

(11.34′)
$$L_u = \frac{(n - 1)MS_S}{\chi_{1-\alpha,n-1}^2} \text{ (upper limit)}$$

For σ_e^2, we have the limits

(11.35)
$$L_l = \frac{n(a - 1)MS_{W.S}}{\chi_{\alpha,n(a-1)}^2}$$

and

(11.35′)
$$L_u = \frac{n(a - 1)MS_{W.S}}{\chi_{1-\alpha,(a-1)}^2}$$

Confidence intervals for other expected mean squares follow the same pattern.

Thus far the development of confidence intervals has followed exactly the discussion of Chapter 3. Because of the difficulty of specifying the distribution of estimates, the problem becomes somewhat more complicated when one tries to develop limits on a single component such as σ_S^2. Numerous papers have been written presenting approximate intervals (i.e., intervals for which the probability is approximately $1 - 2\alpha$); one solution by Scheffé (1959, pp. 231-235) will be considered that provides a reasonable approximation.

Scheffé's derivation, while not mathematically too demanding, is somewhat lengthy and therefore will be omitted. The upper and lower confidence lists are given by

(11.36)
$$L_u = F_{\alpha,\infty,n_1} F - 1 + \frac{1}{F_{\alpha,n_2,n_1} F}\left(1 - \frac{F_{\alpha,n_2,n_1}}{F_{\alpha,\infty,n_1}}\right)$$

and

(11.37)
$$L_l = \frac{F}{F_{\alpha,n_1,\infty}} - 1 - \frac{F_{\alpha,n_1,n_2}}{F}\left(\frac{F_{\alpha,n_1,n_2}}{F_{\alpha,n_1,\infty}} - 1\right)$$

where F is the computed mean square ratio, n_1 and n_2 are the numerator and denominator *df*, respectively, and $F_{\alpha,x,y}$ is the F required for significance at the α level on x (column) and y (row) *df*. Note that the formula for L_u involves seeking the column in Table A–5 corresponding to the denominator *df* and the row corresponding to the numerator *df*, contrary to our usual procedure. This is because the solution for the upper limit involves finding the F significant at the $1 - \alpha$ level; e.g., the .95 level. Such values are not tabled. Fortunately, however,

(11.38)
$$F_{1-\alpha,n_1,n_2} = \frac{1}{F_{\alpha,n_2,n_1}}$$

L_u is taken as zero if $F \leq 1/F_{\alpha,n_2,n_1}$ and L_l is taken as zero if $F < F_{\alpha,n_1,n_2}$.

We next consider the problem of developing confidence limits for ρ_{11}, the population parameter estimated by R_{11}. We assume that the individual difference component, η_i, is a random normally distributed variable with variance σ_S^2. We take Equation (11.17′) as the structural model.

From the developments of Section 3.4.2, we know that $MS_S/(\sigma_e^2 + a\sigma_S^2)$ is distributed as $\chi^2/(n-1)$ and $MS_{W.S}/\sigma_e^2$ is distributed as $\chi^2/n(a-1)$ if ε_{ij} and η_i are normally distributed with variance σ_e^2 and σ_S^2 respectively; $n-1$ and $n(a-1)$ are *df* of Table 11–4. Furthermore, the ratio

$$\frac{MS_S/(\sigma_e^2 + a\sigma_S^2)}{MS_{W.S}/\sigma_e^2}$$

is distributed as *F*. We may rewrite this ratio as

$$\frac{MS_S/MS_{W.S}}{(\sigma_e^2 + a\sigma_S^2)/\sigma_e^2} = \frac{F}{F_{pop}}$$

That is, we have the ratio of the *F* computed from the data to an *F* based on the population from which we are sampling. Then

$$P_r\left(F_{1-\alpha,n-1,n(a-1)} < \frac{F}{F_{pop}} < F_{\alpha,n-1,n(a-1)}\right) = 1 - 2\alpha$$

and

(11.39) $$P_r\left(F_{1-\alpha,n-1,n(a-1)}/F < \frac{1}{F_{pop}} < F_{\alpha,n-1,n(a-1)}/F\right) = 1 - 2\alpha$$

We wish to relate F_{pop} and ρ_{11}. Note that

$$F_{pop} = 1 + a(\sigma_S^2/\sigma_e^2)$$

and

$$\rho_{11} = \frac{\sigma_S^2}{\sigma_e^2 + \sigma_S^2}$$

$$= \frac{1}{1 + (\sigma_e^2/\sigma_S^2)}$$

Some tedious algebra gives

(11.40)
$$\frac{1}{F_{pop}} = \frac{1 - \rho}{a\rho + 1 - \rho}$$

Then

(11.41) $\quad P_r\left(F_{1-\alpha,n-1,n(a-1)}/F < \dfrac{1-\rho}{a\rho+1-\rho} < F_{\alpha,n-1,n(a-1)}/F\right) = 1 - 2\alpha$

Algebraic manipulation of the two inequalities involving F and ρ yields

(11.42)

$$P_r\left(\frac{1 - F_{\alpha,n-1,n(a-1)}/F}{1 + (F_{\alpha,n-1,n(a-1)}/F)(a-1)} < \rho < \frac{1 - F_{1-\alpha,n-1,n(a-1)}/F}{1 + (F_{\alpha,n-1,n(a-1)}/F)(a-1)}\right)$$
$$= 1 - 2\alpha$$

Multiplying numerator and denominator of the limiting expressions by F/F_α (lower limit) and $F/F_{1-\alpha}$ (upper limit) and substituting on the basis of Equation (11.38) gives the final result:

(11.43)

$$P_r\left(\frac{F/F_{\alpha,n-1,n(a-1)} - 1}{F/F_{\alpha,n-1,n(a-1)} + a - 1} < \rho < \frac{F/F_{\alpha,n(a-1),n-1} - 1}{F/F_{\alpha,n(a-1),n-1} + a - 1}\right) = 1 - 2\alpha$$

We turn now to a numerical example of point and interval estimation of ρ_{11}. Consider Table 11–6, which assumes that seven subjects have each been

TABLE 11–6 *Analysis for a reliability example*

SV	df	MS	EMS	F
S	6	32	$\sigma_e^2 + 5\sigma_S^2$	16
Within S	28	2	σ_e^2	

rated by five judges. The implied model is Equation (11.17′). Applying Equation (11.23), we have

$$R = \frac{16 - 1}{16 - 1 + 5}$$
$$= .75$$

In order to obtain confidence limits, we require $F_{\alpha,n-1,n(a-1)}$ and $F_{\alpha,n(a-1),n-1}$; the values are 2.44 and 3.88 (the latter by linear interpolation) respectively, for a 90-percent interval. Then

$$L_l = \frac{(16/2.44) - 1}{(16/2.44) - 1 + 5}$$
$$= \frac{5.56}{10.56}$$
$$= .53$$

$$L_u = \frac{(16)(3.88) - 1}{(16)(3.88) - 1 + 5}$$

$$= \frac{61.08}{66.08}$$

$$= .92$$

EXERCISES

11.1 Discuss in detail the implications of the expected mean square (*EMS*) for the design and analysis of experiments.

11.2 There are three levels of *A* (random variable), four levels of *B* (random), five levels of *C* (fixed), and ten scores in each cell.

SV	df	MS
A	2	3.8
B	3	14.0
C	4	10.0
AB	6	6.0
AC	8	3.6
BC	12	4.5
ABC	24	3.5
Within cells	540	2.0

Pool whenever possible. Provide point estimates for the remaining population variance components.

11.3 Eight rats are run for 25 trials on each of 12 days in a *T* maze. On each day, the subject experiences a different combination of shock intensity (*Sh*) for errors and amount of reward (*R*) for correct responses. There are three levels of *Sh* and four levels of *R*. The 12 combinations are randomly sequenced over days. The experimenter wants to know which is more important: variation in the levels of *Sh* or variation in the levels of *R*.

(a) Should he compare the relative magnitudes of mean squares for *Sh* and *R*? If not, why not?

(b) Should he compare the relative *F* ratios for *Sh* and *R*? If not, why not?

(c) If you reject (a) and (b), describe an alternative procedure that meets your objection.

11.4 Assume a single $a \times a$ Latin square; the additive model is appropriate. Derive the theoretical expression for ω_A^2.

REFERENCES

Bozivich, H., Bancroft, T. A., and Hartley, H. O., "Power of analysis of variance test procedures for certain incompletely specified models," *Annals of Mathematical Statistics*, 27:1017–1043 (1956).

Satterthwaite, F. E., "An approximate distribution of estimates of variance components," *Biometrics Bulletin*, 2: 110–114 (1946).

Scheffé, H. *The Analysis of Variance*. New York: Wiley & Sons, 1959.

Srivastava, S. R, and Bozivich, H., "Power of certain analysis of variance test procedures involving preliminary tests," *Bulletin de l'Institute International Statistique*, 33rd Session (1961).

SUPPLEMENTARY READINGS

The discussion of *EMS* in this and other chapters has been limited to the case in which random effects have been sampled from infinitely large populations. This will usually be a reasonable assumption in psychological research. In those rare instances in which the levels of a variable are a random sample from a finite number of levels, a more general set of rules for establishing *EMS* will be required. Such rules appear in:

Cornfield, J. and Tukey, J. W., "Average values of mean squares in factorials," *Annals of Mathematical Statistics*, 27: 907–949 (1956).

Our brief presentation of the role of analysis of variance in estimating reliability is considerably supplemented by the following book:

Haggard, E. A. *Intraclass Correlation with the Analysis of Variance*. New York: Dryden Press, 1958.

Reliability estimation in a variety of designs is also considered in Chapter 16 of:

Lindquist, E. F. *Design and Analysis of Experiments in Psychology and Education*. Boston: Houghton Mifflin, 1953.

Analysis of Covariance

12.1 INTRODUCTION

As has been indicated, much of the error in experimentation may be traced to those characteristics of individual subjects that correlate highly with the dependent variable. For example, evaluation of the effects of different instructions on concept formation performance is made more difficult by the relationship between this performance and the intelligence of the subject. Variability in intelligence among subjects increases variability in performance within groups. Furthermore, if the average intelligence is higher for some groups than for others, the effects of the independent variable may be either obscured or spuriously enhanced.

Several experimental designs have been presented, each of which provides an alternative approach to this problem of experimental error. In this chapter we take a different tack: we consider a statistical adjustment of each dependent measure (e.g., number of errors in a concept formation task) for the contribution to that measure of a concomitant variable (e.g., intelligence test score). The technique that will be examined is the analysis of covariance.

This approach assumes that some portion of the usual error component is predictable if we have knowledge of the individual's score on some related measure, X, the covariate. Removal of the variance of these predicted error components leaves a smaller error variance and thus a more efficient, more powerful test of treatment effects. The simplest set of computations, and therefore the most frequently used, implies that Y is a linear function of X. In the following discussion, we will assume a knowledge of linear regression and correlation. At a minimum, the reader should be familiar with the equation for a straight line, the concept of linear regression, and the least-squares expression for the regression, or slope, coefficient. If he is not, most introductory statistics texts will provide a quick review.

12.2 THE COMPLETELY RANDOMIZED ONE-FACTOR DESIGN

12.2.1 The covariance model. In Chapter 4, the following structural equation was presented for the simple one-factor design:

$$(12.1) \qquad\qquad Y_{ij} = \mu + \alpha_j + \varepsilon_{ij}$$

We are used to thinking of ε_{ij} as the sum of two components: error due to variation in measurements and error due to individual differences. Individuals differ on many dimensions that may contribute to the difference in their scores. The analysis of covariance provides a way of adjusting for differences on one or more of those dimensions. We assume

$$(12.2) \qquad\qquad Y_{ij} = \mu + \alpha_j + \varepsilon'_{ij} + \beta(X_{ij} - \overline{X}_{..})$$

where β is the regression coefficient that characterizes the rate of change of Y with respect to X in the population sampled; $\beta(X_{ij} - \overline{X}_{..})$ is therefore the error in Y that would be predicted on the basis of knowledge of X, and ε_{ij} is the residual error unaccounted for by X. For example, an individual may differ from the treatment population average because he differs from other individuals in intelligence, in motivation, in amount of previous experience with related tasks, or in any one of a number of other factors. Each of these factors contributes a positive or negative component to the individual's score, Y_{ij}; the sum of these components plus errors in measurements is ε_{ij}. If we have an intelligence test score for the individual, we may estimate that component of the individual's total deviation from μ_j due to intelligence. Removal of components of error predicted on the basis of knowledge of the covariate reduces the variance of the residual error component. Since

$$\varepsilon_{ij} = \varepsilon'_{ij} + \beta(X_{ij} - \overline{X}_{..})$$

and assuming that the predicted and residual components are independent,

$$\sigma_e^2 = \sigma_{e'}^2 + \sigma_{\text{predicted}}^2$$

which demonstrates that $\sigma_{e'}^2$, the residual error variance remaining after adjustment, is less than σ_e^2, the error variance estimated under the usual analysis of variance procedure of Chapter 4. Thus, efficiency may be increased through the covariance procedure.

If the ratio of mean squares computed in the analysis of covariance procedure is to be distributed as F, we require assumptions analogous to those employed in the usual analysis of variance. Specifically, we assume that for any value of the covariate, X_{ij}, within each treatment population, there exists a population of independently and normally distributed scores with variance, $\sigma_{e'}^2$. In addition, we assume that within each treatment population, Y varies as a linear function of X and that the a regression lines characterizing the treatment populations have the same slopes. Finally, we assume that the treatment has no effect upon the covariate. With respect to the last assumption, it is generally wise to measure the covariate *prior* to applying the treatments, although post-experimental attempts to increase efficiency by measuring a covariate are valid if there are logical grounds for concluding that the $\overline{X}_j$ were unaffected by the treatment and if an analysis of variance on the covariate supports this conclusion.

To review briefly, we assume that the adjusted error components, ε'_{ij} (sometimes referred to as the residuals) are distributed within each treatment population:

(a) Independently
(b) Normally
(c) With mean zero and homogeneous variances, $\sigma^2_{e'}$.

Furthermore, we assume:

(d) Linear regression of Y on X within each treatment population
(e) Homogeneous regression coefficients; that is, $\beta_1 = \beta_2 = \cdots = \beta_{j\ldots} = \cdots \beta_a$
(f) The treatment has no effect upon the covariate.

Consequences of violating assumptions. Relative to the analysis of variance, little is known about the effects of violations of assumptions on inferences based on the covariance procedure. However, there is some evidence (Atiqullah, 1964) that (a) nonnormality in Y has little effect if X is normally distributed; (b) nonlinearity results in biased estimates of treatment effects, the magnitude of the bias depending upon the true form of the function relating X and Y, being least severe when subjects are randomly assigned to groups and the dependent variable is normally distributed; and (c) heterogeneity of the β_j results in a conservative F test (i.e., loss of power), at least when $a = 2$.

Psychologists most frequently violate the assumption that the treatment has no effect on the covariate. Indeed, covariance is frequently used not to increase efficiency, but to adjust the treatment effects for differences among groups on a covariate. Generally, such an approach will cause difficulties in interpreting the results of the data analysis. For example, three methods of teaching arithmetic are compared; Y is a performance measure and X is amount of study time. Method 1 is significantly superior to the other two when an analysis of variance test is carried out. However, after adjustment for study time, the effect due to methods is no longer significant. The experimenter concludes that the originally obtained difference in performance was due to differences in study time, and that when study time is held constant, the three methods are equally effective. This interpretation is not necessarily correct. Variation in performance may not be due to variation in study time, but, instead, variability in both measures may be due to a third factor; e.g., differences in motivation resulting from the three methods. If this were the case, and if study time were actually held constant (all subjects were required to study for a set time period), methods might have a significant effect, since the degree of motivation is still free to vary. Another reason why the experimental control of study time may have different results than the statistical control is that the experimental environment has changed. For example, subjects instructed to study for a given amount of time may have a different set from subjects in the covariance experiment, who were given no particular instructions with regard to study time. The point is that *statistically* adjusting for study time is not the same as *experimentally* holding study time constant.

Furthermore, suppose that our interest lies solely in deciding which method is best. How this method achieves its success is irrelevant. In this case, it is of no interest to adjust the mean performance for variability among the mean study times. On the other hand, suppose the interest lies in the influence of study time upon performance. In this case, study time should be systematically manipulated as an independent variable in a factorial design.

12.2.2 Partitioning the adjusted sum of squares. Equation (12.2), which defines Y in terms of population parameters, provides a starting point for our computations. Let $y_{ij} = Y_{ij} - \overline{Y}_{..}$ and $x_{ij} = X_{ij} - \overline{X}_{...}$. Then, the deviation of the adjusted score from the grand mean of the Y values ($\overline{Y}_{.}$) may be expressed as

$$(12.3) \qquad Y_{ij} - b_{\text{tot}}(X_{ij} - \overline{X}_{.}) - \overline{Y}_{..} = y_{ij} - b_{\text{tot}}x_{ij}$$

Since we are concerned with the actual data analysis, we have replaced β of Equation (12.2) with a least-squares estimate based upon the total set of *an* scores. This estimate, b_{tot}, is computed as

$$(12.4) \qquad b_{\text{tot}} = \frac{\sum_j \sum_j x_{ij} y_{ij}}{\sum_i \sum_j x_{ij}^2} = \frac{SP_{\text{tot}}}{SS_{\text{tot}(x)}}$$

Note that the regression coefficient involves (a) a sum of cross-products of deviations about $\overline{X}_{..}$ and $\overline{Y}_{.}$ (SP_{tot}) and (b) the total sum of squares with X as the variable being measured ($SS_{\text{tot}(x)}$). Squaring and summing the right-hand side of Equation (12.3) gives the $SS_{\text{tot}(adj)}$. Our present problem is to determine how this adjusted total sum of squares is to be partitioned to provide the components of our F ratio. The following identity is a step towards the desired partitioning of the $SS_{\text{tot}(adj)}$:

$$(12.5) \qquad y_{ij} - b_{\text{tot}}x_{ij} = (b_j - b_{S/A})(x_{ij} - \bar{x}_{.j}) + [(y_{ij} - \bar{y}_{.j})$$
$$- b_j(x_{ij} - \bar{x}_{.j})] + (\bar{y}_{ij} - b_A\bar{x}_{.j})$$
$$+ [(b_A - b_{S/A})\bar{x}_{.j} - (b_{\text{tot}} - b_{S/A})x_{ij}]$$

where

$$\bar{x}_{.j} = \overline{X}_{.j} - \overline{X}_{..}$$
$$\bar{y}_{.j} = \overline{Y}_{.j} - \overline{Y}_{..}$$
$$b_j = \text{regression coefficient for the best fitting line for } A_j$$

(12.6) $\qquad = \dfrac{\sum_i (x_{ij} - \bar{x}_{.j})(y_{ij} - \bar{y}_{.j})}{\sum_i (x_{ij} - \bar{x}_{.j})^2} = \dfrac{SP_j}{SS_{j(x)}}$

$b_{S/A}$ = a within group regression coefficient based on the pool of the b_j

(12.7) $\qquad = \dfrac{\sum_j SP_j}{\sum_j SS_{j(x)}}$

b_A = regression coefficient for the line that best fits the a pairs of group means

(12.8) $\qquad = \dfrac{n\sum_j \bar{x}_{.j}\,\bar{y}_{.j}}{n\sum_j \bar{x}_{.j}^2} = \dfrac{SP_A}{SS_{A(x)}}$

Squaring both sides of Equation (12.5) and summing yields:

(12.9) $\qquad \sum_j \sum_i (y_{ij} - b_{tot}x_{ij})^2 = \sum_j \left[(b_j - b_{S/A}) \sum_i (x_{ij} - \bar{x}_{.j}) \right]$

$\qquad\quad SS_{tot(adj)} \qquad\qquad = \qquad\qquad SS_1$

$\qquad\qquad\qquad\qquad\qquad + \sum_j \sum_i \left[(y_{ij} - \bar{y}_{.j}) - b_j(x_{ij} - \bar{x}_{.j}) \right]^2$

$\qquad\qquad\qquad\qquad\qquad\qquad SS_2$

$\qquad\qquad\qquad\qquad\qquad + n \sum_j (\bar{y}_{.j} - b_A \bar{X}_{.j})^2$

$\qquad\qquad\qquad\qquad\qquad\qquad SS_3$

$\qquad\qquad\qquad\qquad\qquad + \sum_j \sum_i \left[(b_A - b_{S/A})\bar{x}_{.j} \right.$

$\qquad\qquad\qquad\qquad\qquad\qquad\qquad\left. - (b_{tot} - b_{S/A})x_{ij} \right]^2$

$\qquad\qquad\qquad\qquad\qquad\qquad SS_4$

SS_1 is a measure of the variability of the b_j defined by Equation (12.6). If the slopes were identical for all a groups, all the b_j would equal their average, $b_{S/A}$, and SS_1 would equal zero. Obviously, the group regression coefficients will vary to some extent due to sampling error; the question is whether they vary sufficiently to indicate that the β_j, the regression coefficients for the treatment populations, actually differ. In order to answer this question, we require a measure of error variance. Such a measure is provided by SS_2. This term reduces to a sum of squared deviations of scores (values of Y) about the best fitting straight line for their group. In essence, this is a within-group sum of squares, analogous to $SS_{S/A}$ in the usual analysis of variance; the difference is that we are now taking deviations from the best-fitting straight line rather than from the mean for each group.

The quantities SS_1 and SS_2 provide a basis for testing the assumption that the best fitting regression line has the same slope in all treatment populations. If the F ratio, MS_1/MS_2, is not significant, then it may be concluded that the β_j are essentially homogeneous and that the observed variability in the b_j is attributable to sampling error. In this case, MS_1 and MS_2 are both estimates of error variance, and the two components may be pooled to provide an error term for the test of the adjusted treatment effects.

We next consider the interpretation of SS_3 and SS_4. If the adjusted treatment effects were all zero, then the plot of $\overline{Y}_{.j}$ as a function of $\overline{X}_{.j}$ should resemble the plot of the Y_{ij} as a function of the X_{ij}; in both cases error variance is the only factor contributing to the variability of data points about the best fitting linear function. In deciding that the two plots resemble each other, we are particularly concerned with two aspects of the plots. If the null hypothesis is true, (a) departures of the $\overline{Y}_{.j}$ from the line that best fits the group means should be no greater than one would expect by chance and (b) b_A, the regression coefficient of that line, should not differ significantly from $b_{S/A}$, which represents the regression of individual scores about the average group regression line. SS_3 clearly represents the deviation of group means from the best fitting line. SS_4 is a somewhat more complicated expression but is definitely a function of $b_A - b_{S/A}$.

Generally, SS_3 and SS_4 are pooled and divided by their pooled *df* to provide the numerator of the F test for the adjusted treatment effects. The denominator is the pooled error term based on the first two components of the adjusted total sum of squares.

An artificial numerical example may help clarify relationships among the various regression coefficients as well as the interpretation of SS_3 and SS_4 as measures of treatment effects. Consider a parent population consisting of nine pairs of X and Y scores such that

$$Y = 2X + 3$$

This is errorless data in the sense that all of the variability in Y is attributable to variability in X. Suppose we divide the data into three sets of three pairs each as in the upper half of Table 12–1. Note that there is, at this stage, *no* treatment effect. The variation in the $\overline{Y}_{.j}$ is due solely to the variation in the $\overline{X}_{.j}$. This is readily demonstrated by computing SS_3 and SS_4; both are zero. All the regression coefficients are equal to 2, as can be verified by application of Equations (12.4), (12.6), (12.7), and (12.8).

Suppose we now apply three treatments, one to each group. Let $\alpha_1 = -1$, $\alpha_2 = 2$, $\alpha_3 = -1$. The resulting data set is in the middle of Table 12–1. The b_j are all still 2 as is $b_{S/A}$; the addition of the α_j's has changed only the slope-intercepts of the within-group regression lines. Furthermore, application of Equation (12.8) yields

$$b_A = \frac{3[(0)(2) + (-1)(-3) + (1)(1)]}{3[0^2 + (-1)^2 + (1)^2]} = 2$$

Furthermore, b_{tot} is still 2. Thus, SS_4 must be zero (see Equation (12.9)).

However,

$$SS_3 = 3\{[(10 - 13) - 2(4 - 5)]^2 + [(15 - 13) - 2(5 - 5)]^2$$
$$+ [(14 - 13) - 2(6 - 5)]^2\}$$
$$= 18.$$

In this example, we chose our treatment effects so that they varied in a curvilinear manner with $\overline{X}_{.j}$. Note that $\overline{Y}_{.j}$ at first increases and then decreases with increased $\overline{X}_{.j}$. Thus, the $\overline{Y}_{.j}$ no longer fall on the best-fitting straight line with regression coefficient $b_A = 2$.

SS_4 was unaffected because the straight line that best describes the relation between α_j and $\overline{X}_{.j}$ has a zero regression coefficient. Thus, when the α_j are added to the original $\overline{Y}_{.j}$, b_A is unchanged.

Suppose we chose our α_j in such a way that they did have a linear relationship to $\overline{X}_{.j}$. For example, let $\alpha_j = -1, 0, +1$, respectively. Adding

TABLE 12–1 *An artificial example of relationships in a covariance problem*

$\alpha_1 = \alpha_2 = \alpha_3 = 0$

	A_1		A_2		A_3	
	X	Y	X	Y	X	Y
	3	9	2	7	4	11
	4	11	8	19	9	21
	5	13	5	13	5	13
means =	4	11	5	13	6	15

$\alpha_1 = -1, \alpha_2 = +2, \alpha_3 = -1$

	A_1		A_2		A_3	
	X	Y	X	Y	X	Y
	4	8	2	9	4	10
	4	10	8	21	9	20
	5	12	5	15	11	12
means =	4	10	5	15	6	14

$\alpha_1 = -1, \alpha_2 = 0, \alpha_3 = -1$

	A_1		A_2		A_3	
	X	Y	X	Y	X	Y
	3	8	2	9	4	12
	4	10	8	21	9	22
	5	12	5	15	11	14
means =	4	10	5	15	6	16

these effects to the *original* data set yields the data at the bottom of Table 12–1. Computing b_A again, we now obtain $b_A = 3$. The b_j and $b_{S/A}$ are as usual unchanged by the addition of a constant to the scores within each group. However, b_{tot} is now 55/40, 1.375. Turning to Equation (12.9), it is apparent without further calculation that SS_4 will be greater than zero. However, SS_3 is still zero because the $\overline{Y}_{.j}$ do fall on a straight line when plotted as a function of $\overline{X}_{.j}$.

To summarize: in the analysis of variance, the total sum of squares is partitioned; in the analysis of covariance, the adjusted total sum of squares is partitioned. The first component resulting from this partitioning, SS_1, measures the variability of the group regression coefficients about an average coefficient. The second term, SS_2, measures the variability of scores about each group regression line. The hypothesis of homogeneity of regression may be tested by a ratio of mean squares based on the two terms just described. If the F statistic is not significant, the two terms may be pooled to form a single estimate of error, which will be subsequently used in testing treatment effects. The third component of $SS_{\text{tot}(y')}$, SS_3, reflects the variability of treatment means about the line which gives the predicted value of $\overline{Y}_{.j}$. The fourth component, SS_4, measures the difference between the slope of that line and the slope of the average within-groups regression line. If variability of the means about this best fitting line is significant, then variation among the $\overline{Y}_{.j}$ is attributable to something more than the variation in $\overline{X}_{.j}$ and error variance; the "something more" is presumably treatment effects. If SS_4 is significantly large, the rate of change of $\overline{Y}_{.j}$ as a function of $\overline{X}_{.j}$ is not the same as the rate of change within groups, and again we conclude that the treatments are playing a role. The point is that if either SS_3 or SS_4 is significantly large, then the same function that describes the plot within a group does not adequately describe the plot of group means, and we conclude that the difference is attributable to the presence of treatment effects.

12.2.3 Computational formulas for the analysis of covariance. In order to carry out the analysis of covariance as efficiently as possible, it is desirable to provide raw score computational formulas for the four components of $SS_{\text{tot}(y')}$ which we have just discussed. These four expressions may be obtained by appropriately combining entries in Table 12–2. The entries in the first two columns are simply the usual formulas for the analysis of variance of a completely randomized one-factor design with the addition of formulas for group j; their role will shortly become clearer. The SP (sums of cross-products) terms can be arrived at by analogy to the SS terms. For example,

$$SS_{A(x)} = \frac{\sum_j (\sum_i X_{ij})^2}{n} - \frac{(\sum_i \sum_j X_{ij})^2}{an}$$

$$= \frac{\sum_j (\sum_i X_{ij})(\sum_i X_{ij})}{n} - \frac{(\sum_i \sum_j X_{ij})(\sum_i \sum_j X_{ij})}{an}$$

TABLE 12–2 *Computational formulas for the analysis of covariance*

	Sums of Squares (SS)		Sums of Products (SP)
	Y	X	
Total	$\sum_i \sum_j Y_{ij}^2 - C$	$\sum_i \sum_j X_{ij}^2 - C_x$	$\sum_i \sum_j X_{ij}Y_{ij} - C_{xy}$
A	$\dfrac{\sum_j(\sum_i Y_{ij})^2}{n} - C_y$	$\dfrac{\sum_j(\sum_i X_{ij})^2}{n} - C_x$	$\dfrac{\sum_j(\sum_i X_{ij})(\sum_i Y_{ij})}{n} - C_{xy}$
S/A	$SS_{\text{tot}} - SS_A$	$SS_{\text{tot}} - SS_A$	$SP_{\text{tot}} - SP_A$
Group j	$\sum_i Y_{ij}^2 - \dfrac{(\sum_i Y_{ij})^2}{n}$	$\sum_i X_{ij}^2 - \dfrac{(\sum_i X_{ij})^2}{n}$	$\sum_i X_{ij}Y_{ij} - \dfrac{(\sum_i X_{ij})(\sum_i Y_{ij})}{n}$
	$C_y = \dfrac{(\sum_i \sum_j Y_{ij})^2}{an}$	$C_x = \dfrac{(\sum_i \sum_j X_{ij})^2}{an}$	$C_{xy} = \dfrac{(\sum_i \sum_j X_{ij})(\sum_i \sum_j Y_{ij})}{an}$

Then, replacing one X by Y,

$$SP_A = \frac{\sum_j (\sum_i X_{ij})(\sum_i Y_{ij})}{n} - \frac{(\sum_i \sum_j X_{ij})(\sum_i \sum_j Y_{ij})}{an}$$

Computation of the entries in Table 12–2 is the first step in the analysis of covariance. They will be combined to yield raw score expressions for the four components of $SS_{\text{tot}\,(y)}$. SS_1 and SS_2 will then be used to provide a test of the homogeneity of regression assumption stated in Section 12.2.1. If this assumption appears tenable, SS_1 and SS_2 will provide a pooled error term against which treatment effects SS_3 and SS_4 are tested.

Raw score formulas for the four sums of squares components are readily derived. The terms are defined in Equation (12.9). The regression coefficients are expressed in terms of sums of squares (SS) and sums of cross-products (SP) in Equations (12.4), (12.6), (12.7) and (12.8). Raw score formulas for the SS and SP terms are obtained as indicated earlier. The final equations are:

(12.10) $$SS_1 = \sum_j \frac{SP_j^2}{SS_{j(x)}} - \frac{SP_{S/A}^2}{SS_{S/A(x)}}$$

(12.11) $$SS_2 = SS_{S/A(y)} - \sum_j \frac{SP_j^2}{SS_{j(x)}}$$

(12.12) $$SS_3 = SS_{A(y)} - \frac{SP_A^2}{SS_{A(x)}}$$

(12.13) $$SS_4 = \frac{SP_A^2}{SS_{A(x)}} + \frac{SP_{S/A}^2}{SS_{S/A(x)}} - \frac{SP_{\text{tot}}^2}{SS_{\text{tot}\,(x)}}$$

To complete the analysis of covariance it is also helpful to note that

(12.14) $$SS_{\text{tot}\,(y')} = SS_1 + SS_2 + SS_3 + SS_4$$
$$= SS_{\text{tot}\,(y)} - SS_{\text{lin}}$$
$$= SS_{\text{tot}\,(y)} - \frac{SP_{\text{tot}}^2}{SS_{\text{tot}\,(x)}}$$

The SS_1 is distributed on $(a - 1)$ df, since we are concerned with the variability of a regression coefficients about the pooled coefficient, $b_{S/A}$. Note that this is consistent with Equation (12.10), in which one squared quantity is subtracted from the sum of a squared quantities. The df for SS_2 are $a(n - 2)$; the explanation lies in a closer examination of the meaning of this variability. In each group, the variance of n scores is taken about a group regression line. The estimation of this line involves the loss of two df, one for estimating β_j, the regression coefficient, and the other for estimating $\overline{Y} - \beta\overline{X}$, the slope intercept. Thus there are $(n - 2)$ df pooled over a groups. Referring to Equation (12.11), we note that the relationship of squared quantities to df still holds; we have $a(n - 1) - a\,[= a(n - 2)]$ squared quantities on the right-hand side. The SS_3 measures the variability of a means about a regression line; again the estimation of the line involves the loss of two df, so that SS_3 is distributed on $(a - 2)$ df. Alternatively, we refer to Equation (12.12) and note that we have $(a - 1)$ df for $SS_{A(y)}$, and that one more is lost for the squared cross-product term. Since SS_4 measures the difference between two regression coefficients, it is on 1 df. The computational formula is again consistent with the conclusion.

Before testing treatment effects, we must consider the null hypothesis that

$$\beta_1 = \beta_2 = \cdots = \beta_a$$

The appropriate statistic is

$$F = \frac{SS_1/(a - 1)}{SS_2/a(n - 2)}$$

The logic of this F test should be apparent. The question is whether the variability among treatment regression coefficients is significantly greater than the pooled variability about the group regression lines.

We are now ready to test the adjusted treatment effects. The null hypothesis is that

$$\alpha_1 = \alpha_2 = \cdots = \alpha_a$$

i.e., the adjusted treatment effects are homogeneous. The appropriate statistic is

$$F = \frac{(SS_3 + SS_4)/(a - 1)}{(SS_1 + SS_2)/[a(n - 1) - 1]}$$

The pool of SS_1 and SS_2 requires that the test of homogeneity of regression coefficients does not have a significant result. Table 12-3 summarizes the analysis of covariance and includes the adjusted (adj) sources, their df, the SS formulas for the adjusted terms, the EMS and the F ratios. Note that we obtain the sums of squares for the adjusted A source by subtraction: $SS_{\text{tot}(y')} - SS_{S/A(y')}$. This is equivalent to adding $SS_3 + SS_4$, since

$$
\begin{aligned}
SS_{A(y')} &= SS_{\text{tot}(y')} - SS_{S/A(y')} \\
&= \left(SS_{\text{tot}(y)} - \frac{SP_{\text{tot}}^2}{SS_{\text{tot}(x)}} \right) - \left(SS_{S/A(y)} - \frac{SP_{S/A}^2}{SS_{S/A(x)}} \right) \\
&= SS_{\text{tot}(y)} - SS_{S/A(y)} - \frac{SP_{\text{tot}}^2}{SS_{\text{tot}(x)}} + \frac{SP_{S/A}^2}{SS_{S/A(x)}}
\end{aligned}
$$

TABLE 12-3 *Analysis of covariance for a completely randomized one-factor design*

SV	df	SS	EMS	F
Total(adj)	$an - 2$	$SS_{\text{tot}(y')} = SS_{\text{tot}(y)} - \dfrac{SP_{\text{tot}}^2}{SS_{\text{tot}(x)}}$		
A(adj)	$a - 1$	$SS_{A(y')} = SS_{\text{tot}(y')} - SS_{S/A(y')}$	$\sigma_{e'}^2 + n\theta_{A'}^2$	$\dfrac{MS_{A(\text{adj})}}{MS_{S/A(\text{adj})}}$
S/A(adj)	$a(n - 1) - 1$	$SS_{S/A(y')} = SS_{S/A(y)} - \dfrac{SP_{S/A}^2}{SS_{S/A(x)}}$	$\sigma_{e'}^2$	

and since $SS_{A(y)} = SS_{tot\,(y)} - SS_{S/A(y)}$,

(12.15)
$$SS_{A(y')} = SS_{A(y)} - \frac{SP_{tot}^2}{SS_{tot\,(x)}} + \frac{SP_{S/A}^2}{SS_{S/A(x)}}$$

$$= SS_3 + SS_4$$

summing Equations (12.12) and (12.13).

To summarize the computations briefly:

(a) The total sums of squares for the X and for the Y data are separately analyzed as in the usual analysis of variance; the cross-products sums are similarly treated. The formulas are in Table 12–2.

(b) Substituting into Equations (12.10) and (12.11), a test of homogeneity of regression is carried out.

(c) Assuming the population regression coefficients to be homogeneous, the adjusted sums of squares of Table 12–3 can be calculated, and the F tests of that table carried through.

Relationships between the adjusted sums of squares and correlation coefficients are illuminating; it is important to note that a squared correlation coefficient may be interpreted as a proportion of variance. Consider the adjusted error sum of squares:

$$SS_{S/A(y')} = SS_1 + SS_2$$

$$= SS_{S/A(y)} - \frac{SP_{S/A}^2}{SS_{S/A(x)}}$$

(12.16)
$$= SS_{S/A(y)} - \frac{SP_{S/A}^2}{SS_{S/A(x)}}\left(\frac{SS_{S/A(y)}}{SS_{S/A(y)}}\right)$$

$$= SS_{S/A(y)}\left(1 - \frac{SP_{S/A}^2}{SS_{S/A(x)}SS_{S/A(y)}}\right)$$

$$= SS_{S/A(y)}(1 - r_{xy/A}^2)$$

where $r_{xy/A}$ is the correlation of X and Y scores, pooled over the a treatment groups. Therefore, our adjusted error sum of squares is that proportion of the unadjusted error sum of squares not attributable to the linear relationship between X and Y. It is clear from Equation (12.17) that the efficiency of covariance relative to the usual analysis will depend on the magnitude of the correlation between X and Y; the larger the correlation, the smaller the adjusted error term will be and the greater the profit from performing the covariance analysis.

Manipulations similar to those performed above show that

(12.17)
$$SS_{A(y')} = SS_{A(y)} - [SS_{tot\,(y)}r_{xy/tot}^2 - SS_{S/A(y)}r_{xy/A}^2]$$

where $r_{xy/tot}$ is the correlation of the an X and Y scores, disregarding treatment classifications. Equation (12.17) thus provides an interpretation of the adjusted treatment variability. The adjustment is the difference between the total variability predicted from X and the variability within groups predicted from X.

12.2.4 A numerical example. An analysis of covariance will now be applied to the data of Table 12–4. We first partition the X variability. The total is

$$SS_{\text{tot}(x)} = (12)^2 + (10)^2 + \cdots + (7)^2 + (9)^2$$

$$- \frac{(12 + 10 + 7 + \cdots + 7 + 9)^2}{18}$$

$$= 2{,}385 - 2{,}200.06$$

$$= 184.94$$

Then,

$$SS_{A(x)} = \frac{(12 + 10 + \cdots + 11)^2 + (11 + 12 + \cdots + 11)^2}{6}$$

$$+ \frac{(6 + 13 + \cdots + 9)^2}{6} - C_x$$

$$= 2{,}200.83 - 2{,}200.06$$

$$= .77$$

and

$$SS_{S/A(x)} = 184.94 - .77$$

$$= 184.17$$

We next turn to the Y data:

$$SS_{\text{tot}(y)} = (26)^2 + (22)^2 + \cdots + (30)^2 - \frac{(26 + 22 + \cdots + 30)^2}{18}$$

$$= 17{,}099 - 16{,}260.06$$

$$= 838.94$$

TABLE 12–4 *Data for the analysis of covariance for a one-factor design*

A_1		A_2		A_3	
X	Y	X	Y	X	Y
12	26	11	32	6	23
10	22	12	31	13	35
7	20	6	20	15	44
14	34	18	41	15	41
12	28	10	29	7	28
11	26	11	31	9	30

The components are

$$SS_{A(y)} = \frac{(23 + 22 + \cdots + 26)^2 + (2 + 31 + \cdots + 31)^2}{6}$$
$$+ \frac{(23 + 35 + \cdots + 30)^2}{6} - C_y$$

$$= 16{,}432.17 - 16{,}260.06$$

$$= 172.11$$

and

$$SS_{S/A(y)} = 838.94 - 172.11$$
$$= 666.83$$

Next we obtain the cross-product terms. The total is

$$SP_{tot} = (12)(26) + (10)(22) + \cdots + (9)(30)$$
$$- \frac{(12 + 10 + \cdots + 9)(26 + 22 + \cdots + 30)}{18}$$

$$= 6{,}317 - 5{,}981.06$$

$$= 335.94$$

The treatment term is

$$SP_A = \frac{(12 + 10 + \cdots + 11)(26 + 22 + \cdots + 26)}{6}$$
$$+ \frac{(11 + 12 + \cdots + 11)(32 + 31 + \cdots + 31)}{6}$$
$$+ \frac{(6 + 13 + \cdots + 9)(23 + 35 + \cdots + 30)}{6} - C_{xy}$$

$$= 5{,}978.83 - 5{,}981.06$$

$$= -2.23$$

Note that it is possible to obtain negative SP terms (but not SS), since the SP term is the numerator of a correlation coefficient. The $SP_{S/A}$ is

$$SP_{S/A} = SP_{tot} - SP_A$$
$$= 335.94 - (-2.23)$$
$$= 338.17$$

TABLE 12–5 *Preliminary computations for the covariance analysis for a one-factor design*

	$SS_{(x)}$	$SS_{(y)}$	SP
Total	184.94	838.94	335.94
A	.77	172.11	−2.23
S/A	184.17	666.83	338.17

Table 12–5 summarizes the analysis thus far. The remainder of the analysis of covariance involves the manipulation of the quantities in Table 12–5 according to the formulas of Table 12–3. The adjusted total variability is

$$SS_{\text{tot}\,(y')} = SS_{\text{tot}\,(y)} - \frac{SP_{\text{tot}}^2}{SS_{\text{tot}\,(x)}}$$

$$= 838.94 - \frac{(335.94)^2}{184.94}$$

$$= 228.71$$

The error variability is computed next:

$$SS_{S/A(y')} = SS_{S/A(y)} - \frac{SP_{S/A}^2}{SS_{S/A(x)}}$$

$$= 666.83 - \frac{(338.17)^2}{184.17}$$

$$= 47.89$$

The residual variability accounts for the treatment effects:

$$SS_{A(y')} = SS_{\text{tot}\,(y')} - SS_{S/A(y')}$$

$$= 228.71 - 47.89$$

$$= 180.82$$

TABLE 12–6 *Analysis of covariance for a one-factor design*

SV	df	SS	MS	F
Total	16	228.71		
A	2	180.82	90.41	26.43*
S/A	14	47.89	3.42	
				*$p < .001$

Table 12–6 presents the final results of the analysis of covariance. The *A* main effect is a highly significant source of variance. If the covariance adjustment had not been made, *F* would have had a lower value:

$$F = \frac{SS_{A(y)}/2}{SS_{S/A(y)}/15}$$

$$= \frac{172.11/2}{666.83/15}$$

$$= \frac{86.06}{44.46}$$

$$= 1.94$$

It is clear that the covariance adjustment has resulted in a marked change in the results of the *F* test.

12.3 THE ANALYSIS OF COVARIANCE FOR MULTI-FACTOR DESIGNS

12.3.1 The completely randomized two-factor design. We will first consider a completely randomized two-factor design; the techniques of this section and Section 12.3.2 generalize readily to designs involving more treatment variables.

Calculations of the usual sums of squares for both *X* and *Y* are presented in Chapter 4, and therefore $SS_{A(x)}$, $SS_{A(y)}$, ..., $SS_{AB(y)}$, $SS_{S/AB(x)}$, and $SS_{S/AB(y)}$ require no further comment. We will need calculations of sums of cross-products. The appropriate formulas are

(12.18)
$$C_{xy} = \frac{(\sum_i^n \sum_j^a \sum_k^b X_{ijk})(\sum_i^n \sum_j^a \sum_k^b Y_{ijk})}{abn}$$

(12.19)
$$SP_{\text{tot}} = \sum_i^n \sum_j^a \sum_k^b X_{ijk} Y_{ijk} - C_{xy}$$

(12.20)
$$SP_A = \frac{\sum_j^a (\sum_i^n \sum_k^b X_{ijk})(\sum_i^n \sum_k^b Y_{ijk})}{bn} - C_{xy}$$

(12.21)
$$SP_B = \frac{\sum_k^b (\sum_i^n \sum_j^a X_{ijk})(\sum_i^n \sum_j^a Y_{ijk})}{an} - C_{xy}$$

(12.22)
$$SP_{AB} = \frac{\sum_j^a \sum_k^b (\sum_i^n X_{ijk})(\sum_i^n Y_{ijk})}{n} - C_{xy} - SP_A - SP_B$$

(12.23)
$$SP_{S/AB} = SP_{\text{tot}} - SP_A - SP_B - SP_{AB}$$

Again note the correspondence between the *SP* and *SS* terms. For example,

$$SS_{A(y)} = \frac{\sum_j(\sum_i \sum_k Y_{ijk})^2}{bn} - C = \frac{\sum_j(\sum_i \sum_k Y_{ijk})(\sum_i \sum_k Y_{ijk})}{bn} - C$$

and

$$SP_A = \frac{\sum_j(\sum_i \sum_k X_{ijk})(\sum_i \sum_k Y_{ijk})}{bn} - C_{xy}$$

We are now ready to present exact computational formulas for the adjusted sums of squares. Considering $SS_{A(y')}$, we begin by computing an adjusted pooled sum of squares for A and S/AB:

$$(12.24) \quad SS_{(A+S/AB)(y')} = (SS_{A(y)} + SS_{S/AB(y)}) - \frac{(SP_A + SP_{S/AB})^2}{SS_{A(x)} + SS_{S/AB(x)}}$$

Next, we obtain the adjusted error sum of squares:

$$(12.25) \quad SS_{S/AB(y')} = SS_{S/AB(y)} - \frac{SP_{S/AB}^2}{SS_{S/AB(x)}}$$

The adjusted sum of squares for the treatment variable A is

$$(12.26) \quad SS_{A(y')} = SS_{(A+S/AB)(y')} - SS_{S/AB(y')}$$

$$= SS_{A(y)} - \frac{(SP_A + SP_{S/AB})^2}{SS_{A(x)} + SS_{S/AB(x)}} + \frac{SP_{S/AB}^2}{SS_{S/AB(x)}}$$

Corresponding to Equation (12.26), we have

$$df_{A(adj)} = (a - 1) - 1 + 1 = a - 1$$

The above calculations can be better understood by comparing Equation (12.26) with Equation (12.15). Rewrite Equation (12.15), making the following substitutions:

$$SS_{tot(x)} = SS_{A(x)} + SS_{S/A(x)}$$

and

$$SP_{tot} = SP_A + SP_{S/A}$$

It is now clear that Equation (12.15) and Equation (12.26) are of identical form. Equation (12.26) also describes the pool of two sums of squares quantities, analogous to SS_3 and SS_4 of the preceding section.

The $SS_{B(y')}$ and $SS_{AB(y')}$ are computed in similar manner to $SS_{A(y')}$. We have

$$(12.27) \quad SS_{B(y')} = SS_{B(y)} - \frac{(SP_B + SP_{S/AB})^2}{SS_{B(x)} + SS_{S/AB(x)}} + \frac{SP_{S/AB}^2}{SS_{S/AB(x)}}$$

and

$$(12.28) \quad SS_{AB(y')} = SS_{AB(y)} - \frac{(SP_{AB} + SP_{S/AB})^2}{SS_{AB(x)} + SS_{S/AB(x)}} + \frac{SP_{S/AB}^2}{SS_{S/AB(x)}}$$

The test for homogeneity of regression coefficients also follows that for the one-factor design. We compute

(12.29)
$$SS_1 = \sum_j \sum_k \frac{SP_{jk}^2}{SS_{jk(x)}} - \frac{SP_{S/AB}^2}{SS_{S/AB(x)}}$$

and

(12.30)
$$SS_2 = SS_{S/AB(y)} - \sum_j \sum_k \frac{SP_{jk}^2}{SS_{jk(x)}}$$

To test the null hypothesis that

$$\beta_{11} = \beta_{12} = \cdots = \beta_{ab}$$

we compute

(12.31)
$$F = \frac{SS_1/(ab - 1)}{SS_2/ab(n - 2)}$$

The procedures of this section are readily extended to designs involving a greater number of factors and to designs other than the completely randomized design. All that is necessary is that the transition from *SS* formulas to *SP* formulas be understood, and that the general form for the sum of squares for any main or interaction effect be recognized:

(12.32)
$$SS_{\text{effect}(y')} = SS_{\text{effect}(y)} - \frac{(SP_{\text{effect}} + SP_{\text{error}})^2}{SS_{\text{effect}(x)} + SS_{\text{error}(x)}} + \frac{SP_{\text{error}}^2}{SS_{\text{error}(x)}}$$

In the next section, the application of Equation (12.23) to a design involving between- and within-subjects variability will be illustrated. Following that is a numerical example.

12.3.2 A mixed design. Consider *a* groups of *n* subjects who are given *b* trials on a paired-associate task. The groups differ with respect to the meaningfulness of the material. All *an* subjects have previously been tested for *b* trials on one list of associates which is not included among the *a* experimental lists. Thus, there are *b* pretest scores (*X*) as well as *b* dependent measures (*Y*) for each of the *an* subjects. We may readily apply Equation (12.32) to this experimental design, which involves one between- and one within-subjects variable. The "error" of Equation (12.32) depends upon which treatment effect the adjusted sum of squares is being computed for. Thus, in our example,

$$SS_{A(y')} = SS_{A(y)} - \frac{(SP_A + SP_{S/A})^2}{SS_{A(x)} + SS_{S/A(x)}} + \frac{SP_{S/A}^2}{SS_{S/A(x)}}$$

$$SS_{B(y')} = SS_{B(y)} - \frac{(SP_B + SP_{SB/A})^2}{SS_{B(x)} + SS_{SB/A(x)}} + \frac{SP_{SB/A}^2}{SS_{SB/A(x)}}$$

$$SS_{AB(y')} = SS_{AB(y)} - \frac{(SP_{AB} + SP_{SB/A})^2}{SS_{AB(x)} + SS_{SB/A(x)}} + \frac{SP_{SB/A}^2}{SS_{SB/A(x)}}$$

The error term calculations are

$$SS_{S/A(y')} = SS_{S/A(y)} - \frac{SP_{S/A}^2}{SS_{S/A(x)}}$$

and

$$SS_{SB/A(y')} = SS_{SB/A(y)} - \frac{SP_{SB/A}^2}{SS_{SB/A(x)}}$$

The sum of squares calculations have been previously presented in Chapter 8. The SP calculations follow readily as in the past. For example,

$$SS_{S/A(y)} = \frac{\sum_{j=1}^{a} \sum_{i=1}^{n} (\sum_{k=1}^{b} Y_{ijk})^2}{b} - \frac{\sum_{j=1}^{a} (\sum_{i=1}^{n} \sum_{k=1}^{b} Y_{ijk})^2}{bn}$$

$$= \frac{\sum_j \sum_i (\sum_k Y_{ijk})(\sum_k Y_{ijk})}{b} - \frac{\sum_j (\sum_i \sum_k Y_{ijk})(\sum_i \sum_k Y_{ijk})}{bn}$$

and

$$SP_{S/A} = \frac{\sum_j \sum_i (\sum_k Y_{ijk})(\sum_k X_{ijk})}{b} - \frac{\sum_j (\sum_i \sum_k Y_{ijk})(\sum_i \sum_k X_{ijk})}{bn}$$

12.3.3 A numerical example. Table 12–7 presents X and Y data for an experiment involving four subjects, two at A_1 and two at A_2; all four are tested at all levels of B. The sums of squares for X and for Y are computed as in Chapter 8. Therefore, computational details are omitted and results are merely listed in Table 12–8. The cross-product (SP) calculations parallel those for sums of squares. The total is given by

$$SP_{tot} = (22)(14) + (23)(17) + \cdots + (28)(35)$$

$$- \frac{(22 + 23 + \cdots + 28)(14 + 17 + \cdots + 35)}{12}$$

$$= 5{,}564.00 - 5{,}430.83$$

$$= 133.17$$

As with the sums of squares, the total may be partitioned into a between- and a within-subjects component. Thus, we have

$$SP_{B.S} = \frac{(22 + 23 + 20)(14 + 17 + 22)}{3} + \cdots$$

$$+ \frac{(19 + 26 + 28)(8 + 27 + 35)}{3} - C_{xy}$$

$$= 5{,}452.33 - 5{,}430.83$$

$$= 21.50$$

TABLE 12–7 *Data for the covariance of a mixed design*

		X Data					Y Data		
		B_1	B_2	B_3			B_1	B_2	B_3
	S_{11}	22	23	20		S_{11}	14	17	22
A_1					A_1				
	S_{21}	23	18	26		S_{21}	16	20	24
	S_{12}	22	18	21		S_{12}	6	23	33
A_2					A_2				
	S_{22}	19	26	28		S_{22}	8	27	35

This, in turn, can be partitioned:

$$SP_A = \frac{(22 + 23 + \cdots + 26)(14 + 16 + \cdots + 24)}{6}$$

$$+ \frac{(22 + 19 + \cdots + 28)(6 + 8 + \cdots + 35)}{6} - C_{xy}$$

$$= 5,434.00 - 5,430.83$$

$$= 3.17$$

and

$$SP_{S/A} = SP_{B.S} - SP_A$$

$$= 21.50 - 3.17$$

$$= 18.33$$

TABLE 12–8 *Preliminary computations for the analysis of covariance for a mixed design*

	$SS_{(x)}$	$SS_{(y)}$	SP
Total	115.67	870.92	133.17
Between S	25.00	48.92	21.50
A	.34	30.09	3.17
S/A	24.66	18.83	18.33
Within S	90.67	822.00	111.67
B	15.17	623.17	71.42
AB	8.16	197.16	39.08
SB/A	67.34	1.67	1.17

The within-subjects component is obtained by subtraction:

$$SP_{W.S} = SP_{tot} - SP_{B.S}$$
$$= 133.17 - 21.50$$
$$= 111.67$$

This is now partitioned into SP_B, SP_{AB}, and $SP_{SB/A}$. First,

$$SP_B = \frac{(22 + \cdots + 19)(14 + \cdots + 8)}{4} + \cdots$$
$$+ \frac{(20 + \cdots + 28)(22 + \cdots + 35)}{4} - C_{xy}$$
$$= 5,502.25 - 5,430.83$$
$$= 71.42$$

Next,

$$SP_{AB} = \frac{(22 + 23)(14 + 16) + \cdots + (21 + 28)(33 + 35)}{2}$$
$$- C_{xy} - SP_A - SP_B$$
$$= 5,544.40 - 5,430.83 - 3.17 - 71.42$$
$$= 39.08$$

Finally,

$$SP_{SB/A} = SP_{W.S} - SP_B - SP_{AB}$$
$$= 111.67 - 71.42 - 39.08$$
$$= 1.17$$

These results are also included in Table 12–8.

Using the entries in Table 12–8, we may now proceed to obtain the adjusted sums of squares. The key is the correct application of Equation (12.33). Thus, we have

$$SS_{A(y')} = SS_{A(y)} - \frac{(SP_A + SP_{S/A})^2}{SS_{A(x)} + SS_{S/A(x)}} + \frac{SP_{S/A}^2}{SS_{S/A(x)}}$$
$$= 30.09 - \frac{(3.17 + 18.33)^2}{.34 + 24.66} + \frac{(18.33)^2}{24.66}$$
$$= 25.22$$

The error term is straightforward:

$$SS_{S/A(y')} = SS_{S/A(y)} - \frac{SP_{S/A}^2}{SS_{S/A(x)}}$$
$$= 18.83 - \frac{(18.33)^2}{24.66}$$
$$= 5.21$$

Turning to the within-subjects effects, we have

$$SS_{B(y')} = SS_{B(y)} - \frac{(SP_B + SP_{SB/A})^2}{SS_{B(x)} + SS_{SB/A(x)}} + \frac{SP_{SB/A}^2}{SS_{SB/A(x)}}$$

$$= 623.17 - \frac{(71.42 + 1.17)^2}{15.17 + 67.34} + \frac{(1.17)^2}{67.34}$$

$$= 559.33$$

and

$$SS_{AB(y')} = SS_{AB(y)} - \frac{(SP_{AB} + SP_{SB/A})^2}{SS_{AB(x)} + SS_{SB/A(x)}} + \frac{SP_{SB/A}^2}{SS_{SB/A(x)}}$$

$$= 197.16 - \frac{(39.08 + 1.17)^2}{8.16 + 67.34} + \frac{(1.17)^2}{67.34}$$

$$= 175.72$$

For the error term, we have

$$SS_{SB/A(y')} = SS_{SB/A(y)} - \frac{SP_{SB/A}^2}{SS_{SB/A(x)}}$$

$$= 1.67 - \frac{(1.17)^2}{67.34}$$

$$= 1.65$$

The final analysis is summarized in Table 12–9.

TABLE 12–9 *Analysis of covariance for a mixed design*

SV	df	SS	MS	F
Total	9	767.13		
Between S	2	30.43		
A	1	25.22	25.22	4.83
S/A	1	5.21	5.21	
Within S	7	736.70		
B	2	559.33	279.67	527.68*
AB	2	175.72	87.86	165.77*
SB/A	3	1.65	.53	
				*p < .001

12.4 COMPARISON WITH THE TREATMENTS × BLOCKS DESIGN

Section 12.2.2 suggests that one should generally be interested in covariance as a technique for reducing error variance. In this regard, it is important to compare the analysis with the use of the treatments × blocks design, since both use a concomitant variable to increase precision. Three advantages of the covariance approach are discernible:

(a) The concomitant data may be used after the fact, if the covariance analysis is applied. For example, if problem-solving scores prove to be highly variable, intelligence test data can be collected and an analysis of covariance carried out even though this had not been planned prior to the collection of data.

(b) The establishment of blocks is often impractical. For example, we are interested in the problem-solving behavior of high and low socially cohesive groups. It is most efficient, and possibly more interesting, to work with already established groups, such as Boy Scout troops. However, it may be impossible to find an equal number of high and low intelligence groups at each level of cohesiveness. It is more practical merely to measure intelligence and use it as a covariate rather than as a factor in the design.

(c) The analysis of covariance is more precise than the treatments × blocks design when the true correlation between X and Y is greater than .6 (Feldt, 1958).

Although the covariance approach has the advantages just cited, the treatments × blocks design is superior in several other respects:

(a) The computational labor involved in an analysis of variance performed on treatments × blocks data is approximately one-third to one-half that involved in the covariance analysis, thus partly compensating for the increased experimental labor resulting from establishing blocks of subjects.

(b) The treatments × blocks approach (assuming the optimal number of levels) is more precise than the covariance approach when the true correlation between X and Y is less than .4. This is important, since correlations of less than .4 are more frequent in psychological research than correlations greater than .6.

(c) The treatments × blocks interaction may be of interest. Furthermore, if there is reason to expect such an interaction to be significant, covariance should be avoided. If the block means differ more at one treatment level than at another, then the values of Y_{ij} are changing more rapidly as a function of X_{ij} at one treatment level than at another. In short, if a treatments × blocks interaction is significant, the assumption of homogeneity of regression coefficients is not correct, and the covariance model presented earlier in this chapter is not appropriate to the data.

(d) Perhaps the most important advantage of the experimental over the statistical approach lies in the relative complexity of the covariance model. There are more assumptions, more things that can go wrong, and statisticians have not yet adequately assessed the consequences of violating the model.

Add to the points just cited the inferential problems raised in the preceding section, and one begins to understand why most biometricians and statisticians recommend the experimental over the statistical approach. The author feels that covariance is frequently useful but recommends that the experimenter consider the estimated correlation of X and Y, the probable validity of the covariance model for the data to be collected, and any possible inferential problems. If the choice is approached this way, more often than not the treatments × blocks design will be used.

EXERCISES

12.1 Do the following analysis of covariance:

	A_1		A_2	
	X	Y	X	Y
	23.8	7.9	28.5	25.1
	23.8	7.1	18.5	20.7
	22.6	7.7	20.3	20.3
B_1	22.8	11.2	26.6	18.9
	22.0	6.4	21.2	25.4
	19.6	10.0	24.0	30.0
	27.5	20.1	22.9	19.9
	28.1	17.7	25.2	28.2
	35.7	16.8	20.8	18.1
B_2	27.7	30.5	13.5	13.5
	25.9	21.0	19.1	19.3
	27.9	29.3	32.2	35.1

12.2 The following 3 × 3 Latin square has one S in each sequence and an X (covariate) and Y (dependent) measure in each cell. Compute the adjusted F test for treatments.

	A_1	A_2	A_3
X	4	1	5
Y	8	9	21

	A_2	A_3	A_1
X	3	2	4
Y	5	18	2

	A_3	A_1	A_2
X	1	3	4
Y	10	2	4

12.3 Each of three of eight Ss is run through four problem-solving tasks. Time to solve is recorded. IQ scores are also available.

		P_1	P_2	P_3	P_4	X data (IQ)
	S_{11}	34	46	48	64	108
A_1	S_{21}	36	41	40	60	112
	S_{31}	28	37	35	52	124
	S_{12}	46	60	63	84	116
A_2	S_{22}	40	51	48	74	127
	S_{32}	55	72	73	96	103
	S_{13}	45	70	74	88	106
A_3	S_{23}	41	63	62	70	135
	S_{33}	49	71	70	85	112

Y data (time to solve) above P_1–P_4.

(a) *Set up* the analysis of covariance SS computations.
(b) Comment on the effects of using covariance in the present case.

12.4 The relationship between Y and X for each treatment group can reasonably be described by the nonlinear function:

$$Y_{ij} = K_j + 5X_{ij}^2$$

How would you perform an analysis of covariance?

REFERENCES

Atiqullah, M., "The robustness of the covariance analysis of a one-way classification," *Biometrika*, 51: 365–372 (1964).

Feldt, L. S., "A comparison of the precision of three experimental designs employing a concomitant variable," *Psychometrika*, 23: 335–353 (1958).

SUPPLEMENTARY READING

An excellent summary of the covariance model and the implications of violations of assumptions may be found in:

Elashoff, J. D., "Analysis of covariance: a delicate instrument," *American Educational Research Journal*, 6: 383–401 (1969).
A more technical exposition of the model and its applications may be found in:

Biometrics: 13, No. 3 (September 1957).
In that issue, readers should find particularly useful:

Cochran, W. G., "Analysis of covariance: its nature and uses," *Biometrics*, 13: 261–281 (1957).

Smith, H. F., "Interpretation of adjusted treatment means and regression in analysis of covariance," *Biometrics*, 13: 282–308 (1957).

Additional discussions of covariance analysis may be found in:

Anderson, N. H., "Comparison of different populations: resistance to extinction and transfer," *Psychological Review*, 70: 162–179 (1963).

Evans, S. H. and Anastasio, E. J., "Misuse of analysis of covariance when treatment effect and covariate are confounded," *Psychological Bulletin*, 69: 225–234 (1968).

Lord, F. M., "A paradox in the interpretation of group comparisons," *Psychological Bulletin*, 68: 304–305 (1967).

Sprott, D. A., "Note on Evans and Anastasio on the analysis of covariance," *Psychological Bulletin*, 73: 303–306 (1970).

13

Further Data Analyses: Qualitative Independent Variables

13.1 INTRODUCTION

Chapters 4 to 12 have been concerned with the total variability among treatment population means. The null hypothesis that the variance of the entire set of means is zero has been tested, and estimates, point and interval, of the variance of the entire set of means have been established. This is often only the beginning of a complete analysis of the data. It leaves unanswered questions about the slope and shape of performance curves, and it leaves unexplored differences among means within subsets smaller than the entire treatment set.

If the independent variable is quantitative (e.g., amount of reward, length of time in therapy), the overall F test is often only a preliminary test, useful in determining whether or not to proceed further with the analysis. Knowing that there is significant variability among the means, we may ask whether performance shows a general improvement over the levels of the independent variable, and whether several apparent changes in the direction of the function are significant. In short, the overall F indicates that the means do not fall on a straight line with slope of zero. We then wish to know more about the slope and shape of the best fitting function.

In the case of a qualitative variable (e.g., type of reward, type of therapy), the F test is again only a first step. For example, consider an experiment in which motivation is manipulated. There is one group in which correct responses are rewarded, but errors are not punished (group R), a second group which is punished for errors but not rewarded for correct responses (group P), a third group which is both rewarded for correct responses and punished for incorrect responses (group RP), and a fourth group which receives neither reward for correct responses nor punishment for incorrect responses (NRP). A significant overall F shows that differences exist among the treatment population means. Why? Are reward and punishment different in effect? Is their combination (RP) more effective than either alone?

Does the *RP* effect differ from the average effect for the *R* and *P* groups? Is the effect of the *NRP* treatment significantly different from that of the average of the three incentive groups? These are all reasonable questions, and others could be asked.

Analyses relevant to quantitative independent variables will be considered in the next chapter. For the present, the discussion is restricted to the sort of comparisons among means indicated in the preceding paragraph. These comparisons can also be made with quantitative variables (e.g., Do the effects of one and two food pellets of reward differ?), but for these variables the analyses of Chapter 14 will generally be more fruitful.

To clarify the anlyses that follow, we return to the example of the experiment on reward and punishment. Consider each of the questions raised: Are reward and punishment different in effect? implies the null hypothesis

$$\text{(13.1)} \qquad \mu_R - \mu_P = 0$$

Is their combination (*RP*) more effective than either alone? implies the null hypotheses

$$\text{(13.2)} \qquad \mu_{RP} - \mu_R = 0$$

and

$$\text{(13.3)} \qquad \mu_{RP} - \mu_P = 0$$

Does the *RP* effect differ from the average effect for the *R* and *P* groups? implies the null hypothesis

$$\text{(13.4)} \qquad \mu_{RP} - \tfrac{1}{2}(\mu_R + \mu_P) = 0$$

Is the effect of the *NRP* treatment significantly different from that of the average of the three incentive groups? implies the null hypothesis

$$\text{(13.5)} \qquad \mu_{NRP} - \tfrac{1}{3}(\mu_R + \mu_P + \mu_{RP}) = 0$$

We may rewrite the five null hypotheses:

$$\text{(13.1')} \qquad (1)\mu_R + (-1)\mu_P + (0)\mu_{RP} + (0)\mu_{NRP} = 0$$

$$\text{(13.2')} \qquad (-1)\mu_R + (0)\mu_P + (1)\mu_{RP} + (0)\mu_{NRP} = 0$$

$$\text{(13.3')} \qquad (0)\mu_R + (-1)\mu_P + (1)\mu_{RP} + (0)\mu_{NRP} = 0$$

$$\text{(13.4')} \qquad (-\tfrac{1}{2})\mu_R + (-\tfrac{1}{2})\mu_P + (1)\mu_{RP} + (0)\mu_{NRP} = 0$$

$$\text{(13.5')} \qquad (-\tfrac{1}{3})\mu_R + (-\tfrac{1}{3})\mu_P + (-\tfrac{1}{3})\mu_{RP} + (1)\mu_{NRP} = 0$$

A close examination of Equations (13.1') to (13.5') indicates the general form of the null hypotheses to be considered in this chapter (and, in fact, in the next chapter as well; the difference is in the rationale for selecting the multipliers of the μs). The general null hypothesis is

$$\psi = 0$$

where $\psi = \sum_j w_j \mu_j$, and the w_j are the coefficients which multiply the μ_j; they vary as a function of the specific hypotheses being tested. The sum of these coefficients is always zero for any contrast. It is helpful to note that the analyses will be unchanged if both sides of any of the above equations are multiplied by a constant. Thus, all w_j in Equation (13.5') may be translated into integers by multiplying by 3:

(13.5'') $$-\mu_R - \mu_P - \mu_{RP} + 3\mu_{NRP} = 0$$

The problem of estimation is, in general, the problem of estimating ψ or of obtaining a confidence interval for ψ.

In the next section, we will present a general computational formula for sums of squares for contrasts of the type exemplified by Equations (13.1) to (13.5). Following this, we will examine the inferential problem that exists when several contrasts are simultaneously investigated, and consider some proposed solutions to the problem. The first of these solutions will utilize the computational formulas which are developed below.

13.2 SUMS OF SQUARES FOR MULTIPLE COMPARISONS

13.2.1 Computations. The sum of squares for any estimate of the population contrast, ψ, is given by

(13.6) $$SS_{\hat{\psi}} = \frac{(\sum_j w_j \overline{Y}_{.j})^2}{\sum_j (w_j^2 / n_j)}$$

To clarify the meaning of this sum of squares, consider the contrast in Equation (13.5''). The $SS_{\hat{\psi}}$ is a measure of the variability of two means about the grand mean; one mean is the mean of the combined R, P, and RP groups, and the second mean is the mean of the NRP group. The calculations of Equation (13.6) are equivalent to the calculations involved in comparing two treatment groups, where one group consists of $n_R + n_P + n_{RP}$ scores and the second consists of n_{NRP} scores. If each of the individual treatment means are based on the same n, we have the alternative calculation:

(13.6') $$SS_{\hat{\psi}} = \frac{[\sum_j^q (w_j/n)(\sum_i^n Y_{ij})]^2}{(1/n)\sum_j w_j^2}$$

$$= \frac{[\sum_j w_j (\sum_i Y_{ij})]^2}{n \sum_j w_j^2}$$

As an example of the application of Equation (13.6′), consider the reward and punishment experiment. Assume ten subjects in each of the four treatment groups. The total number of errors in 20 trials for each group is

R	P	RP	NRP
42	34	25	88

To determine whether the average effect of the R and P treatments differs from that of the RP treatment, we compute

$$SS_{\hat{\psi}} = \frac{[42 + 34 - 2(25)]^2}{(10)(6)}$$

$$= \frac{676}{60}$$

$$= 11.27$$

Using Equation (13.6), which would also apply if the cell frequencies were unequal, we have

$$SS_{\hat{\psi}} = \frac{[4.2 + 3.4 - 2(2.5)]^2}{.1 + .1 + .4}$$

$$= \frac{6.76}{.6}$$

$$= 11.27$$

The computation of $SS_{\hat{\psi}}$ is not more complicated for the designs other than the completely randomized one-factor design. It is only necessary to replace n of Equation (13.6′) by the total number of measurements on which each treatment mean is based. For example, if there is a three-factor design with a, b, and c levels of the variables A, B, and C, and if the interest lies in some contrast among the $\overline{Y}_{...m}$ (the means at the levels of C), each mean is based on nab scores:

$$SS_{\hat{\psi}} = \frac{[\sum_{m=1}^{c}(w_m \sum_{i=1}^{n} \sum_{j=1}^{a} \sum_{k=1}^{b} Y_{ijkm})]^2}{nab \sum_{m=1}^{c} w_m^2}$$

That is to say, we find the sum at each level of C, multiply by the appropriate coefficient, add these cross-products (coefficient × sum) together, and square the resulting total; we then divide by the total number of measures in a level of C times the sum of squared coefficients.

13.2.2 A general single *df* formula. It may help the reader to note the relationship between the calculations just presented and other calculations previously encountered in this text. The quantity

$$\frac{[\sum_j w_j (\sum_i Y_{ij})]^2}{n \sum_j w_j^2}$$

and its equivalent for unequal n [Equation (13.6)] are general formulas for any sum of squares distributed on 1 df. For example, there is no real difference between the above formula and the quantity on the right of Equation (5.51). In Chapter 5 we had a 2×2 design, and we provided a shortcut formula for SS_{AB}:

$$SS_{AB} = \frac{(T_{11} + T_{22} - T_{12} - T_{21})^2}{4n}$$

This can be rewritten

$$SS_{AB} = \frac{[(1)(\sum_i Y_{i11}) + (1)(\sum_i Y_{i22}) + (-1)(\sum_i Y_{i12}) + (-1)(\sum_i Y_{i21})]^2}{[(1)^2 + (1)^2 + (-1)^2 + (-1)^2]n}$$

which is clearly of the form of the right-hand quantity of Equation (13.6). In general, the formula for any sum of squares on 1 df may be represented by

(13.7) $$SS_{1df} = \frac{(\text{sum of the weighted cell totals})^2}{(\text{sum of the squared weights}) \times (\text{cell frequencies})}$$

and if the ns are unequal,

(13.8) $$SS_{1df} = \frac{(\text{sum of the weighted cell means})^2}{\text{sum of the ratios of squared weights to cell frequencies}}$$

13.3 NULL HYPOTHESIS TESTS AND CONFIDENCE INTERVALS FOR MULTIPLE COMPARISONS

Contrasts are tested under one or both of two sets of circumstances: (a) prior to the collection of data, the experimenter raises certain questions that dictate specific sets of contrasts; (b) after the data have been collected, the usual F test has proven significant and the investigator wishes to examine the results further to detect those contrasts that cause the overall null hypothesis to be rejected. In either case, we have a situation that might be likened to a series of tosses of a coin. Although the probability of a head on any one toss is .5, the probability of *at least one head* in a series of tosses is greater than .5, and the magnitude of this probability is a direct function of the total number of tosses. Similarly, any single test of a comparison has probability α of a Type I error. However, as the number of comparisons made increases, the probability of *at least one Type I error* increases. For example, if we did 10 independent significance tests, each at the .05 level, the probability of at least one Type I error would be $1 - (.95)^{10} \approx .40$. Note that the same problem exists if the largest observed contrast is selected for testing. This is equivalent to testing all contrasts, since the probability that the largest observed contrast is significant is the probability that at least one contrast is significant.

At this point, it will be helpful to distinguish between two types of error rates. The usual α level, the probability that a single comparison results in a Type I error, will be referred to as the *error rate per comparison* (*EC*). The probability that an entire set of comparisons contains at least one Type I error will be referred to as the *error rate experimentwise* (*EW*). In the coin toss example, $EC = .05$ but $EW = .40$. The distinction may be clarified by conceptualizing 2,000 replications of an experiment. Five contrasts, all of which are true for the population, are tested in each experiment. Thus, we have 2,000 sets of five significance tests. A single experiment can result in 0-5 erroneous rejections of H_0. The actual results might look like:

Number of experiments with X errors	X
1,548	0
402	1
43	2
5	3
2	4
0	5
2,000	

The total number of Type I errors is $402 + (2)(43) + (3)(5) + (4)(2) = 511$. Thus, the *EC* equals $511/10,000 \approx .05$. However, 452 experiments contain one or more Type I errors. Thus, the *EW* equals $452/2,000 = .226$.

There are two extreme viewpoints with respect to the discrepancy between *EW* and *EC*. On the one hand, we might ignore the *EW* and test every contrast at a fixed α level that is chosen independently of the total number of contrasts to be tested. The objection to this approach lies in the fact that the likelihood of a Type I error increases with the number of contrasts tested. Significant findings that are not readily replicable, because they are erroneous, are more likely to be obtained by investigators who test relatively many contrasts. If the experiment is a pilot study, not to be published and be used as a source of leads for subsequent studies, the damage may be slight. But if the results are to be published, or to be put to practical applications, the use of the *EC* can be dangerous.

Of course, one could move to another extreme. The energetic researcher tests thousands of null hypotheses in his lifetime. Should he guard against some lifetime-wise error rate, carrying out each significance test at some infinitesimally low α level? If he did, the Type II error rate would soar to undesirable heights.

In selecting the critical region of our test statistic, we must always consider both Type I and Type II errors. Our proposed compromise is to use the set of contrasts involved in a single experiment as the unit for controlling error rate. We will choose a criterion for significance such that the *EW* is constant regardless of the number of treatment groups. In effect, a rule is required for adjusting the *EC* downwards as the total number of comparisons increases, and adjusting in such a way that the change in the number of comparisons does not alter the *EW*. Several such rules will be considered in Sections 13.3.1–13.3.5.

13.3.1 Planned comparisons. We first consider the general case in which the experimenter designates k contrasts of interest prior to the experiment. This involves two situations, orthogonal and nonorthogonal contrasts.

Orthogonal contrasts. There are many possible contrasts of the form

$$\psi = \sum w_j \mu_j$$

if the sole restriction is that $\sum w_j = 0$. Many of these contrasts will be correlated with each other; they will carry redundant information. To take an extreme example, suppose:

$$\overline{Y}_3 - \overline{Y}_2 = 3 \quad \text{and} \quad \overline{Y}_2 - \overline{Y}_1 = 7$$

Then, the value of $\overline{Y}_3 - \overline{Y}_1$ must be 10 and, if the first two contrasts are significant, then the third must be. However, sets of independent contrasts can be obtained; each are called *orthogonal sets*. Assuming a treatment means are contrasted, each set of orthogonal contrasts will have $a - 1$ members. The sum of squares for each term will be distributed on 1 *df*, and the total of the $a - 1$ sums of squares will equal the SS_A distributed on $(a - 1)$ *df*.

Two contrasts, ψ_p and $\psi_{p'}$, are orthogonal if (a) the sum of the coefficients for each contrast is zero; i.e., $\sum_j w_{jp} = 0$ and $\sum_j w_{jp'} = 0$; and (b) the sum of cross-products of coefficients is zero; i.e., $\sum_j w_{jp} w_{jp'} = 0$. To illustrate the concept of orthogonality, we again return to the example of the experiment on reward and punishment. One possible set of three orthogonal contrasts is

	R	P	RP	NRP
$w_{j1} - 1$	$+1$	0	0
$w_{j2} - 1$	-1	$+2$	0
$w_{j3} - 1$	-1	-1	$+3$

To verify that the three contrasts are independent, we note that

$$\sum_j w_{j1} w_{j2} = (-1)(-1) + (+1)(-1) + (0)(+2) + (0)(0) = 0$$

$$\sum_j w_{j1} w_{j3} = (-1)(-1) + (+1)(-1) + (0)(-1) + (0)(+3) = 0$$

$$\sum_j w_{j2} w_{j3} = (-1)(-1) + (-1)(-1) + (+2)(-1) + (0)(+3) = 0$$

Using the data of Section 13.2.1,

$$SS_A = \frac{(42)^2 + (34)^2 + (25)^2 + (88)^2}{10} - \frac{(189)^2}{40}$$

$$= 1{,}128.9 - 893.025$$

$$= 235.875$$

For the three contrasts, we have

$$SS_{\psi_1} = \frac{(42 - 34)^2}{(2)(10)}$$

$$= 3.2$$

$$SS_{\psi_2} = \frac{[42 + 34 - (2)(25)]^2}{(6)(10)}$$

$$= 11.267$$

$$SS_{\psi_3} = \frac{[42 + 34 + 25 - (3)(88)]^2}{(12)(10)}$$

$$= \frac{26{,}569}{120}$$

$$= 221.408$$

Adding,

$$\sum_p SS_{\psi_p} = 3.2 + 11.267 + 221.408$$

$$= 235.875$$

$$= SS_A$$

It is not necessary to test all members of an orthogonal set. Suppose that there were six levels of A and only two contrasts of interest, orthogonal to each other. We might divide the SS_A into three quantities corresponding to the two contrasts and a residual $(SS_A - SS_{\psi_1} - SS_{\psi_2})$.

If the contrasts are orthogonal, control of the EW is fairly simple. Note that under these conditions,

$$EW = 1 - (1 - EC)^k$$

where EC is the significance level for each contrast and k is the number of contrasts to be tested. We choose EC equal to EW/k. Then,

$$1 - \left(1 - \frac{EW}{k}\right)^k = 1 - \left[1 - k\left(\frac{EW}{k}\right) + (k_2)\left(\frac{EW}{k}\right)^2 - \cdots - \left(\frac{EW}{k}\right)^k\right]$$

If EW is not too close to 1, $(EW/k)^j (j > 1)$ will be very small and therefore

$$1 - \left(1 - \frac{EW}{k}\right)^k \approx 1 - \left[1 - k\left(\frac{EW}{k}\right)\right] = EW$$

Thus, for orthogonal sets of k contrasts, we can achieve any desired EW by testing each contrast at the EW/k level; the df are 1 and $a(n - 1)$ for the one-factor design. Frequently, the desired significance level will not be available in Table A–5; for example, if $EW = .10$ and $k = 8$, each contrast must be tested at the .0125 α level. Table A–12 provides values of $t(= \sqrt{F}$ when $df_1 = 1)$ for various combinations of k and error df at $EW = .01, .05,$ and .10. If other combinations are involved in an experiment, find z_α, the normal deviate significant at the EW/k level, using a two-tailed test. The quantity,

$$\left[z_\alpha + \frac{z_\alpha^3 + z_\alpha}{4(df_{\text{error}} - 2)}\right]^2$$

is distributed approximately as F; if its value is exceeded, reject H_0. Be careful to note that the required z-score is that which is exceeded with probability equal to $(\frac{1}{2})(EW/k)$.

The use of $\frac{1}{2}$ in the preceeding calculation is related to the general point that all of the tests of contrasts described in this section are two-tailed: differences in either direction that exceed some critical magnitude will be rejected.

A further point in considering significance tests of orthogonal contrasts is that they need not be equally weighted. Let us assume that we desire an EW of .10 and that k equals 3. One of the contrasts is of greater interest than the other two and we are willing to have a higher Type I error rate in testing it in order to achieve greater power. We might carry out this test at the .05 level and the other two at the .025 level. As long as the sum of the ECs is EW, the actual experimentwise error rate will be approximately equal to EW.

We can readily obtain confidence intervals on the magnitudes of the contrasts. However, we first need some estimate of the variance of the sampling distribution of $\hat\psi$. We know from Section 13.1 that $\hat\psi$ s defined as $\sum_j w_j\mu_j$. Therefore, its estimate is

(13.9) $$\hat\psi = \sum_j w_j \overline{Y}_{.j}$$

Consider $w_j\overline{Y}_{.j}$. The weight, w_j, will not vary over repeated samples of the jth treatment group. Therefore, the variance of $w_j\overline{Y}_{.j}$ is just $w_j^2 \text{var}(\overline{Y}_{.j})$. The variance of a mean is the variance of scores divided by n. Then

(13.10) $$\sigma_{\hat\psi}^2 = \sum_j w_j^2 \frac{\sigma_Y^2}{n_j}$$

In the completely randomized one-factor design, we have the estimate

(13.11)
$$S_{\hat{\psi}}^2 = MS_{S/A} \sum_j \frac{w_j^2}{n_j}$$

In general,

(13.12)
$$S_{\hat{\psi}}^2 = MS_{error} \sum_j \frac{w_j^2}{N_j}$$

where N_j is the number of observations upon which the jth mean is based; for example, if we were contrasting means for different levels of B in a two-factor design with equal cell frequencies, the squared weights would be divided by an, the number of scores at each level of B.

We are now ready to return to the problem of establishing confidence intervals for orthogonal planned contrasts. The probability is $1 - \alpha$ that the values of all contrasts simultaneously lie within the confidence intervals of the form

(13.13)
$$\hat{\psi} - S_{\hat{\psi}}\sqrt{F_{\alpha/k}} \leq \hat{\psi} \leq \hat{\psi} + S_{\hat{\psi}}\sqrt{F_{\alpha/k}}$$

where $S_{\hat{\psi}}$ is the square root of the expression of Equation (13.12) and $F_{\alpha/k}$ is the F required for significance at the α/k level with 1 df for the numerator.

Such confidence intervals, based on the EW concept, are somewhat different in interpretation than those previously encountered in the text, the latter being based on the EC concept. When EC is our error rate, we are concerned with whether a single interval does or does not contain the parameter; confidence is the proportion of intervals that, in the long run, would contain the parameter. In the present context, in which we wish to control EW, the experiment falls into one of two categories: (a) either the intervals computed for all k contrasts contain the parameters being estimated or (b) at least one interval does not contain its parameter; confidence is the proportion of experiments that, in the long run, fall in the first category.

The reward and punishment experiment referred to earlier may clarify the preceding discussion. We are interested in whether average performances differ under reward and punishment ($\hat{\psi}_1$) and whether the average of subjects who are rewarded for correct responses and punished for incorrect responses differ from the average of the two single incentive groups ($\hat{\psi}_2$). The contrasts are:

$$\psi_1 = \mu_R - \mu_P$$

$$\psi_2 = (\tfrac{1}{2})(\mu_R + \mu_P) - \mu_{RP}$$

The means for the three groups are

$$\overline{Y}_R = 4.2 \qquad \overline{Y}_P = 3.4 \qquad \overline{Y}_{RP} = 2.5$$

There are 10 subjects in each group. $MS_{S/A}$ is equal to 2.0. We have

$$\hat{\psi}_1 = 4.2 - 3.4 = .8$$
$$\hat{\psi}_2 = (\tfrac{1}{2})(4.2 + 3.4) - 2.5 = 1.3$$

From Equation (13.11) we have

$$S_{\hat{\psi}_1} = (200)(2)\left(\frac{2}{10}\right) = .4$$

$$S_{\hat{\psi}_2} = (2)\left(\frac{1.5}{10}\right) = .3$$

Assume that $EW = .05$. Then we require $F_{.025,1,36}$, which is approximately 5.48. Inserting these values into Equation (13.13) gives the 95 percent experimentwise confidence limits on $\mu_R - \mu_P$,

$$.8 - (2.33)\sqrt{(.4)} \leq \psi_1 \leq .8 + (2.33)\sqrt{(.4)} = -.67 < \psi_1 < 2.27$$

If we are also interested in testing

$$H_0: \mu_R - \mu_P = 0$$

we note that the interval contains zero, the value under the null hypothesis. Therefore, with $EW = .05$, the hypothesis cannot be rejected. The confidence interval for ψ_2 is

$$1.3 - (2.33)\sqrt{(.3)} \leq \psi_2 \leq 1.3 + (2.33)\sqrt{(.3)} = .02 \leq \psi_2 \leq 2.58$$

In this instance, H_0 would be rejected.

Nonorthogonal contrasts. Orthogonal contrasts are easily interpreted because they are not redundant and a close approximation to the desired EW is readily obtained. However, in many and perhaps most cases the contrasts of interest will not be orthogonal. After all, contrasts are tested because they are of psychological import, not because they are independent of each other.

It can be proven from elementary probability theory that the probability of at least one Type I error for a set of k nonindependent tests, each carried out at the EW/k level, is less than EW. Therefore, we recommend the procedures described above for the orthogonal case, even when the contrasts are not orthogonal. Note that with correlated contrasts the approach is conservative with respect to Type I error rate; the probability of one or more Type I errors over the set of k contrasts is at most EW. Furthermore, the test tends to be more conservative as k increases. Therefore, one should realize that testing nonorthogonal contrasts and adding more contrasts involves a compromise. We gain more information in the sense of investigating more questions, or questions that are not part of a restricted orthogonal set. However, the quality of the information is somewhat degraded in the sense that our test becomes conservative, more so with more comparisons; we lose power and have wider confidence intervals. In any specific instance, it may be worth trying to divide the questions of interest into two sets in terms of level of importance. Frequently, certain contrasts are basic to the study; others would be nice to know about but are not of fundamental importance. Then the total error rate might be divided so that the *EC* is higher for the primary contrasts. These will then be tested with somewhat more power and the confidence intervals will be narrower.

13.3.2 Post hoc analysis: Scheffé's multiple comparison method.

Despite careful planning, certain differences among means that had not been considered will attract attention when the data have been collected. It would be foolish to ignore these merely because they are unanticipated. The contrasts in question might well provide insights into the processes under investigation. On the other hand, as noted earlier, picking out the largest contrasts after the experiment has been completed is tantamount to investigating the universe of possible contrasts. We are obligated to pay a price in power and confidence interval width for this additional information, for we are controlling error rate over a larger set of contrasts than would be the case if the contrasts had been planned. In this section, we will consider one approach to controlling *EW* for post hoc comparisons.

Assume *a* levels of the treatment variable *A* in a completely randomized one-factor design. To test the null hypothesis that the *p*th contrast is zero; i.e., $\sum_j w_{jp}\mu_j = 0$, Scheffé (1959) proposes that the obtained *F* statistic be evaluated against $(a - 1)F_{\alpha,a-1,a(n-1)}$, where the criterion *F* is that required for significance at the α level on $a - 1$ and $a(n - 1)$ *df*. (With unequal *n*, the denominator *df* are $\sum n_j - a$.) The basis for this null hypothesis test is the following theorem (proved by Scheffé in Section 3.5 of his text): the probability is $1 - \alpha$ that the values of all contrasts simultaneously lie within the confidence intervals of the form,

$$(13.14)\quad \hat{\psi}_P - S_{\hat{\psi}_P}\sqrt{(a - 1)F_{\alpha,a-1,a(n-1)}} < \psi_P < \hat{\psi}_P$$
$$+ S_{\hat{\psi}_P}\sqrt{(a - 1)F_{\alpha,a-1,a(n-1)}}$$

where $\hat{\psi}_P = \sum_j w_{jp}\overline{Y}_{.j}$ and $S_{\hat{\psi}_P}$ is defined in Equations (13.11) and (13.12). We are again controlling the experimentwise error rate, this time for the set of all possible contrasts.

Experimenters who have used the Scheffé procedure (or, for that matter, the Tukey procedure, which will be considered next), have sometimes been perplexed to find that a significant test of the overall main effect has not been followed by at least one significant contrast. If there is some difference within the set of μ_j, why is this difference not reflected within one of the subsets of μ_j subsequently considered? The answer lies in the fact that if the overall test is significant at the α level, at least the maximum possible contrast will also be significant at the α level. Unfortunately, the maximum possible contrast may have been of little interest and, therefore, may not have been computed. There is no guarantee that the obvious contrasts (e.g., differences within pairs of means) or the contrasts most interesting for the psychologist will be significant when the overall F test is.

Related to the point just discussed is the fact that the power of the Scheffé test is equal to that of the overall F test only when the detection of the maximum possible contrast is at issue. The power to detect the significance of other contrasts is lower than that for the main effect test because the EW is held constant over the entire set of possible contrasts. Scheffé, recognizing the power problem, suggests that the EW be set at 10 percent. This may strike some as heresy, but it should be noted that EW and EC are different concepts, and there is no reason to demand that traditional EC levels of significance be applied to the EW. Even with the EW at 10 percent, the EC will generally be quite low.

A numerical example. Consider the contrasts $\mu_{NRP} - \frac{1}{3}(\mu_N + \mu_R + \mu_P)$ based on the illustrative experiment presented earlier in this chapter. For the sample data of Section 13.2.1, with $n = 10$, $\hat{\psi}_3 = 8.8 - (\frac{1}{3})(4.2 + 3.4 + 2.5) = 5.43$. Assume that $MS_{S/A} = 15$. Then the 90 percent confidence interval based on Equation (13.14) is

$$5.43 - \sqrt{(15)\left(\frac{12/9}{10}\right)}\sqrt{(3)(2.25)} \leq \psi_3 \leq 5.43 + \sqrt{(15)\left(\frac{12/9}{10}\right)}\sqrt{(3)(2.25)}$$

$$= 1.76 < \psi_3 < 9.10$$

Since the interval does not contain zero, the null hypothesis will be rejected.

13.3.3 Post hoc analysis: Tukey's multiple comparison method. The statistical literature contains an overabundance of proposals for dealing with the multiple comparison-error rate problem. Of this variety of approaches, only the Scheffé and the Tukey HSD (honestly significant difference) (1953) approaches hold the EW at α for the entire possible set of contrasts. Tukey's method is based on the distribution of q, the studentized range. This distribution is defined by first taking the range (R) for a set of a independent, normally distributed values, group means in our applications. R is then divided by S_Y, the estimate of the standard deviation of the values whose range is being considered. The sampling distribution of q is the sampling distribution of R/S_Y and depends upon a (the number of means ranged over) and upon the df associated with S_Y. Significant values are presented in Table A–9 in the Appendix. Assuming a completely randomized one-factor design and that the estimates of the treatment population means are independent, normally distributed, and have homogeneous variances, the probability is $1 - a$ that

$$\textbf{(13.15)} \qquad \psi_P - qS_{\bar{Y}}(\tfrac{1}{2}\sum_j |w_{jp}|) \leq \psi_p \leq \hat{\psi}_p + qS_{\bar{Y}}(\tfrac{1}{2}\sum_j |w_{jp}|)$$

for all values of p (i.e., for all possible contrasts), where $q = q_{\alpha;a,a(n-1)}$, the q required for significance at the α level when there are a means within the range and the error *df* are $a(n-1)$, $S_{\bar{Y}} = \sqrt{MS_{S/A}/n}$, and $|w_{jp}|$ is the absolute value of the jth weight for the pth contrast. To test the null hypothesis that

$$\psi = 0$$

we note whether

(13.16)
$$\frac{\hat{\psi}}{S_{\bar{Y}}(\frac{1}{2}\sum_j |w_j|)} > q$$

or, as is more commonly done, whether

(13.17)
$$\hat{\psi} > S_{\bar{Y}}(\tfrac{1}{2} \sum_j |w_j|)q$$

A numerical example. We will apply Equation (13.15) to the data previously analyzed by the Scheffé method. The 90 percent confidence interval is

$$5.43 - (3.36)\sqrt{\frac{15}{10}}(\tfrac{1}{2})(2) \leq \psi_B \leq 5.43 + (3.36)\sqrt{\frac{15}{10}}(\tfrac{1}{2})(2) = 1.33 \leq \psi_3 \leq 9.53$$

Again, we reject H_0 with $EW = .10$.

13.3.4 Comparison of the Scheffé and Tukey methods. Note that the interval computed for the Tukey procedure is slightly wider than that computed earlier, for the same data, for the Scheffé method. In view of our discussion of confidence intervals and hypothesis tests (Section 3.8), it follows that the Scheffé procedure also provides a more powerful significance test. This will generally be true of contrasts involving more than two treatment means. The situation is somewhat different when each side of the contrast involves exactly one treatment. For example, suppose we contrast the R and NRP treatments. Then $\psi_p = 8.8 - 4.2 = 4.6$. The Scheffé 90 percent interval is

$$4.60 - \sqrt{(15)\left(\frac{2}{10}\right)}\sqrt{(3)(2.25)} \leq \psi_p \leq 4.60 + \sqrt{(15)\left(\frac{2}{10}\right)}\sqrt{(3)(2.25)}$$

or $.1 < \psi_p < 9.1$. The Tukey 90 percent interval is

$$4.60 - \sqrt{\frac{15}{10}}(\tfrac{1}{2})(2)(3.36) \leq \psi_p \leq 4.60 + \sqrt{\frac{15}{10}}(\tfrac{1}{2})(2)(3.36)$$

or $.50 \leq \psi_p \leq 8.70$. The Tukey approach provides a narrower interval and consequently a more powerful test of the contrast. This will generally be true when the contrast involves only two treatments.

In attempting to evaluate the merits of the Scheffé and Tukey procedures, we note the following:

(a) The Scheffé procedure is less sensitive to violations of normality and homogeneity of variance assumptions.

(b) The Tukey procedure requires equal n, while the Scheffé procedure does not.

(c) Tables of the F statistic are more widely available than tables of the q statistic.

(d) As noted above, the Tukey procedure is more powerful for contrasts of the type, $\mu_j - \mu_{j'}$, while the Scheffé procedure is more powerful for more complex contrasts involving more than two treatment means.

If the ns are equal, and if the normality and homogeneity of variance assumptions appear reasonable, (d) should determine the choice of approach. When the experimenter is only interested in comparing two means at a time, the Tukey procedure is preferred; if more complex processes are also to be investigated, the Scheffé procedure should be used.

13.3.5 Alternatives to the Scheffé and Tukey methods, the Newman-Keuls procedure. When only contrasts between two treatments (that is, the w_j are $1, -1, 0, 0, \ldots$) are of interest, the Scheffé and Tukey procedures described are overly conservative since they control EW for infinitely large sets of contrasts. The Newman-Keuls test, like the Tukey test, uses Table A–9; however, the procedure is somewhat different. For simplicity, we label the group means $\overline{X}_1, \overline{X}_2, \ldots, \overline{X}_a$, from smallest to largest. Then, we first compare $\overline{X}_a - \overline{X}_1$ with $S_{\overline{Y}} q_{\alpha;a,df_2}$ and reject $H_a(\mu_a = \mu_1)$ if the difference between means exceeds the critical value. If the result is not significant, we stop testing; if the result is significant, we evaluate $\overline{X}_{a-1} - \overline{X}_1$ and $\overline{X}_a - \overline{X}_2$, each against $S_{\overline{Y}} q_{\alpha;a-1, df_2}$. Note that the q-value is now obtained from the $(a - 1)$st column of Table A–9. Testing proceeds as in the first stage. If we reject $H_{a-1}: \mu_{a-1} = \mu_1$, we now evaluate $\overline{X}_{a-2} - \overline{X}_1$ and $\overline{X}_{a-1} - \overline{X}_2$, this time using $q_{\alpha;a-2, df_2}$.

The Newman-Keuls test controls EW, yet provides greater power than other tests we have considered because (a) it is based only on the set of $\binom{a}{2}$ possible contrasts between pairs of individual treatments and (b) successive tests involving smaller subsets of means are carried out as though only those treatments existed in the experiment.

Setting the significance level equal to EW/k. In discussing planned contrasts, we suggested controlling EW by testing each contrast at the EW/k level. It is worth noting that even when many contrasts are planned, the confidence interval for the Bonfonerri t (so named because the approach is based on Bonfonerri's inequality) will frequently be shorter and power consequently greater than that for the Scheffé or Tukey tests. Comparisons of entries in A–12 (Bonfonerri t values) with $\sqrt{(a - 1)F}$, the corresponding critical value for the Scheffé test, will permit the experimenter to determine which procedure to use, assuming that he has planned a large number of contrasts. Dunn (1961) has published explicit comparisons of the Bonfonerri, Scheffé, and Tukey procedures, as well as a further discussion.

13.3.6 Dunnett's test: comparing a control group with experimental groups.
If one of the a groups is a control group, there are $a - 1$ nonindependent
comparisons which may be of interest. In order to keep the error rate at α
for the entire set of comparisons, the statistic,

$$\frac{\overline{Y}_{.j} - \overline{Y}_c}{\sqrt{MS_{S/A}[(1/n_j) + (1/n_c)]}}$$

is evaluated against the statistic d, whose distribution depends upon the value
of a and the error df. Significant values of d are presented in Table A–10 in
the Appendix. The null hypothesis test can also be carried out by determining whether

$$(13.18) \qquad \overline{Y}_{.j} - \overline{Y}_c > \left[\sqrt{MS_{S/A}\left(\frac{1}{n_j} + \frac{1}{n_c}\right)}\right] d_{\alpha;a,a(n-1)}$$

where $\overline{Y}_c$ is the mean of the control group, n_j and n_c are the numbers of observations in the jth experimental group and the control group, and $d_{\alpha;a,a(n-1)}$
is the value of d required for significance at the α level when there are a
groups (including group C), with n measurements in a group. For designs
other than the completely randomized one-factor, $MS_{S/A}$ and $a(n - 1)$ are
replaced by the appropriate error term and error df, and the ns are always
the numbers of observations on which the two means are based.

Note that Table A–10 presents one-tailed probabilities. Therefore, if
we are concerned with detecting differences in either direction at the .05 level,
we require the value of d significant at the .025 level. Similarly, if we require
a 95 percent confidence interval, the d to use would be the value appropriate
to the .025 level of significance.

The probability is $1 - \alpha$ that all $a - 1$ confidence intervals include the
true difference $(\mu_{.j} - \mu_c)$ if the confidence interval is computed as

(13.19)

$$(\overline{Y}_{.j} - \overline{Y}_c) - dS\sqrt{\frac{1}{n_j} + \frac{1}{n_c}} \leq \mu_j - \mu_c \leq (\overline{Y}_{.j} - \overline{Y}_c) + dS\sqrt{\frac{1}{n_j} + \frac{1}{n_c}}$$

where $S^2 = MS_{\text{error}}$ and d is the d required for significance at the α level.

13.4 THE ANALYSIS OF INTERACTION

Suppose that we have selected a group of subjects who score high on the
MMPI obsessive-compulsive scale (OC), a group that scores high on the
psychopathic deviant (PD) scale, and a control (C) group consisting of subjects who are close to the mean on both measures. A second variable in
the study is the type of motivation: gains and losses of money following

correct and incorrect decisions in a gambling task (GL), gains for correct decisions but no losses for incorrect decisions (G), and losses for incorrect decisions but no gains for correct ones (L). We might attempt to determine whether the difference between the GL mean and the average of the combined G and L data varies with personality type. That is, we are interested in variability in the contrast $2\mu_{GL} - \mu_G - \mu_L$ as a function of personality type. This suggests that the interaction of personality and motivation is of interest, which is true but not sufficient to describe our goal. A significant interaction merely suggests that the spread among motivation effects is a function of personality; we want to know if the particular motivational contrast cited varies with personality. We might push our inquiry into the nature of the interaction still further. For example, it is possible to ask whether the contrast cited above is different for the PD population as compared to the same contrast for the OC population. In this case, we are interested in a contrast among contrasts, specifically,

$$(2\mu_{GL,OC} - \mu_{G,OC} - \mu_{L,OC}) - (2\mu_{GL,PD} - \mu_{G,PD} - \mu_{L,PD})$$

The first question involves the interaction of personality type with the pth component of motivation; i.e., pers $\times$ p(mot). The general form of the null hypothesis when we are considering contrasts among the means at the levels of A as a function of level of B (in our example, *motivation* would be A, *personality*, B) is

$$H_0 : \sum_j w_j \mu_{j1} = \sum_j w_j \mu_{j2} = \cdots = \sum_j w_j \mu_{jb}$$

That is, the null hypothesis is that the pth contrast among the a treatment means is the same at all levels of B.

The second question raised above involves the interaction of the qth component of personality with the pth component of motivation; i.e., q(pers) $\times$ p(mot). The general form of the null hypothesis is

$$H_0 : \sum_j \sum_k w_{j.} w_{.k} \mu_{jk} = 0$$

and $\sum_j w_j = \sum_k w_k = 0$.

In this section we will present computations for significance tests of the two types of null hypotheses.

13.4.1 Calculations for $SS_{p(A) \times B}$ and its error terms. We are interested in whether the magnitude of the pth contrast among the means of the various levels of A changes as a function of the level of B. Computations for this sort of contrast are very similar to those for the sum of squares for a main effect. In the present case, we measure the variability of b contrasts about an average contrast; in the case of the main effect, we measure the variability of b means about a grand mean. This analogy suggests that the computations for $SS_{p(A) \times B}$ should involve $b - 1$ squared quantities, and that this sum of squares will be distributed on $(b - 1)$ *df*. We can develop a set of rules for establishing computational formulas for contrasts on the basis of *df*, just as we did with main and interaction effects in previous chapters. Knowing that there are $b - 1$ squared quantities, we begin with

$$\sum_k^b (\quad)^2 - (\quad)^2$$

Next, we place all summation signs that are not outside the parentheses inside them, yielding

$$\sum_k^b \left(\sum_i^n \sum_j^a \quad \right)^2 - \left(\sum_i^n \sum_j^a \sum_k^b \quad \right)^2$$

The weights and the data values are next considered, yielding

$$\sum_k^b \left(\sum_i^n \sum_j^a w_{jp} Y_{ijk} \right)^2 - \left(\sum_i^n \sum_j^a \sum_k^b w_{jp} Y_{ijk} \right)^2$$

We divide by the product of the sum of the squared weights times the number of scores that are summed at a level of *A* prior to squaring. Thus, the final result is

(13.20) $$SS_{p(A) \times B} = \frac{\sum_k^b (\sum_i^n \sum_j^a w_{jp} Y_{ijk})^2}{n \sum_j w_{jp}^2} - \frac{(\sum_i^n \sum_j^a \sum_k^b w_{jp} Y_{ijk})^2}{nb \sum_j w_{jp}^2}$$

Note that the "correction term" is actually $SS_{p(A)}$.

The application of Equation (13.20) is illustrated using the data of Table 13–1. The sum of squares will be calculated for $p(A) \times B$, where 2, -1, and -1 are the coefficients for A_1, A_2, and A_3, respectively. We first calculate $SS_{p(A)}$, multiplying the sum of all A_1 scores by 2, the sum of all A_2 scores by -1, and the sum of all A_3 scores by -1, squaring the sum of these three cross-products, and dividing by 72 ($= bn \sum_j w_j^2$):

$$SS_{p(A)} = [(2)(4 + 5 + \cdots + 5 + 8) - (6 + 9 + \cdots + 17 + 15)$$
$$- (12 + 14 + \cdots + 20 + 28)]^2 / (12)(6)$$
$$= 715.68$$

To complete the sum of squares calculations, we carry out the above calculations at each level of *B*, and then sum the results:

$$\frac{\sum_k^b (\sum_i^n \sum_j^a w_{jp} Y_{ijk})^2}{n \sum_j w_j^2} = \frac{[(2)(4 + \cdots + 8) - (6 + \cdots + 8) - (12 + \cdots + 16)]^2}{(4)(6)}$$

$$+ \cdots$$

$$+ \frac{[(2)(6 + \cdots + 8) - (14 + \cdots + 15) - (24 + \cdots + 28)]^2}{(4)(6)}$$

$$= 812.71$$

TABLE 13–1 *Data for a numerical example*

	A_1	A_2	A_3	$\sum_j Y_{ijk}$	
	4	6	12	22	
	5	9	14	28	
B_1	6	8	13	27	$w_{1_q} = 0$
	8	8	16	32	
$\sum_i Y_{ij1} = 23$		31	55	$\sum_i \sum_j Y_{ij1} = 109$	
	5	14	15	34	
	7	12	18	37	
B_2	6	16	15	37	$w_{2_q} = +1$
	4	16	17	37	
$\sum_i Y_{ij2} = 22$		58	65	$\sum_i \sum_j Y_{ij2} = 145$	
	6	14	24	44	
	4	14	22	40	
B_3	5	17	20	42	$w_{3_q} = -1$
	8	15	28	51	
$\sum_i Y_{ij3} = 23$		60	94	$\sum_i \sum_j Y_{ij3} = 177$	
$w_{jp} =$	2	-1	-1		

Combining terms,

$$SS_{p(A) \times B} = 812.71 - 715.68$$

$$= 97.03$$

In order to test the significance of the $p(A) \times B$ term, we form an F ratio, dividing the sum of squares by the associated df ($b - 1$, in our example), and then dividing this by an error mean square. The answer to the question of what is the appropriate error term depends upon the design and the structural model. We will consider several cases in turn.

(a) *Completely randomized; repeated measurements, additive model.* The appropriate error term for $p(A) \times B$ is the error term for the AB interaction effect. This would be $MS_{S/AB}$ in the completely randomized design and MS_{ABS} in the repeated measurements design, assuming no interactions involving subjects in the population.

(b) *Repeated measurements, nonadditive model.* Assuming the existence of σ^2_{ABS}, we have

$$E(MS_{p(A) \times B}) = \sigma^2_e + \sigma^2_{S \times p(A) \times B} + \theta^2_{p(A) \times B}$$

which would be tested against a mean square with expectation,

$$E(MS_{S \times p(A) \times B}) = \sigma_e^2 + \sigma_{S \times p(A) \times}^2$$

We require a computational formula for $MS_{S \times p(A) \times B}$. The nature of the computations will be clearer if we better understand the nature of the interaction. What is involved is the variability of the contrasts among the a treatment means over the nb combinations of S and B, adjusting for the variability in the contrast that is due to S $(p(A) \times S)$ and the variability in the contrast as a function of B $(p(A) \times B)$. This is quite similar to the SB interaction effect; there our concern is with variability among means; here, it is with variability among contrasts. This analogy suggests $(n - 1)(b - 1)$ df. Using the df approach suggested earlier, we arrive at the formula

(13.21)

$$SS_{S \times p(A) \times B} = \frac{\sum_k^b \sum_i^n (\sum_j^a w_{jp} Y_{ijk})^2}{\sum_j^a w_{jp}^2} - SS_{p(A)} - SS_{p(A) \times B} - SS_{p(A) \times S}$$

We have already presented formulas for the $p(A) \times B$ and $p(A)$ terms. For the subject interaction, we have

(13.22)
$$SS_{p(A) \times S} = \frac{\sum_i^n (\sum_j^a \sum_k^b w_{jp} Y_{ijk})^2}{b \sum_j^a w_{jp}^2} - SS_{p(A)}$$

Let us assume for Table 13–1 that the scores $\langle 4, 6, 12; 5, 14, 15; 6, 14, 24 \rangle$ all come from one subject, $\langle 5, 9, \ldots, 14, 22 \rangle$ from a second, and so on. Then we have scores under all nine treatment combinations for each of four subjects. Then,

$$SS_{p(A) \times S} = \frac{[2(4 + 5 + 6) - (6 + 14 + 14) - (12 + 15 + 24)]^2}{(3)(6)} + \cdots$$

$$+ \frac{[2(8 + 4 + 8) - (8 + 16 + 15) - (16 + 17 + 28)]^2}{(3)(6)} - SS_{p(A)}$$

$$= 720.61 - 715.68$$

$$= 4.93$$

In order to complete the calculations for $S \times p(A) \times B$, we require a measure of the variability in $p(A)$ over SB combinations, which is

$$SS_{\overline{S \times p(A) \times B}} = \frac{[(2)(4) - 6 - 12]^2 + \cdots + [(2)(8) - 15 - 28]^2}{6} - SS_{p(A)}$$

$$= 826.50 - 715.68$$

$$= 110.82$$

Finally, we have

$$SS_{S \times p(A) \times B} = SS_{\overline{S \times p(A) \times B}} - SS_{p(A) \times B} - SS_{p(A) \times S}$$

$$= 110.82 - 97.03 - 4.93$$

$$= 8.86$$

Suppose we desire an F test of the $p(A) \times B$ effect. Then,

$$MS_{p(A) \times B} = \frac{SS_{p(A) \times B}}{b - 1}$$

$$= \frac{97.03}{2}$$

$$= 48.52$$

and

$$MS_{S \times p(A) \times B} = \frac{SS_{S \times p(A) \times B}}{(n - 1)(b - 1)}$$

$$= \frac{8.86}{6}$$

$$= 1.48$$

Then,

$$F = \frac{48.52}{1.48}$$

$$= 32.78$$

(c) *One between- and one within-subjects variable.* Suppose B is a between-subjects variable and A is a within-subjects variable. (If A, the variable whose means are contrasted, is a between-subjects variable, the error term is just $MS_{S/A}$; partitioning of the overall error term is sensible only when there are repeated measurements on the variable whose means are contrasted.) Reasoning similar to that for the repeated measurements design leads us to

$$E(MS_{p(A) \times B}) = \sigma_e^2 + \sigma_{S \times p(A)/B}^2 + n\theta_{p(A) \times B}^2$$

and

$$E(MS_{S \times p(A)/B}) = \sigma_e^2 + \sigma_{S \times p(A)/B}^2$$

Essentially, we want to measure the variability of the contrast among subjects in a particular level of B, do this for each level of B, and then pool the results. Thus, there will be n contrasts corrected for their average contrast within a level of B. In short, there are $b(n - 1)$ squared quantities in the formula for $SS_{S \times p(A)/B}$, and this sum of squares is distributed on $b(n-1)$ *df*. According to the approach to sums of squares developed earlier in this section, we begin with

$$\sum_k^b \sum_i^n (\quad)^2 - \sum_k^b (\quad)^2$$

Next, we place the remaining summations within the parentheses:

$$\sum_k^b \sum_i^n \left(\sum_j^a \quad Y_{ijk} \right)^2 - \sum_k^b \left(\sum_i^n \sum_j^a \quad Y_{ijk} \right)^2$$

The weights are still missing within the parentheses. Therefore, we have

$$\sum_k^b \sum_i^n \left(\sum_j^a w_{jp} Y_{ijk} \right)^2 - \sum_k^b \left(\sum_i^n \sum_j^a w_{jp} Y_{ijk} \right)^2$$

Division by $\sum_j^a w_{jp}^2$ times the number of scores within the parentheses at a level of j (the index of the weights) yields the final result:

$$(13.23) \qquad SS_{S \times p(A)/B} = \frac{\sum_k^b \sum_i^n (\sum_j^a w_{jp} Y_{ijk})^2}{\sum_j w_{jp}^2} - \frac{\sum_k^b (\sum_i^n \sum_j^a w_{jp} Y_{ijk})^2}{n \sum_j w_{jp}^2}$$

The application of Equation (13.23) is illustrated by calculations on the data of Table 13–1, again assuming that we wish to contrast the effect of A_1 with the average A_2 and A_3 effect. In this case, we assume that there is a total of 12 subjects, four at each level of B. We have

$$SS_{S \times p(A)/B} = \frac{[(2)(4) - 6 - 12]^2 + \cdots + [(2)(8) - 15 - 28]^2}{6}$$

$$- \left\{ \frac{[(2)(4 + \cdots + 8) - (6 + \cdots + 8) - (12 + \cdots + 16)]^2}{(4)(6)} + \cdots \right.$$

$$\left. + \frac{[(2)(6 + \cdots + 8) - (14 + \cdots + 15 - (24 + \cdots + 28)]^2}{(4)(6)} \right\}$$

$$= 826.50 - 812.71$$

$$= 13.79$$

To test the $p(A) \times B$ effect, we have

$$F = \frac{SS_{p(A) \times B}/(b-1)}{SS_{p(A) \times S/B}/[b(n-1)]}$$

$$= \frac{48.52}{13.79/9}$$

$$= 31.71$$

13.4.2 Single *df* components of interaction and their error terms. It was suggested earlier that contrasts among contrasts might be of interest in some studies. For example, does

$$(2\mu_{GL,OS} - \mu_{G,OS} - \mu_{L,OS}) - (2\mu_{GL,PD} - \mu_{G,PD} - \mu_{L,PD})$$

equal zero? We might speak of the pth contrast among motivation level and the qth contrast among personality types, where

$$w_{1p} = 2, \qquad w_{2p} = -1, \qquad w_{3p} = -1$$

$$w_{1q} = 0, \qquad w_{2q} = 1, \qquad w_{3q} = -1$$

In general, the null hypothesis is

$$H_0: \sum_{j}^{a} \sum_{k}^{b} w_{jp} w_{kq} \mu_{jk} = 0$$

The appropriate sum of squares is

(13.24) $$SS_{p(A) \times q(B)} = \frac{\left(\sum_{j}^{a} \sum_{k}^{b} \sum_{i}^{n} w_{jp} w_{kq} Y_{ijk}\right)^2}{n \sum_{j} \sum_{k} (w_{jp} w_{kq})^2}$$

This formula conforms to our discussion of computational formulas in the preceding section. To see this, consider the product of $w_{jp} w_{kq}$ as a single weight; i.e., $w'_{jk,pq} = w_{jp} w_{kq}$. Then, since the quantity is on 1 *df*, we have first

$$(\quad)^2$$

All summations not outside the parentheses should be within the parentheses, yielding

$$\left(\sum_{j}^{a} \sum_{k}^{b} \sum_{i}^{n} \quad\right)^2$$

As before, we insert the weight and the score within the parentheses, yielding

$$\left(\sum_{j}^{a} \sum_{k}^{b} \sum_{i}^{n} w'_{jp,kq} Y_{ijk}\right)^2$$

We divide by the sum of the squared weights times the number of scores at each level of the index, jk. The final result is

(13.25) $$SS_{p(A) \times q(B)} = \frac{\left(\sum_{j}^{a} \sum_{k}^{b} \sum_{i}^{n} w'_{jp,kq} Y_{ijk}\right)^2}{n \sum_{j} \sum_{k} w'^2_{jp,kq}}$$

which is algebraically identical to that of Equation (13.24).

As an illustrative example, we apply Equation (13.24) to the data of Table 13–1:

$$SS_{p(A) \times q(B)} = \frac{[(0)(2)(23) + (0)(-1)(31) + \cdots + (-1)(-1)(94)]^2}{(4)(12)}$$

$$= 17.52$$

We next consider the error terms for this contrast among contrasts. If there are n different subjects in each of the ab cells, $MS_{S/AB}$ is the appropriate error term. If there are n subjects at each level of b, all of whom go through all levels of A, the error SS is given by Equation (13.23). If A is the between-subjects variable and B is the within-subjects variable, we modify Equation (13.23) by interchanging $\sum_j^a$ and $\sum_k^b$. Thus,

$$(13.23')\quad SS_{S \times q(B)/A} = \frac{\sum_j^a \sum_i^n (\sum_k^b w_{kq} Y_{ijk})^2}{\sum_k w_{kq}^2} - \frac{\sum_j^a (\sum_i^n \sum_k^b w_{kq} Y_{ijk})^2}{n \sum_k w_{kq}^2}$$

If all subjects go through all ab combinations, we require $SS_{S \times p(A) \times q(B)}$ for the error sum of squares. This is a measure of the variability of the $p(A) \times q(B)$ contrast over subjects. Then, we have n contrasts, one for each subject, deviated about their average. The result is

$$(13.26)\quad SS_{S \times p(A) \times q(B)} = \frac{\sum_i^n (\sum_j^a \sum_k^b w_{jp} w_{kq} Y_{ijk})^2}{\sum_j \sum_k (w_{jp} w_{kq})^2} - SS_{p(A) \times q(B)}$$

We apply this formula to the data of Table 13–1, again assuming four subjects with nine scores on each; e.g., $\langle 4, 6, 12; 5, 14, 15; 6, 14, 24 \rangle$ are assumed to come from S_1. We have

$$SS_{S \times p(A) \times q(B)} = \frac{[(0)(2)(4) + (0)(-1)(6) + \cdots + (-1)(-1)(24)]^2}{12} + \cdots$$

$$+ \frac{[(0)(2)(8) + (0)(-1)(8) + \cdots + (-1)(-1)(28)]^2}{12} - 17.52$$

$$= 21.75 - 17.52$$

$$= 4.23$$

The F test of $p(A) \times q(B)$ would be

$$F = \frac{17.52}{4.23/3}$$

$$= \frac{17.52}{1.41}$$

$$= 12.43$$

13.5 CONCLUDING REMARKS

Much time and space has been devoted to some rather complex analyses of interaction that provide inferences about highly specific null hypotheses. This was done because it is felt that the fact that the AB interaction is significant is only a first step toward understanding the interaction. A significant F for a main effect merely tells us that some treatment population means differ, not which ones; similarly, a precise understanding of the nature of interaction only *begins* with the overall test.

The analysis of interaction may result in the same error rate problems that were discussed in Section 13.3. When comparisons are planned, the developments of Section 13.3.1 again apply. For post hoc comparisons, we may readily generalize to tests for null hypotheses of the form

$$H_0: \sum_j \sum_k w_{jp} w_{jk} \mu_{jk} = 0$$

The criterion F will be $(a - 1)(b - 1)$ times the F required for significance at the α level, distributed on $(a - 1)(b - 1)$ and the error df. Note that contrasts of this form do not exhaust the universe of possible contrasts embedded in the set of ab means. For example, $\overline{Y}_{11} - \overline{Y}_{32}$ is not of this form. If one is interested in this wider universe of contrasts in which we are essentially treating the two-factor design as consisting of one factor with ab levels, then the appropriate factor by which the critical F is to be multiplied should be $ab - 1$. We do not have a specific solution when the null hypothesis is of the form

$$H_0: \sum_j w_j \mu_{j1} = \cdots = \sum_j w_j \mu_{jb} = 0$$

We can only note that such contrasts should be regarded with even more than the usual caution until further data are available.

EXERCISES

13.1   Consider the following anova table and group means (each mean is the sum of scores for the group divided by 75—15 subjects, five measurements each).

SV	df	SS	MS	F
Total	374	32,086.80		
Between S	74	27,938.15		
Groups	4	7,117.16	1,779.29	5.98*
S/G	70	20,820.99	297.44	
Within S	300	4,148.65		
Time	4	1,846.20	461.55	60.49*
T × G	16	166.24	10.39	1.36
S × T/G	280	2,136.21	7.63	

*p < .001

A	B	C	D	E
20.546	14.560	13.332	11.160	7.200

Perform both the Scheffé and Tukey tests, making all comparisons of the form *A* versus *B*, *A* versus *C*, etc.

13.2 (a) *GSR* measures are taken on parachutists. Measures are obtained on groups two weeks before the jump (*BJ*-2), one week before (*BJ*-1), on the day of the jump prior to jumping (*DJ*-*P*), and on the day of the jump after jumping (*DJ*-*A*). In addition, a control group of normal (nonparachuting) cowards is tested (*C*). The mean scores for the groups are:

BJ-2	*BJ*-1	*DJ*-*A*	*DJ*-*P*	*C*
5	5	7	9	2

One hypothesis is that parachutists two weeks before the jump behave similarly to the controls. Determine whether these two groups differ significantly from the other three groups. There are six subjects in each group, and the MS_W was 4.0.

(b) Suppose the above contrast was one of a number being tested. What would your criterion of significance then be? Find both the Scheffé and Tukey confidence intervals.

13.3 Compute the *SS* for numerator and denominator of each of the following contrasts, assuming that all subjects go through all levels of *B*:

(a) the mean for treatments B_1 and B_2 versus the mean for treatment B_3,

(b) the variability in the above contrast over the levels of *A*.

		B_1	B_2	B_3
A_1	S_{11}	4	5	3
	S_{21}	1	6	2
A_2	S_{12}	5	4	2
	S_{22}	3	7	4

13.4

	A_1	A_2	A_3
B_1	21	16	9
B_2	4	12	7

Each cell contains a group total based on five subjects. Compute the SS_{AB} by making use of the fact that the *AB* interaction can be viewed as a pool of two orthogonal sums of squares.

13.5 Five rats are each tested in a Skinner Box under treatments: four drugs and a placebo. Tests are separated in time to minimize carryover effects and the orders of presentation are Latin squared. The residual *MS* is 10.0. The means are

D_1	D_2	D_3	D_4	*C*
1	6	8	13	7

Carry out both Dunnett's test for the D_j versus *C* and Tukey's test for all pairs at the 5 percent level. What conclusions do you reach on the relative power to test D_j versus *C*? Why do you think this happens?

13.6 Post Ph.D. productivity measures are obtained on random samples of size 10 of clinicians, experimentalists, and social psychologists from the University of California (Berkeley), Stanford, Minnesota, Northwestern, Penn State, and Yale. Lay out a data matrix for the appropriate design. Then set up contrasts to test the following hypotheses:

H_1: Midwestern Ph.D.s are less likely to perish than are non-Midwestern Ph.D.s.

H_2: This regional difference is more marked for the public than for the private institutions.

H_3: This regional difference is more marked among clinicians than among nonclinicians.

H_4: Minnesota clinicians outpublish all other clinicians.

H_5: Minnesota's superiority is more marked in clinical than in other areas.

REFERENCES

Dunn, O. J., "Multiple comparisons among means," *Journal of the American Statistical Association*, 56: 52–64 (1961).

Scheffé, H. *The Analysis of Variance*, New York: Wiley, 1959.

Tukey, J. W., "The problem of multiple comparisons," Unpublished manuscript.

SUPPLEMENTARY READINGS

Ryan has discussed the rationale underlying multiple comparison tests and has suggested some extension of available tests in the following two articles:

Ryan, T. A., "Multiple comparisons in psychological research," *Psychological Bulletin*, 56: 26–47 (1959).

Ryan, T. A., "Significance tests for multiple comparisons of proportions, variances, and other statistics," *Psychological Bulletin*, 57: 318–328 (1960).

Wilson has criticized Ryan's approach. Ryan expanded his original discussion in a reply to Wilson. The references are:

Wilson, W., "A note on the inconsistency inherent in the necessity to perform multiple comparisons," *Psychological Bulletin*, 59: 296–300 (1962).

Ryan, T. A., "The experiment as the unit for computing rate of error," *Psychological Bulletin*, 59: 301–305 (1962).

A general review of a variety of techniques for making multiple comparisons may be found in:

Miller, R. G. Jr., *Simultaneous Statistical Inference*. New York: McGraw-Hill, 1966.

Further Data Analyses: Quantitative Independent Variables

14.1 INTRODUCTION

In Chapter 13 various procedures for comparing the means of treatment groups were discussed. The primary concern was answering questions that seem most appropriate when the independent variable is qualitative. Certainly it is possible to investigate the contrasts of the previous chapter even when the independent variable is quantitative; however, with quantitative variables it will usually be more interesting to consider the overall trend in the treatment group means rather than to make the specific comparisons among means, which were the chief concern in Chapter 13. The formulas are basically the same for quantitative and qualitative contrasts. This being the case, the main topic of the present chapter will be development of the rationale for the evaluation of trend.

Consider an experimental study of generalization. A mild shock is paired with a 1,000-Hz tone. Subjects are then divided into several groups, each of which experiences a different tonal frequency but no shock. Galvanic skin responses (GSR) are measured in this test phase for each individual. Let us assume that there are two independent processes at work that sum together to produce the individual scores. One process is the consequence of stimulus frequency; the effect upon GSR increases linearly as a function of the frequency of the test tone. A second process is generalization; the effect upon GSR decreases symmetrically as a function of distance of the test tone from the training stimulus of 1,000 Hz. Figure 14.1 depicts this hypothetical situation. The dashed lines represent the effects of the two processes underlying the data. Summing these yields the α_j, represented by the solid line.

In actual fact, all the experimenter really has is a set of observed deviations from the means, the xs of Figure 14.1. Assuming that the F test of the *frequency* main effect is significant, we are left with the inferential problem of determining whether there is a stimulus frequency process, a generalization process, or both, or other processes contributing to the significant variation

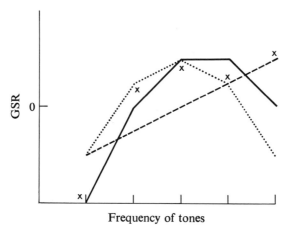

FIGURE 14-1 *Assumed effects of two processes, the sum of effects, and observed data points*

within the observed set of means. With respect to other possible processes, note the slight upturn in the last data point. If this is more than chance variability, the straight line and inverted U representing stimulus frequency and generalization could not account for it. There is the possibility that some S-shaped process is also at work.

Drawing inferences about the true shape of the function relating treatment population means and levels of the independent variable is the subject matter of the present chapter. The procedures involved are usually referred to as trend, or orthogonal polynomial, analysis. Such procedures can attack additional questions as well. For example, suppose we carried out the experiment with several different groups of subjects drawn from populations defined by various degrees of severity of schizophrenia. Several personality theorists have considered the possibility that schizophrenics generalize more than normal subjects. We would therefore expect that the underlying generalization gradient would become flatter as severity increased, the very schizophrenic subject showing little variation in *GSR* over a wide range of test tones. If this hypothesis is true, we should have a personality × frequency interaction. However, that in itself does not prove the case because such an interaction could be due to other variations in underlying processes. For example, if the linear function relating *GSR* and intensity varied in regression coefficient over clinical categories, this too would give rise to a significant interaction. In this chapter, we will consider analyses that attack questions about variations in the slope and curvature of several population functions.

Before developing the rationale and computations involved in trend analysis, it might help to consider a second instance of its application, this time an experiment on signal detection. A subject sits a fixed distance from a 25-sq. ft. screen. Every ten seconds, a small circle of light appears on the screen, and the subject must report the location ot this target. Detection time is recorded by the experimenter. One independent variable is the intensity of the target illumination. A second variable is the structure of the screen, which may be one open area, or which may be divided by a vertical line into two equal segments or by two vertical lines into three equal segments, and so on. The experimenter hypothesizes that a certain amount of segmenting of the screen facilitates the search. However, there will be some point at which further segmenting will actually result in deterioration of performance; the subject will become confused and will repeatedly search the same segment, thus losing time. It is further hypothesized that this optimal number of segments will be a function of intensity of illumination. The higher the illumination, the higher will be the optimal number of segments. Figure 14–2 illustrates one way the data might look if the experimenter's hypotheses were correct.

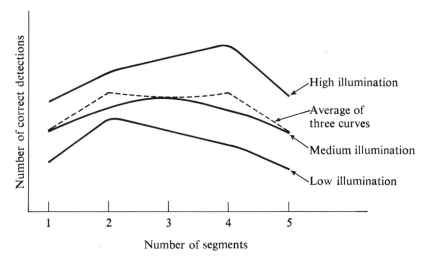

FIGURE 14–2 *Hypothesized functions for a signal detection experiment*

In Figure 14–2 the behavioral hyptheses mentioned above have been translated into hypotheses about performance trends over numbers of segments. With regard to the average curve—the plot of the main effect of segments—one might note that (a) if a straight line were fitted to this curve, its slope would be zero, and (b) the function appears to be better fitted by a curved than by a straight line. The behavioral hypotheses also lead to definite inferences about the three intensity curves. If a straight line were fitted to each, the three lines would have different slopes. Furthermore, it appears that the three functions differ in their shapes.

In view of the initial hypotheses, and of Figure 14–2, which provides one plausible representation of the hypotheses, it seems that tests are required to answer such questions as

(a) Does the plot of mean detection time as a function of number of segments have a slope other than zero?
(b) Is the plot of mean detection time as a function of segments adequately fitted by a straight line?
(c) Do the three plots of detection time against segments differ in slope?
(d) Do the three plots of detection time against segments differ in shape?

The computations involved in answering questions such as these about the shapes and slopes of functions come under the general heading of trend. As in the example of the generalization experiment, the answer to our questions lies within the context of trend analysis.

14.2 ORTHOGONAL POLYNOMIALS

14.2.1 Rationale for testing null hypotheses. We begin our discussion of the complete trend analysis by noting that any a data points may be described by an equation having the general form

(14.1) $\qquad Y = b_0 + b_1 X + b_2 X^2 + \cdots + b_p X^p + \cdots + b_{a-1} X^{a-1}$

Equations of this form are referred to as *polynomial functions of order a* $-$ *1*. Figure 14–3 presents several such functions, each labeled by the appropriate equation. Note the restriction that if there are a points, the order of the polynomial is at most $a - 1$ (it can be less, since b_{a-1}, b_{a-2}, etc., can be zero). To understand why this is so, consider a first order polynomial, the straight line

(14.2) $\qquad\qquad\qquad Y = b_0 + b_1 X$

At least two data points are required in order to estimate the parameters, b_0 and b_1. Similarly, three data points are required in order to fit a second-order polynomial, otherwise known as a quadratic function,

(14.3) $\qquad\qquad\qquad Y = b_0 + b_1 X + b_2 X^2$

since three parameters are to be estimated. In general, a polynomial of order $a - 1$ involves the estimation of a parameters, and at least that many data points are required.

Suppose that the a data points are treatment population means and the values of X are levels of some quantitative independent variable; e.g., stimulus intensity, magnitude of reward, amount of practice. In this context, Equation (14.1) becomes

(14.1′) $\quad \mu_j = \beta_0 + \beta_1 X_j + \beta_2 X_j^p + \cdots + \beta_p X_j^p + \cdots + \beta_{a-1} X_j^{a-1}$

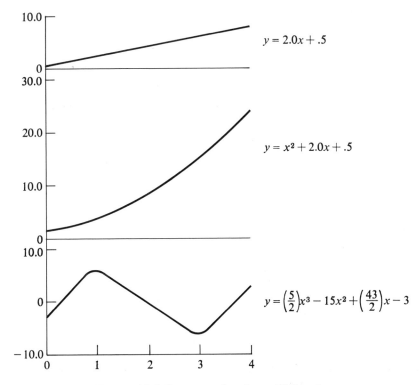

FIGURE 14–3 *Some sample polynomial functions*

which may be rewritten as

$$(14.4) \quad \mu_j = \beta'_0 + \beta'_1(a_1 + X_j) + \beta'_2(a_2 + b_2X_j + X_j^2) + \cdots$$
$$+ \beta'_{a-1}(a_{a-1} + b_{a-1}X_j + c_{a-1}X_j^2 + d_{a-1}X_j^3 + \cdots + X_j^{a-1})$$

where

$$\beta'_0 = \mu$$
$$\beta'_0 + a_1\beta'_1 + a_2\beta'_2 + \cdots + a_{a-1}\beta'_{a-1} = \beta_0$$
$$\beta'_{a-1} = \beta_{a-1}$$

Given the above definitions of β'_p, Equations (14.1′) and (14.4) are algebraically identical. The presentation will now be considerably simplified if we rewrite Equation (14.4) as

$$(14.5) \quad \mu_j = \beta'_0 + \beta'_1\xi_{j1} + \beta'_2\xi_{j2} + \cdots + \beta'\xi_{j,a-1}$$

where the definitions of the ξ_{jp} follow from a comparison of Equations (14.4) and (14.5). We have taken the trouble to recast Equation (14.1′) as (14.5) because, if the ξ_{jp} are properly chosen, the terms being summed are independent. The term, $\beta'_1\xi_{j1}$, represents an underlying linear process. $\beta'_2\xi_{j2}$ represents an underlying independent quadratic process, and so on; the critical point is that in this form, with ξ_{jp} properly chosen, the $a - 1$ functions are independent of each other. We generally refer to *orthogonal* (independent) *polynomial* functions.

Let us now reconsider some of the questions raised in Section 14.1. In the generalization example, we desired to determine whether a linear process due to stimulus frequency contributed to the overall main effect. In the detection example, the experimenter's hypotheses led him to expect that the underlying linear component would have a slope of zero. In terms of Equation (14.5), the relevant null hypothesis is

$$H_0: \beta'_1 = 0$$

In both experiments, there is reason to believe that the plot of the main effect against levels of the independent variable (frequency, number of segments) would have a quadratic component. The relevant null hypothesis is

$$H_0: \beta'_2 = 0$$

In general, tests of hypotheses about polynomial components that may contribute to main effects are of the form

$$H_0: \beta'_p = 0$$

In short, Equation (14.5) states that the plot of the μ_j against the levels of the independent variable may be viewed as resulting from the summing of several independent polynomial functions of different orders; Figure 14.1 presented a graphic analog to Equation (14.5). The statistical problem is to determine which of the possible $a - 1$ polynomial functions make a significant contribution to the plot of the μ_j, which of the regression coefficients (β') are greater than zero.

How do we test null hypotheses of the sort described? First, note that β'_p is the coefficient for the regression of μ_j on ξ_{jp}. In Chapter 12, we pointed out that in dealing with the regression of Y on X the appropriate expression for the regression coefficient is

$$\beta = \frac{\sum_j(Y_j - \bar{Y})(X_j - \bar{X})}{\sum_j(X_j - \bar{X})^2}$$

By analogy,

(14.6)
$$\beta'_p = \frac{\sum_j(\mu_j - \mu)(\xi_{jp} - \bar{\xi}_p)}{\sum_j\xi_{jp}^2}$$

We will show shortly that $\bar{\xi}_p = 0$.

From Equation (14.6) it follows that the null hypothesis $\beta'_p = 0$, is true if

$$\sum_j (\mu_j - \mu)\xi_{jp} = 0$$

However,

$$\sum_j (\mu_j - \mu)\xi_{jp} = \sum_j \xi_{jp}\mu_j - \mu \sum_j \xi_{jp} = \sum_j \xi_{jp}\mu_j$$

But this is the exact form of the null hypotheses introduced in Section 13.1 except that there we used w_j instead of ξ_j. It therefore follows that the computations to test for contributions of orthogonal polynomial functions are identical to those employed to test contrasts in Chapter 13 and that, furthermore, each null hypothesis of the class now under discussion will be tested on a single *df*.

Since Chapter 13 provides the computational formulas for sums of squares, all we really require is a method for obtaining the weights—in the case of trend, the ξ_{jp}. We will consider this in the next section and follow with a simple numerical example. But before proceeding, it might be well to briefly reconsider the application of trend analysis.

We apply the analyses of this chapter when we are interested in testing the contributions of various polynomial functions to the variability among the treatment effects. We may, as in the generalization example, actually conceptualize the data as the sum of several independent processes represented by different orders of polynomial functions. Alternatively, as in the signal detection example, we may be unable to label separate component processes but have reason to believe that the observed data function will be of a particular nature; this belief is consistent with viewing the data as if it were the sum of certain underlying components and proceeding to test for these components. In both the generalization and detection examples, we have reason to be interested in the linear and quadratic components. Each of these should be tested. Since the components are orthogonal, these tests will have accounted for two of the $a - 1$ *df*; we would then test the residual, $SS_A - SS_{\text{lin}} - SS_{\text{quad}}$, on $(a - 1) - 2$ *df*. Significance of the residual term would suggest the presence of processes that had not been expected prior to the experiment. If the residual is significant, it may be desirable to partition it into its polynomial components, testing them separately to better specify the unlooked-for process.

Frequently, the investigator's sole interest is in possible deviations from linearity. For example, a common paradigm in the study of memory involves presenting subjects with displays of digits and then measuring the latency with which a subject reports whether a test digit was or was not in the original display. The most common theoretical position holds that the subject has a representation of the display in memory and scans that representation one digit at a time, comparing each with the test digit. If the test digit had not been in the display, and if it requires t milliseconds to scan each digit in memory, latency is a linear function of K, the display size. In this example, we would perform two tests: linearity on 1 *df* and residual (deviations from linearity) on $(a - 1) - 1$ *df*. Significance of the latter term would pose a problem for the theory.

14.2.2 Deriving values of ξ. Table A–11 presents values of ξ' for various values of a under the assumptions that there are equal numbers of measures for each level of A and that the levels of A are equally spaced. The derivational technique frequently gives noninteger values of ξ. These values were multiplied by λ to give the integers, ξ', of Table A–11; remember that multiplication of a set of weights by a constant (λ) will not change the sum of squares based on those weights.

For several reasons, it is worthwhile to illustrate the method by which the tabled polynomial coefficients are derived. Hopefully, we will be more comfortable with the analyses if we have some feeling for where the coefficients come from. More important, there may be occasions in which ns are not equal or, more frequently, the levels of A are not equally spaced. Suppose we have a study of delay of reward with four delay intervals. The values of X, delay magnitude, and numbers of subjects are:

X_j	0	1	3	4
n_j	4	5	5	4

From Equations (14.4) and (14.5), we know that the linear coefficients, ξ_{j1}, are of the form

$$\xi_{j1} = a_1 + X_j$$

Substituting the actual delay intervals, we have

$$\xi_{11} = a_1 + 0$$
$$\xi_{21} = a_1 + 1$$
$$\xi_{31} = a_1 + 3$$
$$\xi_{41} = a_1 + 4$$

In order to achieve orthogonality, we now impose the restriction:

$$\sum_j n_j \xi_{jp} = 0$$

Then,

$$(4)(a_1) + (5)(a_1 + 1) + 5(a_1 + 3) + 4(a_1 + 4) = 0$$

and upon solving this, we have

$$18a_1 + 36 = 0$$
$$a_1 = -2$$

Then,

$$\xi_{11} = -2$$
$$\xi_{21} = -2 + 1 = -1$$
$$\xi_{31} = -2 + 3 = 1$$
$$\xi_{41} = -2 + 4 = 2$$

The quadratic coefficients are of the form

$$\xi_{j2} = a_2 + b_2 X_j + X_j^2$$

Then,

$$\xi_{12} = a_2 + (0)(b_2) + 0^2$$
$$\xi_{22} = a_2 + b_2 + 1$$
$$\xi_{32} = a_2 + 3b_2 + 9$$
$$\xi_{42} = a_2 + 4b_2 + 16$$

We have two unknowns, a_2 and b_2. However, because of the orthogonality requirement, we also have two simultaneous equations:

$$\sum_j n_j \xi_{j2} = 0$$
$$\sum_j n_j \xi_{j1} \xi_{j2} = 0$$

The second equation represents the orthogonality requirement. Applying these restrictions yields

$$(4)(a_2) + (5)(a_2 + b_2 + 1) + (5)(a_2 + 3b_2 + 9) + (4)(a_2 + 4b_2 + 16) = 0$$

$$(4)(-2)(a_2) + (5)(-1)(a_2 + b_2 + 1) + (5)(1)(a_2 + 3b_2 + 9)$$
$$+ (4)(2)(a_2 + 4b_2 + 16) = 0$$

Simplifying, we have

$$18a_2 + 36b_2 + 114 = 0$$
$$42b_2 + 168 = 0$$

Then,

$$b_2 = \frac{-168}{42} = -4$$

and

$$18a_2 + (36)(-4) + 114 = 0$$
$$a_2 = \tfrac{5}{3}$$

Substituting into the original expressions for ξ_{j2},

$$\xi_{12} = \tfrac{5}{3}$$
$$\xi_{22} = \tfrac{5}{3} + (-4) + 1 = -\tfrac{4}{3}$$
$$\xi_{32} = \tfrac{5}{3} + (3)(-4) + 9 = \tfrac{4}{3}$$
$$\xi_{42} = \tfrac{5}{3} + (4)(-4) + 16 = \tfrac{5}{3}$$

Let $\lambda = 3$. Then, we have integer values:

$$\xi'_{12} = 5$$
$$\xi'_{22} = -4$$
$$\xi'_{32} = -4$$
$$\xi'_{42} = 5$$

The cubic coefficients will involve three unknowns: a_3, b_3, c_3. However, we have three simultaneous equations:

$$\sum n_j \xi_{j3} = 0$$
$$\sum n_j \xi_{j1}\xi_{j3} = 0$$
$$\sum n_j \xi_{j2}\xi_{j3} = 0$$

We leave the solution as an exercise.

Usually, the values of X are equally spaced or can be transformed to a scale on which they are equally spaced; for example, if $A = 1, 2, 4, 8$, $\log_2 X = 0, 1, 2, 3$. When spacing is equal, and observations at each value of X occur equally frequently, Table A–11 immediately provides the desired coefficients.

14.2.3 A numerical example. As an example of the calculations, assume that we have the following treatment totals:

A_1	A_2	A_3	A_4
22	22	20	36

Further assume that $n = 10$ and $MS_{S/A} = 1.2$. We have

$$SS_A = \frac{(22)^2 + \cdots + (36)^2}{10} - \frac{(22 + \cdots + 36)^2}{40}$$
$$= 16.4$$

Applying Equation (13.6) with w_{jp} first equal to the ξ'_{j1}, we have

$$SS_{\text{lin }(A)} = \frac{[(-3)(22) + (-1)(22) + (1)(20) + (3)(36)]^2}{(10)(20)}$$
$$= 8.0$$

Using the ξ'_2 next, we have

$$SS_{\text{quad }(A)} = \frac{(-22 + 22 + 20 - 36)^2}{(10)(4)}$$
$$= 6.4$$

The cubic contrast yields

$$SS_{\text{cub }(A)} = \frac{[-22 + (3)(22) + (-3)(20) + 36]^2}{(10)(20)}$$
$$= 2.0$$

Note that

$$SS_A = SS_{\text{lin }(A)} + SS_{\text{quad }(A)} + SS_{\text{cub }(A)}$$

which must be true if our calculations are correct. Only the linear and quadratic components are significant; we conclude that the cubic sum of squares reflects chance variability.

The upper panel of Figure 14–4 contains plots of b_0', $b_1'\xi_1'$, and $b_2'\xi_2'$ for the data set just analyzed. The parameter b_0' is merely $\overline{Y}_{..}$; other coefficients are obtained from the general expression

(14.6) $$b_p' = \frac{\sum_j \xi_{jp}\overline{Y}_{.j}}{\sum_j \xi_{jp}^2}$$

Our numerical coefficients are

$$b_0' = \frac{22 + 22 + 20 + 36}{40} = 2.5$$

$$b_1' = \frac{(-3)(2.2) - 2.2 + 2.0 + (3)(3.6)}{20} = .20$$

$$b_2' = \frac{2.2 - 2.2 - 2.0 + 3.6}{4} = .40$$

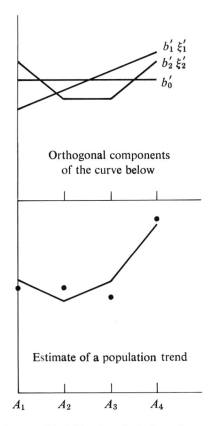

FIGURE 14–4 *Trend analysis for a data set*

14.3 MULTI-FACTOR DESIGNS

Since the cubic coefficient was not significant, we assume the population value of b_3' to be zero and consequently do not estimate a cubic component. As required by Equation (14.5), we literally sum the orthogonal components plotted in the upper panel of Figure 14–4. The result is the second-order polynomial plotted in the bottom panel. This function is not the function that best fits the data points; a perfect fit could be obtained by adding in $b_3'\xi_3'$. However, we do not wish to fit the data points (the $\overline{Y}_{.j}$); our goal is to describe the plot of the population values (the μ_j). Since our analysis has led us to conclude that the cubic component is not significant, the curve in the lower panel of Figure 14–4 is our estimate of the population function. The variability of data points about that function is assumed to be error variability, rather than the result of a cubic component in the population function.

14.3.1 The analysis of interaction. We have thus far considered the orthogonal components of a main effect. This analysis enables us to answer several specific questions, such as, Is the slope of the best fitting straight line significantly different from zero? Is there quadratic curvature? Other questions arise when the design involves more than one independent variable. If there is a significant interaction between the variables A and B, and if A is a quantitative variable, we might wish to investigate further the source of the interaction. Why are the b curves not parallel? Do they differ in their slopes? In their quadratic components? In their cubic components? These equations imply tests of the null hypothesis that the β_p' are the same for all values of k (levels of B). For example, the comparison of slopes involves testing the null hypothesis

$$H_0: \beta_{11}' = \beta_{21}' = \cdots = \beta_{k1}' = \cdots = \beta_{b1}'$$

Let us consider some examples. In the generalization experiment, A might be the intensity of the test stimulus and B might represent clinical populations. Our theory is that generalization increases with degree of severity of schizophrenic symptoms. In terms of trend analysis, we believe that the quadratic coefficient, β_{k2}', decreases (the quadratic component becomes flatter) as schizophrenia becomes more pronounced. The corresponding null hypothesis is that β_{k2}' is the same at all levels of B.

Consider a second example. In the memory experiment described earlier, it was hypothesized that response time would increase linearly with size of display, each added item in the display resulting in a constant increment in time to scan memory of the display. It has been further hypothesized that the time to scan each item would decrease with amount of prior practice with the experimental task. Let B represent several groups having had different numbers of previous trials and let A still be display size. Now consider a plot of response time as a function of display size with a separate curve for each level of practice. Our theory holds that the linear coefficient, β'_{k1}, will decrease as practice increases; that is, response time will change less rapidly as a function of display size for more practiced subjects. The null hypothesis is that β'_{k1} will be the same at all levels of B.

Consider carefully what is implied. There are b curves, each plotted over the levels of A. We wish to compare the b linear components of A, then the b quadratic components of A, and so on. There are $a - 1$ possible comparisons of this type, i.e., as many as there are orthogonal components of the A effect. In general, we are interested in testing the interaction of the pth component of A with the levels of B, in determining whether the pth component of A varies as a function of the level of B. The relevant source of variance is labeled $p(A) \times B$, following our practice in Section 13.4.1. The calculations and df are exactly those presented in Chapter 13, with the w_{jp} of the sum of squares formula replaced by the appropriate ξ'_{jp}. The error terms are also those presented in Section 13.4.1 and are dictated by the design and analysis of variance model.

In Section 13.4.2 we considered the complete analysis of interaction into $(a - 1)(b - 1)$ components, each distributed on a single df. Replacing the ws by ξs, the same breakdowns are possible in trend analysis. For example, suppose that there are four performance curves, each obtained under a different training method (A), plotted over five blocks of trials (B). In our analysis of the interaction of trend components, we would first test the source, lin (B) $\times$ A, which represents the variability of the slopes of the four curves. If this is significant, the slopes differ as a function of training method. Additional hypotheses may now be considered. For example, the average slope of the A_1 and A_2 curves may differ from the average slope of the A_3 and A_4 curves. In this case, we are concerned with lin (B) $\times$ $p(A)$ (p represents the contrast of A_1 and A_2 against A_3 and A_4), whose sum of squares is distributed on a single df. Two sets of weights are involved:

$$\xi_{11} = -2 \qquad w_{1p} = +1$$
$$\xi_{21} = -1 \qquad w_{2p} = +1$$
$$\xi_{31} = 0 \quad \text{and} \quad w_{3p} = -1$$
$$\xi_{41} = +1 \qquad w_{4p} = -1$$
$$\xi_{51} = +2$$

Given these weights, we proceed to calculate $SS_{\text{lin}(B) \times p(A)}$ as we calculated $SS_{q(B) \times p(A)}$ in Section 13.4.2. Calculation of error terms again follows directly from the developments in Chapter 13.

Suppose that in the experiment just described, A is also a quantitative variable, for example, amount of practice. Then, there are two sets of ξs, enabling us to obtain such terms as lin (A) $\times$ lin (B) and quad (A) $\times$ cub (B). The calculations are the same as for any single df component of interaction. The interpretation is somewhat more complicated than before.

Reconsider the example of the memory experiment in which A is display size and B is amount of practice. We earlier hypothesized that B'_{k1} would decrease as practice increases. In other words, we are interested in the variation in lin (A) as a function of level of B. Suppose that we had a still more precise theory; we expect that, as practice increases, B'_{k1} will decrease rapidly at first and then more slowly. This implies hypotheses about the plot of B_{k1}, the linear component of A at B_k, as a function of level of B. Our theory implies that the observed function will have a linear component (lin (A) × lin (B)) because the overall trend is for β'_{k1} to decrease as B_k increases. Given three levels of B, it also implies that the plot of the β'_{k1} as a function of level of B will have a quadratic component (lin (A) × quad (B)) because we expect that β'_{k1} will not decrease at a constant rate as practice is increased.

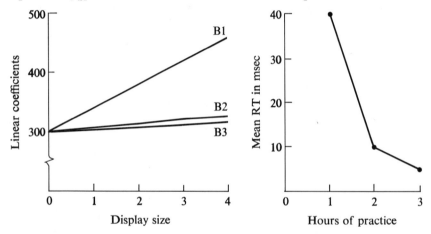

FIGURE 14–5 *Cell means and linear coefficients for a two-factor experiment*

Figure 14–5 presents a set of population means for the memory experiment described above. In the upper panel, mean reaction times are plotted as a function of display size (A) for each of three levels of practice (B). Note that each set of four means falls on a straight line; if an analysis of variance were carried out, the only significant component of SS_A should be $SS_{\text{lin}(A)}$. However, the linear regression coefficients differ; they are 40, 10, and 5. If we partitioned the SS_{AB}, $SS_{\text{lin }(A) \times B}$ would reflect this variation in linear regression as a function of level of B. Now consider the lower panel. We have plotted the linear regression coefficients as a function of level of B. The best-fitting straight line for this set of three points would have a slope different from zero. In other words, we expect a significant $SS_{\text{lin }(A) \times \text{lin}(B)}$ component of SS_{AB}. However, the three linear regression coefficients would not fall on the best-fitting straight line. Consequently, there should be a significant lin (A) × quad (B) component of the overall interaction.

Before proceeding to a numerical example, it may be wise to summarize the analysis of the interaction source. Assuming that B is a quantitative scaled variable, we may examine the variation in the qth component of B as a function of the levels of A. Therefore, our first breakdown of the AB source yields

SV	df
AB	$(a-1)(b-1)$
lin $(B) \times A$	$a-1$
quad $(B) \times A$	$a-1$
$\vdots$	$\vdots$
$q(B) \times A$	$a-1$
$\vdots$	$\vdots$
$(b-1)(B) \times A$	$a-1$

Each of the above sources may be further analyzed by assigning weights to the levels of A. If A is also a quantitative variable, we obtain

SV	df
AB	$(a-1)(b-1)$
lin $(B) \times A$	$a-1$
lin $(B) \times$ lin (A)	1
lin $(B) \times$ quad (A)	1
$\vdots$	$\vdots$
lin $(B) \times p(A)$	1
$\vdots$	$\vdots$
lin $(B) \times (a-1)(A)$	1
quad $(B) \times A$	$a-1$
quad $(B) \times$ lin (A)	1
$\vdots$	$\vdots$
quad $(B) \times p(A)$	1
$\vdots$	$\vdots$
quad $(B) \times (a-1)(A)$	1
$q(B) \times A$	$a-1$
$q(B) \times$ lin (A)	1
$\vdots$	$\vdots$
$q(B) \times p(A)$	1
$\vdots$	$\vdots$
$q(B) \times (a-1)(A)$	1
$(b-1)(B) \times A$	$a-1$
$(b-1)(B) \times$ lin (A)	1
$\vdots$	$\vdots$
$(b-1)(B) \times p(A)$	1
$\vdots$	$\vdots$
$(b-1)(B) \times (a-1)(A)$	1

The sum of squares formulas and error terms follow from Chapter 13. However, to ensure that the application is clear, we proceed to a numerical example involving two quantitative variables.

TABLE 14–1 *Data for a trend analysis*

	A_1	A_2	A_3	$\sum_j Y_{ijk}$	ζ'_{k1}	ζ'_{k2}	ζ'_{k3}	ζ'_{k4}
	20	18	16	54				
B_1	18	17	15	50				
	19	18	17	54	-2	2	-1	1
	16	16	15	47				
$\sum_i Y_{ij1} =$	73	69	63	$\sum_i \sum_j Y_{ij1} = 205$				
	18	15	14	47				
B_2	18	16	13	47				
	17	14	14	45	-1	-1	2	-4
	16	13	13	42				
$\sum_i Y_{ij2} =$	69	58	54	$\sum_i \sum_j Y_{ij2} = 181$				
	16	12	11	39				
B_3	18	13	14	45				
	17	12	13	42	0	-2	0	6
	16	10	12	38				
$\sum_i Y_{ij3} =$	67	47	50	$\sum_i \sum_j Y_{ij3} = 164$				
	15	5	6	26				
B_4	18	8	8	34				
	17	7	9	33	1	-1	-2	-4
	17	5	5	27				
$\sum_i Y_{ij4} =$	67	25	28	$\sum_i \sum_j Y_{ij4} = 120$				
	17	7	6	30				
B_5	18	10	9	37				
	18	9	8	35	2	2	1	1
	15	8	7	30				
$\sum_i Y_{ij5} =$	68	34	30	$\sum_i \sum_j Y_{ij5} = 132$				
$\sum_i \sum_k Y_{ijk} =$	344	233	225	$\sum_i \sum_j \sum_k Y_{ijk} = 802$				
ζ_{j1}	-1	0	1					
ζ_{j2}	1	-2	1					

14.3.2 A numerical example. We assume three equally spaced levels of the independent variable A and five equally spaced levels of the independent variable B with four scores in each cell. The data, together with various totals and subtotals, are presented in Table 14–1. The appropriate sets of orthogonal coefficients, taken from Table A–11 are also present. The total variability is calculated in the usual way:

$$SS_{tot} = (20)^2 + (18)^2 + \cdots + (7)^2 - \frac{(802)^2}{60}$$

$$= 11,792.000 - 10,720.067$$

$$= 1,071.933$$

For the A main effect, we have

$$SS_A = \frac{(344)^2 + (233)^2 + (225)^2}{20} - 10,720.067$$

$$= 442.433$$

and for B,

$$SS_B = \frac{(205)^2 + \cdots + (132)^2}{12} - 10,720.067$$

$$= 405.433$$

We next compute the variability due to polynomial components according to Equation (13.6):

$$SS_{lin\,(A)} = \frac{[-344 + (0)(233) + 225]^2}{(20)(2)}$$

$$= 354.025$$

and

$$SS_{quad\,(A)} = \frac{[344 + (-2)(233) + 225]^2}{(20)(6)}$$

$$= 88.408$$

The sum of the above two terms is 442.433, which equals SS_A, as it should.
Partitioning SS_B, we obtain

$$SS_{lin\,(B)} = \frac{[(-2)(205) - 181 + (0)(164) + 120 + (2)(132)]^2}{(12)(10)}$$

$$= \frac{(207)^2}{120}$$

$$= 357.075$$

$$SS_{quad\,(B)} = \frac{[(2)(205) - 181 + (-2)(164) - 120 + (2)(132)]^2}{(12)(14)}$$

$$= \frac{(45)^2}{168}$$

$$= 12.054$$

$$SS_{cub\,(B)} = \frac{[-205 + (2)(181) + (0)(164) + (-2)(120) + 132]^2}{(12)(10)}$$

$$= \frac{(49)^2}{120}$$

$$= 20.008$$

and

$$SS_{\text{quart}(B)} = \frac{[205 + (-4)(181) + (6)(164) + (-4)(120) + 132]^2}{(12)(70)}$$

$$= \frac{(117)^2}{840}$$

$$= 16.296$$

Summing the four components, we obtain SS_B.

The interaction sum of squares is obtained as in previous chapters; thus

$$SS_{AB} = \frac{(73)^2 + (69)^2 + \cdots + (30)^2}{4} - C - SS_A - SS_B$$

$$= 151.067$$

There are two ways to partition the AB variability; the choice depends on the questions that are asked of the data. We may compare the five curves (one at each level of B) with respect to the linear and quadratic components of A; then we are concerned with $SS_{p(A) \times B}$. Alternatively, we may compare the three curves (one at each level of A) with respect to each of the four components of B; then we are concerned with $SS_{q(B) \times A}$. The choice of approach does not affect the subsequent further analysis of interaction into $(a - 1)(b - 1)$ single df components of the form $p(A) \times q(B)$. For simplicity, we will assume that our primary interest lies in plotting a curve for each level of A over the five B_k data points and, therefore, in assessing $q(B) \times A$. Then, according to Equation (13.20),

$$SS_{\text{lin}(B) \times A} = \frac{[(-2)(73) - 69 + (0)(67) + 67 + (2)(68)]^2}{(4)(10)}$$

$$+ \frac{[(-2)(69) - 58 + (0)(47) + 25 + (2)(34)]^2}{(4)(10)}$$

$$+ \frac{[(-2)(63) - 54 + (0)(50) + 28 + (2)(30)]^2}{(4)(10)} - SS_{\text{lin}(B)}$$

$$= 123.350$$

Calculations of $SS_{\text{quad}(B) \times A}$, $SS_{\text{cub}(B) \times A}$, and $SS_{\text{quart}(B) \times A}$ are parallel to the above. We merely use a different set of ξ'. For example,

$$SS_{\text{quad}(B) \times A} = \frac{[(2)(73) - 69 + (-2)(67) - 67 + (2)(68)]^2}{(4)(14)}$$

$$+ \frac{[(2)(69) - 58 + (-2)(47) - 25 + (2)(34)]^2}{(4)(14)}$$

$$+ \frac{[(2)(63) - 54 + (-2)(50) - 28 + (2)(30)]^2}{(4)(14)} - SS_{\text{quad}(B)}$$

$$= 5.821$$

Similar manipulations yield

$$SS_{\text{cub}(B) \times A} = 13.067$$

and

$$SS_{\text{quart}(B) \times A} = 8.829$$

Each of the above four sum of squares components is distributed on 2 *df*, indicating the possibility of further partitioning. For example,

$$SS_{\text{lin}(B) \times A} = SS_{\text{lin}(B) \times \text{lin}(A)} + SS_{\text{lin}(B) \times \text{quad}(A)}$$

We compute these single *df* components as follows:

$$SS_{\text{lin}(B) \times \text{lin}(A)} = \frac{[(-2)(-1)(73) + (-2)(0)(69) + \cdots + (2)(1)(30)]^2}{(4)(20)}$$
$$= 80$$

and

$$SS_{\text{lin}(B) \times \text{quad}(A)} = \frac{[(-2)(1)(73) + (-2)(-2)(69) + \cdots + (2)(1)(30)]^2}{(4)(60)}$$
$$\doteq 43.350$$

The complete breakdown of the interaction sum of squares is presented in Table 14–2.

TABLE 14–2* *Trend analyses for three two-factor designs*

SV	df	SS
A	2	442.433
lin (A)	1	354.025
quad (A)	1	88.408
B	4	405.433
lin (B)	1	357.075
quad (B)	1	12.054
cub (B)	1	20.008
quart (B)	1	16.296
AB	8	151.067
lin (B) × A	2	123.350
lin (B) × lin (A)	1	80.000
lin (B) × quad (A)	1	43.350
quad (B) × A	2	5.821
quad (B) × lin (A)	1	.571
quad (B) × quad (A)	1	5.250
cub (B) × A	2	13.067
cub (B) × lin (A)	1	5.000
cub (B) × quad (A)	1	8.067
quart (B) × A	2	8.829
quart (B) × lin (A)	1	7.779
quart (B) × quad (A)	1	1.050

* Table 14–2 is continued on the following page.

Table 14-2 (*Continued*)

Error Terms

Design	SV	df	SS	Error Term for
Completely randomized	S/AB	45	73.000	all sources
Repeated measurements	$S \times A$	6	3.567	A
	$S \times \text{lin}(A)$	3	2.275	$\text{lin}(A)$
	$S \times \text{quad}(A)$	3	1.292	$\text{quad}(A)$
	$S \times B$	12	21.767	B
	$S \times \text{lin}(B)$	3	15.692	$\text{lin}(B)$
	$S \times \text{quad } B$	3	3.065	$\text{quad}(B)$
	$S \times \text{cub}(B)$	3	2.892	$\text{cub}(B)$
	$S \times \text{quart}(B)$	3	.118	$\text{quart}(B)$
	$S \times A \times B$	24	12.912	AB
	$S \times A \times \text{lin}(B)$	6	1.183	$A \times \text{lin}(B)$
	$S \times \text{lin}(A) \times \text{lin}(B)$	3	1.000	$\text{lin}(A) \times \text{lin}(B)$
	$S \times \text{quad}(A) \times \text{lin}(B)$	3	.183	$\text{quad}(A) \times \text{lin}(B)$
	$S \times A \times \text{quad}(B)$	6	5.418	$A \times \text{quad}(B)$
	$S \times \text{lin}(A) \times \text{quad}(B)$	3	3.000	$\text{lin}(A) \times \text{quad}(B)$
	$S \times \text{quad}(A) \times \text{quad}(B)$	3	2.418	$\text{quad}(A) \times \text{quad}(B)$
	$S \times A \times \text{cub}(B)$	6	2.734	$A \times \text{cub}(B)$
	$S \times \text{lin}(A) \times \text{cub}(B)$	3	2.500	$\text{lin}(A) \times \text{cub}(B)$
	$S \times \text{quad}(A) \times \text{cub}(B)$	3	.234	$\text{quad}(A) \times \text{cub}(B)$
	$S \times A \times \text{quart}(B)$	6	3.600	$A \times \text{quart}(B)$
	$S \times \text{lin}(A) \times \text{quart}(B)$	3	2.349	$\text{lin}(A) \times \text{quart}(B)$
	$S \times \text{quad}(A) \times \text{quart}(B)$	3	1.251	$\text{quad}(A) \times \text{quart}(B)$
Mixed	S/B	15	56.500	B and all $q(B)$
	$S \times A/B$	30	16.500	AB, all $q(B) \times A$
	$S \times \text{lin}(A)/B$	15	11.130	all $q(B) \times \text{lin}(A)$
	$S \times \text{quad}(A)/B$	15	5.360	all $q(B) \times \text{quad}(A)$

We will now consider appropriate error terms for three experimental designs: completely randomized, repeated measurements ($S \times A \times B$), and a mixed design in which B is a between-subjects variable.

Completely randomized. Assume that each of the 60 scores in Table 14–1 represents a different subject. Then, all terms in the analysis of variance are tested against $MS_{S/AB}$. The results are presented in Table 14–2.

Repeated measurements. Assume that the first row at each level of B in Table 14–1 represents the performance of a single subject, that the second row represents a second subject, etc. Then we have four subjects going through 15 combinations of levels of A and B. Assuming a nonadditive model, the selection of error terms follows the developments of Chapters 7 and 13. The A and B main effects are tested against the $S \times A$ and $S \times B$ terms, respectively. The $p(A)$ term is tested against the $S \times p(A)$ term. For example, if we wish to test whether there is a significant linear component in the plot of the A main effect, we compute $SS_{S \times \text{lin}\,(A)}$, which is distributed on 3 df. Calculation of this error term is aided if we pool the data over levels of B, establishing a table containing totals for each $S \times A$ combination. Table 14–3 accomplishes this for the data of Table 14–1. Then we calculate

$$SS_{S \times \text{lin}\,(A)} = \frac{[-86 + (0)(57) + 53]^2 + \cdots + [-80 + (0)(52) + 52]^2}{(5)(2)}$$
$$- SS_{\text{lin}\,(A)}$$

$$= 356.300 - 354.025$$

$$= 2.275$$

The calculations of $SS_{S \times \text{quad}\,(A)}$ and $SS_{S \times q(B)}$ ($q = \text{lin, quad, cub, quart}$) are left as an exercise for the reader. The results which should be obtained are presented in Table 14–2.

We next consider an error term for the F test of the significance of the general interaction component, $q(B) \times A$. The appropriate error sum of squares, $SS_{S \times q(B) \times A}$, which is distributed on $(n - 1)(a - 1)$ df, will be most easily calculated if the data are regrouped into four (in general, n) $S \times A$ tables, as has been done in Table 14–3. Then, for the specific test of lin $(B) \times A$, we require

$$SS_{S \times \text{lin}\,(B) \times (A)} = \frac{[(-2)(20) - 18 + (0)(16) + 15 + (2)(17)]^2}{10} + \cdots$$

$$+ \frac{[(-2)(15) - 13 + (0)(12) + 5 + (2)(7)]^2}{10}$$

$$- SS_{\text{lin}\,(B)} - SS_{S \times \text{lin}\,(B)} - SS_{A \times \text{lin}\,(B)}$$

$$= 1.182$$

Values for sums of squares for other error terms of the form $S \times q(B) \times A$ are presented in Table 14–2.

TABLE 14–3 *Reorganization of Table 14–1 to facilitate calculations of interaction effects involving subjects*

		A_1	A_2	A_3	$\sum_j \xi'_{j1} Y_{ijk}$	$\sum_j \xi'_{j2} Y_{ijk}$
	B_1	20	18	16	-4	0
	B_2	18	15	14	-4	2
S_1	B_3	16	12	11	-5	3
	B_4	15	5	6	-9	11
	B_5	17	7	6	-11	9
	$\sum_k Y_{1jk} = 86$		57	53	$\sum_k \sum_j \xi'_{j1} Y_{1jk} = -33$	$\sum_k \sum_j \xi'_{j2} Y_{1jk} = 25$
	B_1	18	17	15	-3	-1
	B_2	18	16	13	-5	-1
S_2	B_3	18	13	14	-4	6
	B_4	18	8	8	-10	10
	B_5	18	10	9	-9	7
	$\sum_k Y_{2jk} = 90$		64	59	$\sum_k \sum_j \xi'_{j1} Y_{2jk} = -31$	$\sum_k \sum_j \xi'_{j2} Y_{2jk} = 21$
	B_1	19	18	17	-2	0
	B_2	17	14	14	-3	3
S_3	B_3	17	12	13	-4	6
	B_4	17	7	9	-8	12
	B_5	18	9	8	-10	8
	$\sum_k Y_{3jk} = 88$		60	61	$\sum_k \sum_j \xi'_{j1} Y_{3jk} = -27$	$\sum_k \sum_j \xi'_{j2} Y_{3jk} = 29$
	B_1	16	16	15	-1	-1
	B_2	16	13	13	-3	3
S_4	B_3	16	10	12	-4	8
	B_4	17	5	5	-12	12
	B_5	15	8	7	-8	6
	$\sum_k Y_{4jk} = 80$		52	52	$\sum_k \sum_j \xi'_{j1} Y_{4jk} = -28$	$\sum_k \sum_j \xi'_{j2} Y_{4jk} = 28$

To test $p(A) \times q(B)$ we require $SS_{S \times p(A) \times q(B)}$, which is distributed on $n - 1$ *df*. For example, if we are interested in testing lin $(A) \times$ lin (B), we compute

$$SS_{S \times \text{lin} (A) \times \text{lin} (B)}$$

$$= \frac{[(-2)(-1)(20) + (-2)(0)(18) + \cdots + (2)(0)(7) + (2)(1)(6)]^2}{20} + \cdots$$

$$+ \frac{[(-2)(-1)(16) + (-2)(0)(16) + \cdots + (2)(0)(8) + (2)(1)(7)]^2}{20}$$

$$- SS_{\text{lin} (A) \times \text{lin} (B)}$$

$$= 1.00$$

The general expression is provided by Equation (13.26).

Mixed design. Suppose that there are different subjects at each level of *B*, but all subjects are tested at all levels of *A*. Then each row of scores in Table 14–1 represents a different subject. The *q*th component of the *B* main effect is tested against $MS_{S/B}$, the between-subjects error term. To test the *p*th component of *A*, we require $SS_{S \times p(A)/B}$, which is distributed on $b(n - 1)$ *df*. For example, if we wish to test lin (*A*), we compute

$$SS_{S \times \text{lin}(A)/B} = \frac{[-20 + (0)(18) + 16]^2 + \cdots + [(-15) + (0)(8) + 7]^2}{2}$$

$$- SS_{\text{lin}(A)} - SS_{\text{lin}(A) \times B}$$

$$= 11.130$$

The $MS_{S \times p(A)/B}$ is also the appropriate error term for testing all terms of the form $p(A) \times q(B)$. Thus $S \times \text{lin}(A)/B$ would be the appropriate component for a test of either lin (*A*) $\times$ lin (*B*) or lin (*A*) $\times$ quad (*B*) or lin (*A*) $\times$ cub (*B*) or lin (*A*) $\times$ quart (*B*). This is analogous to the test of the interaction of a between-subjects and a within-subjects variable; the error term is the error term for testing the within-subjects main effect. In the present instance, the error term for the interaction of a between- and a within-subjects polynomial component is the error term for testing the within-subjects component.

14.4 CONCLUDING REMARKS

The analyses discussed in this chapter should never be routinely applied whenever one or more independent variables are quantitative. Any set of *a* data points can be fit by a polynomial of order $a - 1$, but if the population function is not polynomial (e.g., a sine curve), the polynomial analysis will be misleading. It is also dangerous to identify statistical components freely with psychological processes. It is one thing to postulate a cubic component of *A*, to test for it, and to find it significant, thus substantiating the theory. It is another matter to assign psychological meaning to a significant component which has not been postulated on a priori grounds. An unexpected significant component would be of interest and should alert the experimenter to the possible need to revise his behavioral hypotheses. However, remembering that the calculation of several polynomial *F* tests will increase the overall Type I error rate, significant results established on an a posteriori basis should require subsequent experimental validation before they are drawn into the body of scientific conclusions. Furthermore, when several trend tests are carried out on a set of data, the problem of inflation of the experimentwise error rate arises. For planned tests, the approach developed in Section 13.3.1 applies. For post hoc trend tests, *EW* may be controlled by setting the significance level for the individual tests equal to the *EW* divided by the total number of orthogonal tests. With these caveats in mind, trend analysis can be a powerful tool for establishing the true shapes of data functions. As such, these methods of analyses should go hand in hand with the development of precise quantitative behavioral theories.

EXERCISES

14.1 In the following problem assume that the levels of A are equally spaced. Test the orthogonal polynomial components of A and AB (including single df components of AB). Assume that $n = 5$ and $MS_{S/AB} = .40$.

	A_1	A_2	A_3	A_4
B_1	15	5	5	15
B_2	24	18	12	26

14.2 A conflict theorist has scaled TAT cards and has chosen seven that are equally spaced along a sexual content continuum. He predicts that low-guilt subjects will give increasing numbers of sexual responses as sexual content increases, and that high-guilt subjects will show the same number of responses to the low-sex-content cards but will inhibit responses to the high-sex-content cards. Assuming 20 subjects in each group (no counterbalance required), give the SV, df, and error terms, stating explicitly what terms should be significant according to the hypotheses and why.

14.3 There are three training methods (T) and four amounts of practice (P) in a completely randomized design. There are five subjects in each cell. The total number of errors for each of the 12 groups are given below.

	P_1	P_2	P_3	P_4
T_1	9	5	3	4
T_2	6	8	5	4
T_3	5	4	6	3

The levels of P are so chosen as to be equally spaced (i.e., 1, 2, 3, and 4 hr).
(a) Find the sum of squares for the linear components of P and $P \times T$.
(b) We are interested in comparing the rate of learning under T_1 with the average of T_2 and T_3. Compute the appropriate sum of squares.
(c) Suppose the subjects at each level of T went through all levels of P (P might be the stage of practice). Give computational formulas for the error terms for the effects to be tested in (a) and (b).

14.4 Three classes of 20 patients each undergo group therapy over a 2-year period. They are individually rated by the clinical staff every six months (five times). According to one theory of personality (by Freud out of Estes), one group (A) should show improvement, then regression, then improvement but with overall trend to improve; A_2 should improve and then regress to the initial level; A_3 should show a steady rate of improvement. List the terms that should be significant.

14.5 In a study of short-term memory, 12 groups of 10 Ss are run. They are all required to tell whether a comparison tone is the same as or different from a standard presented several seconds earlier. The groups differ with respect to D, duration of the interstimulus interval: $D = 1, 2,$ or 4. They also differ with respect to I, the stimulus presented during the interval: blank, noise, tone 15 Hz above standard, tone 30 Hz above standard. It is hypothesized that

H_1: D will not influence memory when the interval is blank.

H_2: All other I levels will result in a nonlinear (exponential drop) with increased D.

H_3: The rate of decay (drop in memory over D) will be most pronounced if $I =$ noise, next when $I = +30$ Hz, next when $I = +15$ Hz.

Describe the statistical tests that you would make to test these Hs.

14.6 A large-scale study of programmed instruction is carried out with three variables: Method (linear program, branching program, material is just read), Material (math, English, social studies), and time/day (15, 30, or 45 min.). There are 20 students in each cell. The hypotheses are:

H_1: Math scores are higher than scores on the other materials

H_2: Programmed instruction is superior to nonprogrammed instruction.

H_3: Performance improves overall with instruction time but there is a leveling off between 30 and 45 minutes.

H_4: The superiority of math performance is more marked under the two programming methods than under the straight reading method.

H_5: The superiority of math is more marked under branching than under linear programming.

H_6: The superiority of mathematics performance increases as instructional time increases.

Set up the appropriate contrasts for each hypothesis.

14.7 The levels of A are

$$X_1 = 0 \qquad X_2 = 1 \qquad X_3 = 2$$

The corresponding means are

$$\bar{Y}_1 = 2 \qquad \bar{Y}_2 = 20 \qquad \bar{Y}_3 = 8$$

(a) Find the numerical values of b_0', b_1', and b_2' for the equations

$$\bar{Y}_j = b_0' + b_1'\xi_1' + b_2'\xi_2'$$

(b) Derive the values of ξ that are tabled in A–11 for the three groups.

(c) Using the values obtained in the course of doing (b), find the numerical values of b_0, b_1, and b_2 for: $\bar{Y}_j = b_0 + b_i X_j + b_2 X_j^2$.

Answers to Exercises

CHAPTER 2

2.1 $\displaystyle\sum_{i=3}^{6} Y_i$

2.2 (a)

$$\sum_{i=1}^{n} (Y_i - \bar{Y}) = \sum_{i=1}^{n} Y_i - \sum_{i=1}^{n} \bar{Y} \qquad \text{(Rule 3)}$$

$$= \sum Y_i - n\bar{Y} \qquad \text{(Rule 2)}$$

$$= \sum Y_i - n\frac{\sum Y_i}{n} \qquad \text{(Substitution)}$$

$$= 0$$

(b)

$$\sum_{i=1}^{n} (Y_i + k - n) = \sum_{i=1}^{n} Y_i + nk - n^2$$

2.3

$$\frac{1}{k} \sum_{i=1}^{k} (k - kY_i) = \frac{1}{k}\left(\sum_{i=1}^{k} k + \sum_{i=1}^{k} kY_i\right) \qquad \text{(Rule 3)}$$

$$= \frac{1}{k}\left(k^2 + k\sum_{i=1}^{k} Y_i\right) \qquad \text{(Rules 1 and 2)}$$

$$= k + \sum_{i=1}^{k} Y_i$$

2.4

$$S_{y+c}^2 = \frac{\sum_{i=1}^{n}[(Y_i + C) - \overline{(Y + C)}]^2}{n - 1} \qquad \text{(by definition of a variance)}$$

$$\overline{(Y + C)} = \frac{\sum_{i=1}^{n}(Y_i + C)}{n}$$

$$= \frac{\sum Y_i}{n} + \frac{\sum C}{n}$$

$$= \bar{Y} + C$$

Then,

$$S_{y+c}^2 = \frac{\sum_{i=1}^{n}[(Y_i + C) - (\bar{Y} + C)]^2}{n - 1}$$

$$= \frac{\sum(Y_i - \bar{Y})^2}{n - 1} = S_y^2$$

2.5
$$S_{cy}^2 = \frac{\sum_{i=1}^n [CY_i - (\overline{CY})]^2}{n-1}$$

$$(\overline{CY}) = \frac{\sum_{i=1}^n CY_i}{n}$$

$$= \frac{C\sum Y_i}{n}$$

$$= C\bar{Y}$$

Then,

$$S_{cy}^2 = \sum_{i=1}^n \frac{(CY_i - C\bar{Y})^2}{n-1}$$

$$= \frac{\sum [C(Y_i - \bar{Y})]^2}{n-1}$$

$$= C^2 \frac{\sum (Y_i - \bar{Y})^2}{n-1} = C^2 S_y^2$$

2.6   (a)
$$\bar{Z} = \sum_{i=1}^n \frac{(Y_i - \bar{Y})/n}{S_y}$$

$$= \frac{1}{S_y} \frac{\sum (Y_i - \bar{Y})}{n} \qquad \text{(since } S_y \text{ is a constant)}$$

$$= \frac{1}{S_y} \cdot 0 \qquad \text{(see Exercise 2.2)}$$

(b)
$$S_z^2 = \frac{\sum_{i=1}^n (Z_i - \bar{Z})^2}{n-1} \qquad \text{(by definition)}$$

$$\bar{Z} = 0$$

Therefore,

$$S_z^2 = \frac{\sum_{i=1}^n Z_i^2}{n-1}$$

$$Z_i^2 = \frac{(Y_i - \bar{Y})^2}{S_y^2}$$

Then,

$$\sum_{i=1}^n Z_i^2 = \frac{\sum_{i=1}^n (Y_i - \bar{Y})^2}{S_y^2}$$

Dividing by $n-1$,

$$S_z^2 = \frac{\sum (Y_i - \bar{Y})^2/(n-1)}{S_y^2}$$

$$= \frac{S_y^2}{S_y^2}$$

$$= 1$$

2.7 (a)
$$\bar{D} = \frac{\sum_{i=1}^{n}(X_i - Y_i)}{n}$$

$$= \frac{\sum X_i - \sum Y_i}{n}$$

$$= \bar{X} - \bar{Y}$$

(b)
$$S_d^2 = \frac{\sum_{i=1}^{n}(D_i - \bar{D})^2}{n - 1}$$

$$= \frac{\sum[(X_i - Y_i) - (\bar{X} - \bar{Y})]^2}{n - 1}$$

Rearrange terms inside the brackets. Then

$$S_d^2 = \frac{\sum[(X_i - \bar{X}) - (Y_i - \bar{Y})]^2}{n - 1}$$

$$= \frac{\sum[(X_i - \bar{X})^2 + (Y_i - \bar{Y})^2 - 2(X_i - \bar{X})(Y_i - \bar{Y})]}{n - 1}$$

$$= \frac{\sum(X_i - \bar{X})^2}{n - 1} + \frac{\sum(Y_i - \bar{Y})^2}{n - 1} - \frac{2\sum(X_i - \bar{X})(Y_i - \bar{Y})}{n - 1}$$

$$= S_x^2 + S_y^2 - \frac{2\sum(X_i - \bar{X})(Y_i - \bar{Y})}{n - 1}$$

Note:

$$r_{xy} = \frac{\sum_{i=1}^{n}(X_i - \bar{X})(Y_i - \bar{Y})/(n - 1)}{S_x S_y}$$

Then

$$\frac{\sum(X_i - \bar{X})(Y_i - \bar{Y})}{n - 1} = r_{xy}S_x S_y$$

Therefore,

$$S_d^2 = S_x^2 + S_y^2 - 2r_{xy}S_x S_y$$

2.8 $i = 1, 2, \ldots, n$ (subjects)
$j = 1, 2, \ldots, a$ (A)
$k = 1, 2, \ldots, d$ (D)
$m = 1, 2, \ldots, t$ (T)

(a)
$$\sum_{j=1}^{a}\left[\sum_{i=1}^{n}\sum_{k=1}^{d}\sum_{m=1}^{t}Y_{ijkm}\right]^2$$

(b)
$$\sum_{k=1}^{d}\sum_{m=1}^{t}\left[\sum_{i=1}^{n}\sum_{j=1}^{a}Y_{ijkm}\right]^2$$

2.9 $\displaystyle\sum_{j=1}^{a}\sum_{i=1}^{n}(\bar{Y}_{i.} - \bar{Y}_{..})^2 = \sum_{j}^{a}\sum_{i}^{n}(\bar{Y}_{i.}^2 + \bar{Y}_{..}^2 - 2\bar{Y}_{i.}\bar{Y}_{..})$

$$= \sum_{i}(a\bar{Y}_{i.}^2 + a\bar{Y}_{..}^2 - 2a\bar{Y}_{i.}\bar{Y}_{..})$$

$$= a\sum_{i}\bar{Y}_{i.}^2 + an\bar{Y}_{..}^2 - 2a\bar{Y}_{..}\sum_{i}\bar{Y}_{i.}$$

$$= a\sum_{i}\frac{(\sum_{j}Y_{ij})^2}{a^2} + an\frac{(\sum_{i}\sum_{j}Y_{ij})^2}{a^2n^2}$$

$$- 2a\left(\frac{\sum_{i}\sum_{j}Y_{ij}}{an}\right)\sum_{i}\left(\frac{\sum_{j}Y_{ij}}{a}\right)$$

$$= \frac{\sum_{i}(\sum_{j}Y_{ij})^2}{a} + \frac{(\sum_{i}\sum_{j}Y_{ij})^2}{an} - \frac{2(\sum_{i}\sum_{j}Y_{ij})^2}{an}$$

$$= \frac{\sum_{i}(\sum_{j}Y_{ij})^2}{a} - \frac{(\sum_{i}\sum_{j}Y_{ij})^2}{an}$$

2.10 $\displaystyle na\sum_{k=1}^{b}(\bar{Y}_{.k.} - \bar{Y}_{...})^2 = na\sum_{k}(\bar{Y}_{.k.}^2 + \bar{Y}_{...}^2 - 2\bar{Y}_{.k.}\bar{Y}_{...})$

$$= na\sum_{k}\bar{Y}_{.k.}^2 + nab\bar{Y}_{...}^2 - 2na\bar{Y}_{...}\sum_{k}\bar{Y}_{.k.}$$

$$= na\sum_{k}\frac{(\sum_{i}\sum_{j}Y_{ijk})^2}{n^2a^2} + nab\frac{(\sum_{i}\sum_{j}\sum_{k}Y_{ijk})^2}{n^2a^2b^2}$$

$$- 2na\frac{(\sum_{i}\sum_{j}\sum_{k}Y_{ijk})}{nab}\sum_{k}\frac{(\sum_{i}\sum_{j}Y_{ijk})}{na}$$

$$= \sum_{k}\frac{(\sum_{i}\sum_{j}Y_{ijk})^2}{na} - \frac{(\sum_{i}\sum_{j}\sum_{k}Y_{ijk})^2}{nab}$$

2.11 (a) $(14 + 12 + 2 + 8)^2 + (3 + 8 + 1 + 9)^2 + \cdots + (2 + 3 + 6 + 3)^2$

(b) $(14 + 2)^2 + (12 + 8)^2 + \cdots + (3 + 3)^2$

(c) $\displaystyle \bar{Y}_{...2} = \frac{\sum_{i}^{2}\sum_{j}^{3}\sum_{k}^{3}Y_{ijk2}}{18} = \frac{2 + 8 + 1 + 9 + \cdots + 7 + 6 + 3}{18}$

(d) $\displaystyle \bar{Y}_{..2.} = \frac{\sum_{i}^{2}\sum_{j}^{3}\sum_{m}^{2}Y_{ij2m}}{12} = \frac{3 + 6 + 4 + 7 + \cdots + 7 + 6}{12}$

2.12 (a) $(42 + 93)(70 + 80) + (60 + 99)(77 + 91)$

(b) $(27 + 30)^2 + (19 + 20)^2 + (24 + 30)^2 + \cdots + (33 + 27)^2$

(c) $(11 + 21)(27 + 22) + (15 + 16)(19 + 22) + \cdots$

$$+ (27 + 32)(30 + 27)$$

(d) $(4 + 1 + 8 + 6 + \cdots + 9 + 14)^2 + (3 + 9 + \cdots + 6 + 7)^2$

$$+ (4 + 5 + \cdots + 13 + 11)^2$$

CHAPTER 3

3.2 (b) The result is not in conflict with the statement that these are 95 percent confidence intervals. *In the long run*, we expect 95 percent of the intervals computed to contain μ. Alternatively, if we compute many sets of 100 intervals, the average proportion of intervals containing μ should be .95.

3.3 From Equation (3.1)

$$\left(\bar{Y} + \frac{1.96}{\sqrt{n}}\right) - \left(\bar{Y} - \frac{1.96}{\sqrt{n}}\right) = 10$$

Then

$$\frac{(2)(1.96)(15)}{\sqrt{n}} = 10$$

$$\sqrt{n} \approx \tfrac{60}{10}$$

$$n \approx 36$$

3.4 With this unusual critical region, the area under the H_1 distribution corresponding to power will be a small segment in the tail of the distribution.

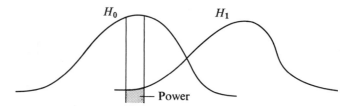

In fact, as the distributions become more separate, power will actually decrease.

3.5 (a) The critical region is $z \geq 1.645$.

$$\sigma_{\bar{x}} = \frac{\sigma}{\sqrt{n}} = 4$$

$$\frac{\mu_1 - \mu_0}{\sigma_{\bar{x}}} = \frac{10}{4} = 2.5$$

Thus μ_1 lies 2.5 standard deviation units above μ_0. The critical value of z lies $2.5 - 1.645 = .855\sigma$ below μ_1. Then power is approximately .80, the proportion of the μ_1 distribution above $-.855$.

(b) The critical region consists of values of $z \geq 1.96$ or ≤ -1.96. The upper value, 1.96, lies $.54\sigma$ below μ_1 ($2.5 - 1.96 = .54$) and this is exceeded approximately 71 percent of the time under H_1. The lower critical value, -1.96, is at 4.46σ below μ_1. Power is the proportion of the H_1 distribution $> -.54$ or < -4.46 under H_1, or .71 plus (essentially) .00. Note that we have lost power against one alternative in switching from a one- to a two-tailed test.

3.6 $\quad Y - \mu = Y - \bar{Y} + \bar{Y} - \mu$

$$\frac{\Sigma(Y - \mu)^2}{n} = \frac{\Sigma(Y - \bar{Y})^2}{n} + \frac{\Sigma(\bar{Y} - \mu)^2}{n}$$

Noting that $\bar{Y} - \mu$ is a constant, and rearranging terms.

$$\frac{\Sigma(Y - \bar{Y})^2}{n} = \frac{\Sigma(Y - \mu)^2}{n} - (\bar{Y} - \mu)^2$$

$$E\left(\frac{\Sigma(Y - \bar{Y})^2}{n}\right) = \Sigma\frac{E(Y - \mu)^2}{n} - E(\bar{Y} - \mu)^2$$

$$= \frac{n\sigma^2}{n} - \frac{\sigma^2}{n}$$

$$= \left(\frac{n - 1}{n}\right)\sigma^2$$

CHAPTER 4

4.1 Since ε_{ij} is a random variable, and a given experimental group contains only n of the infinite number of values in the treatment population, its sum will not generally be zero. The α_j, on the other hand, are fixed-effect variables; all deviations are represented in the experiment, and the sum will therefore be zero. If the α_j were a random sample from a larger population of such values, they would no longer sum to zero, although $E(\Sigma_j \alpha_j) = 0$.

4.2 $\quad (\Sigma_i \Sigma_j \varepsilon_{ij})^2 = \Sigma_i \Sigma_j \varepsilon_{ij}^2 + 2 \text{ (Sum of all cross products)}$

$$E\left[\frac{n \Sigma_j(\Sigma_i \Sigma_j \varepsilon_{ij})^2}{a^2 n^2}\right] = \frac{n \Sigma_j E(\Sigma_i \Sigma_j \varepsilon_{ij})^2}{a^2 n^2}$$

The cross-product terms have zero expectations if we assume independence. We now have:

$$\frac{n \Sigma_j E(\Sigma_i \Sigma_j \varepsilon_{ij}^2)}{a^2 n^2} = \frac{n \Sigma_j \Sigma_i \Sigma_j E(\varepsilon_{ij}^2)}{a^2 n^2}$$

Assuming homogeneity of variance, we have:

$$\frac{n \Sigma_j \Sigma_i \Sigma_j \sigma_e^2}{a^2 n^2} = \frac{nana\sigma_e^2}{a^2 n^2}$$

4.3 (a)

SV	df	S	MS	F
A	2	3,886.9	1,943.5	31.55
S/A	27	1,664.1	61.6	

(b)

A	4	3,541.9	885.5	1.30
S/A	26	17,727.5	681.8	

4.4
$$\frac{\sum(\mu_j - \mu)^2}{3} = \frac{200}{3}$$

$$\phi' = \sqrt{\frac{200}{(3)(225)}} \, n \approx .55\sqrt{n}$$

if $n = 16$, $df_2 = 45$ and $\phi = 2.2$. Then, *Power* $\approx .90$.

CHAPTER 5

5.1 The significant effects are: B, AB, AC

5.2

SV	S
A	108.00
B	84.25
C	97.20
AB	28.75
AC	35.20
BC	16.85
ABC	5.75

5.3
$$\sum_j n_j(\bar{Y}_j - \bar{Y}_{..}) = \sum_j n_j \left[\frac{\sum_i \sum_k Y_{ijk}}{n_j} - \frac{\sum_i \sum_j \sum_k Y_{ijk}}{\sum_j n_j} \right]$$

$$= \sum_i \sum_j \sum_k Y_{ijk} - \frac{(\sum_j n_j)(\sum_i \sum_j \sum_k Y_{ijk})}{\sum_j n_j} = 0$$

5.4 These effects should be significant: A, D, AD, CD.

5.5 $E[\sum_i \sum_j \sum_k \hat{\varepsilon}_{ijk}^2] = E\left[\sum_i \sum_j \sum_k \left(\varepsilon_{ijk} - \frac{\sum_i \sum_k \varepsilon_{ijk}}{bn} - \frac{\sum_i \sum_j \varepsilon_{ijk}}{an} \right. \right.$

$$\left. \left. + \frac{\sum_i \sum_j \sum_k \varepsilon_{ijk}}{abn} \right)^2 \right]$$

Expanding the right-hand side, we have the following terms:

(a) $\qquad\qquad\qquad \varepsilon_{ijk}^2$

(b) $\qquad\qquad\qquad \dfrac{(\sum_i \sum_k \varepsilon_{ijk})^2}{b^2 n^2}$

(c) $\qquad\qquad\qquad \dfrac{(\sum_i \sum_j \varepsilon_{ijk})^2}{a^2 n^2}$

(d) $\qquad\qquad\qquad \dfrac{(\sum_i \sum_j \sum_k \varepsilon_{ijk})^2}{a^2 b^2 n^2}$

(e) $\qquad\qquad\qquad -\dfrac{2\varepsilon_{ijk} \sum_i \sum_k \varepsilon_{ijk}}{bn}$

(f) $\qquad\qquad\qquad -\dfrac{2\varepsilon_{ijk} \sum_i \sum_j \varepsilon_{ijk}}{an}$

(g)
$$+ \frac{2\varepsilon_{ijk} \sum_i \sum_j \sum_k \varepsilon_{ijk}}{abn}$$

(h)
$$+ \frac{2 \sum_i \sum_k \varepsilon_{ijk} \sum_i \sum_j \varepsilon_{ijk}}{abn^2}$$

(i)
$$- \frac{2 \sum_i \sum_k \varepsilon_{ijk} \sum_i \sum_j \sum_k \varepsilon_{ijk}}{ab^2n^2}$$

(j)
$$- \frac{2 \sum_i \sum_j \varepsilon_{ijk} \sum_i \sum_j \sum_k \varepsilon_{ijk}}{a^2bn^2}$$

Next, we consider $E(\sum_i \sum_j \sum_k)$ for each term. Remember that the expectation of the cross-product of two different εs is zero under the independence assumption and that $E(\varepsilon^2)$ is σ_e^2 for all i, j, and k under the homogeneity of variance assumption.

(a)
$$E(\sum_i \sum_j \sum_k \varepsilon_{ijk}^2) = abn\sigma_e^2$$

(b)
$$E\left[\frac{\sum_i \sum_j \sum_k(\sum_i \sum_k \varepsilon_{ijk})^2}{b^2n^2} \right]$$

Expanding $(\sum_i \sum_k \varepsilon_{ijk}^2)^2 = \sum_i \sum_k \varepsilon_{ijk}$ + cross-products. Placing the "E" before the ε^2's and dropping the cross-products, we have

$$\frac{\sum_i \sum_j \sum_k[\sum_i \sum_k(E\varepsilon_{ijk}^2)]}{b^2n^2} = a\sigma_e^2$$

(c) Following the steps in (b) we obtain

$$b\sigma_e^2$$

(d) Similarly, we obtain

$$\sigma_e^2$$

(e)
$$E\frac{\sum_i \sum_j \sum_k(-2\varepsilon_{ijk} \sum_i \sum_k \varepsilon_{ijk})}{bn} = -2 \sum_j E\frac{(\sum_i \sum_k \varepsilon_{ijk})^2}{bn}$$

$$= -2a\sigma_e^2$$

(f) Similarly, we obtain

$$-2b\sigma_e^2$$

(g) and also

$$2\sigma_e^2$$

(h)

$$2E\frac{\sum_i \sum_j \sum_k(\sum_i \sum_k \sum_{ijk} \sum_i \sum_j \sum_{ijk})}{abn^2} = 2E\frac{\sum_i(\sum_i \sum_j \sum_k \sum_{ijk} \sum_i \sum_j \sum_k \sum_{ijk})}{abn^2}$$

$$= 2 \sum_i \frac{nba\sigma_e^2}{abn^2}$$

$$= 2\sigma_e^2$$

(i) Similarly, we obtain

$$-2\sigma_e^2$$

(j) and also

$$-2\sigma_e^2$$

Collecting terms, we have

$$(abn - a - b + 1)\sigma_e^2$$

The analysis of variance table is

SV	df
A	$a - 1$
B	$b - 1$
Residual	$abn - a - b + 1$
Total	$abn - 1$

The potential advantage lies in the increased error *df* and the consequent greater power. This would be a noticeable gain in power only if $(a - 1)(b - 1)$ were large relative to $ab(n - 1)$. However, if $\theta_{AB}^2 > 0$, the $E(MS_{res})$ will be larger than σ_e^2 and $E(MS_a)/E(MS_{res})$ and $E(MS_b)/E(MS_{res}) < 1$ under H_0. The ratio of mean squares would not be distributed as *F* and there will be a loss of power. We will return to the question of pooling terms in later chapters (see, particularly, Chapter 11). For the present, note that alternative models are possible, they may dictate alternative analyses, and there are possible advantages and disadvantages.

5.6 (a) $df_{pool} = (abcen - 1) - (b - 1) - (c - 1) - (b - 1)(c - 1)$
$$-abce(n - 1)$$
$$= bc(ae - 1)$$

Note that this result can be arrived at more directly. Within each *BC* cell, there are *ae* treatment combinations on $ae - 1$ *df*. Summing this variability over *BC* combinations yields $bc(ae - 1)$ *df*.

Expanding the *df* immediately yields the *SS*:

$$abce - bc \rightarrow \sum^a \sum^b \sum^c \sum^e \frac{(\sum^n Y)^2}{n} - \sum^b \sum^c \frac{(\sum^a \sum^e \sum^n Y)^2}{aen}$$

(b) If any of the effects involving *A* or *E* are not negligible (e.g., $\theta_{AB}^2 > 0$ or $\theta_E^2 > 0$), the residual error term will be spuriously inflated (see Exercise 5.5). The *df* are $abce(n - 1) + bc(ae - 1)$.

5.7
$$SS_D = \frac{[(52 + 80 + 84 + 81) - (107 + 58 + 60 + 115)]^2}{(8)(24)}$$

$$SS_{AC} = \frac{[(52 + 107) + (81 + 115) - (108 + 58) - (84 + 60)]^2}{(8)(24)}$$

$$SS_{ACD} = \frac{[(52 + 81) - (84 + 80) - (107 + 115) + (160 + 58)]^2}{(8)(24)}$$

CHAPTER 6

6.1 Let C designate blocks. To equate the sizes of the two designs, we will have cn scores in each of the ab cells of the completely randomized (c.r) design. For that design,

$$E(SS_{tot}) = (a - 1)(\sigma^2_{e_{c.r.}} + nbc\theta^2_A) + (b - 1)(\sigma^2_{e_{c.r.}} + nac\theta^2_B)$$
$$+ (a - 1)(b - 1)(\sigma^2_{e_{c.r.}} + nc\theta^2_{AB}) + ab(cn - 1)\sigma^2_{e_{c.r.}}$$

For the blocks design ($t \times b$), we have

$$E(SS_{tot}) = (a - 1)(\sigma^2_{e_{t \times b}} + nbc\theta^2_A) + (b - 1)(\sigma^2_{e_{t \times b}} + nac\theta^2_B)$$
$$+ (a - 1)(b - 1)(\sigma^2_{e_{t \times b}} + nc\theta^2_{AB}) + E(SS_C) + E(SS_{AC})$$
$$+ E(SS_{BC}) + E(SS_{ABC}) + abc(n - 1)\sigma^2_{e_{t \times b}}$$

Equating the two $E(SS_{tot})$ and cancelling the θ^2_A, θ^2_B, and θ^2_{AB} terms, we have

$$(abcn - 1)\sigma^2_{e_{c.r.}} = [abc(n - 1) + (ab - 1)]\sigma^2_{e_{t \times b}} + E(SS_C)$$
$$+ E(SS_{AC}) + E(SS_{BC}) + E(SS_{ABC})$$

Relative efficiency of the $t \times b$ design to c.r. design is

$$\text{R.E.} = \frac{[abc(n - 1) + (ab - 1)]MS_{S/ABC} + SS_{AC} + SS_{BC} + SS_{ABC} + SS_C}{(abcn - 1)MS_{S/ABC}}$$

$$= 1 - \frac{ab(c - 1)}{abcn - 1} + \frac{SS_{AC} + SS_{BC} + SS_{ABC} + SS_C}{(abcn - 1)MS_{S/ABC}}$$

6.2

SV	df	SS	MS
A	2	136.8	63.4
B	3	25.5	8.5
AB	6	153.9	25.7
S/AB	12	62.5	5.2

$$\text{R.E.} = \frac{25.5 + 153.9 + 62.5}{62.5}$$
$$= 3.87$$

The AB variability is considerable.

6.3 Consider $p = .2$, $N = 70$. Then $a = 4$, by linear interpolation, yields $b = 3.7$. If $N = 100$, $b = 5$ then, for $N = 80$, we want a value of b one-third the distance between the values at $N = 70$ ($b = 3.7$) and $N = 100$ ($b = 5$); $b = 4$ is a reasonable approximation. Similarly, for $p = .4$, we arrive at $b = 7$. Our estimates of p range from .25 to .35. Taking $p = .30$, its squared value, .09, lies a little less than midway between $(.2)^2$ and $(.4)^2$; $b = 5$ would seem a reasonable approximation.

6.4 (a) Taking $MS_{S/AB}$ as σ^2_e,

$$\phi^2 = \frac{(12)(32)/3}{16} = 8$$

Then $\phi = 2.8$ and, with 2 and 18 df,

$$Power \approx .91$$

(b) We must re-estimate σ_e^2. From Equation (6.5),

$$MS_{S/A} = \left(1 - \frac{15}{35}\right)(16) + \frac{800 + 220}{35} = 9.1 + 29.1 \approx 38$$

Now

$$\phi^2 = \frac{(12)(32)/3}{38}$$

and

$$\phi \approx 1.84$$

with $df = 2$ and 33

$$Power \approx .52$$

CHAPTER 7

7.1 The interaction term is now the appropriate error term. Note that in this case the presence of an AB interaction poses the problems generally associated with nonadditivity.

7.2 We again multiply all scores by 1,000.

$SS_{nonadd} = 298.5$

$SS_{bal} = 969.7 - 298.5 = 671.2$

$F = 1.33$

7.3 Only set 3 should be sensitive to the Tukey test.

7.4
$$(\eta\alpha)_{ij} = \mu_{ij} - \mu_i - \mu_j + \mu$$
$$\sum_j (\eta\alpha)_{ij} = \sum_j \mu_{ij} - a\mu_i - \sum_j \mu_j + a\mu$$
$$= a\mu_i - a\mu_i - a\mu + a\mu$$
$$= 0.$$

7.5 The variance-covariance matrix is

$$\begin{bmatrix} 2.917 & 4.083 & .750 \\ 4.083 & 10.917 & 3.583 \\ .750 & 3.583 & 4.250 \end{bmatrix}$$

$$\bar{V}_{..} = 3.88$$
$$\bar{V}_{jj} = 6.03$$
$$\sum_j \sum_k \bar{V}_{jk}^2 = 205.89$$
$$\sum \bar{V}_j^2 = 53.23$$

$$= \frac{a^2(\bar{V}_{jj} - \bar{V}_{..})^2}{(a-1)(\sum_j \sum_k V_{jk}^2 - 2a \sum_j \bar{V}_j^2 + a^2 \bar{V}_{..}^2)}$$

$$= \frac{9(6.03 - 3.88)^2}{2[205.89 - (6)(53.23) + (9) 3.88^2]}$$

$$= .94$$

7.6	SV	EMS Components
	A	e, A, AB
	B	e, B, AB
	C	e, AC, BC, ABC
	AB	e, AB
	AC	e, AC, ABC
	BC	e, BC, ABC
	ABC	e, ABC

CHAPTER 8

8.1	SV	df	SS	MS	F
	A	2	90.70	45.35	5.94
	S/A	6	45.78	7.63	
	B	1	93.35	93.35	10.73
	AB	2	84.93	42.46	4.88
	SB/A	6	52.22	8.70	
	C	2	667.70	333.85	53.25
	AC	4	10.07	2.52	.40
	SC/A	12	75.22	6.27	
	BC	2	89.93	44.96	5.77
	ABC	4	13.63	3.41	.44
	SBC/A	12	93.44	7.79	

8.2

SV	EMS
A	$\sigma_e^2 + c\sigma_{AB}^2 + b\sigma_{AC}^2 + \sigma_{ABC}^2 + bc\sigma_A^2$
B	$\sigma_e^2 + c\sigma_{AB}^2 + a\sigma_{BC}^2 + \sigma_{ABC}^2 + ac\sigma_B^2$
C	$\sigma_e^2 + a\sigma_{BC}^2 + b\sigma_{AC}^2 + \sigma_{ABC}^2 + ab\sigma_C^2$
AB	$\sigma_e^2 + \sigma_{ABC}^2 + c\sigma_{AB}^2$
AC	$\sigma_e^2 + \sigma_{ABC}^2 + b\sigma_{AC}^2$
BC	$\sigma_e^2 + \sigma_{ABC}^2 + a\sigma_{BC}^2$
ABC	$\sigma_e^2 + \sigma_{ABC}^2$

(a) $E(MS_A)$, $E(MS_B)$, and $E(MS_C)$ contain extra terms so that, under H_0, no appropriate error term is immediately available.

(b) If σ_{AB}^2 is zero, under H_0

$$\frac{E(MS_A)}{E(MS_{AC})} = 1$$

$$\frac{E(MS_B)}{E(MS_{BC})} = 1$$

8.3 Two runs through the computer suffice. *First run:* First factor = Ss (*an* levels). Second factor = B (*b* levels) we obtain :

Between Ss	$an - 1$
B	$b - 1$
Between Ss × B	$(an - 1)(b - 1)$

Second run: First factor = A. Second factor = B. We obtain A, B, and AB sums of squares. Subtracting by hand:

$$SS_{B.Ss} - SS_A = SS_{S/A}$$

$$SS_{B.Ss} \times B - SS_{AB} = SS_{SB/A}$$

The analysis is complete. The approach is readily extended to other variables and to other computer limitations (e.g., if there is a limit on the total number of scores to be read in, much of the analysis can still be carried out automatically by analyzing in parts, say first at A_1, then at A_2 in the next run, etc. This would yield S/A, SB/A. Further reduction in number of observations might be obtained by computing SS' for cell means, then multiplying by appropriate number of observations to obtain SS.)

CHAPTER 9

9.1

SV	df	SS	MS	F
Between G	5	1,422.25		
A	2	1,324.67	662.33	20.36
G/A	3	97.58	32.53	.57
S/G/A	30	1,706.50	56.88	

9.2

SV	df	SS	MS	F
Between S	15	1,277.25		
Between G	3	1,210.42		
A	1	1,160.33	1,160.33	46.34
G/A	2	50.08	25.04	4.49
S/G/A	12	66.83	5.57	
Within S	32	1,006.00		
B	2	919.63	459.81	50.86
AB	2	17.04	8.52	0.94
GB/A	4	36.16	9.04	6.55
SB/G/A	24	33.26	1.38	

9.3

SV	df	EMS
Between S	119	
Between Classes (C)	11	
Between Schools (Sc)	5	
Machines (M)	1	$\sigma_e^2 + 2\sigma_{S/C/Sc/M}^2 + 20\sigma_{C/Sc/M}^2 + 40\sigma_{Sc/M}^2 + 120\theta_M^2$
Sc/M	4	$\sigma_e^2 + 2\sigma_{S/C/Sc/M}^2 + 20\sigma_{C/Sc/M}^2 + 40\sigma_{Sc/M}^2$
C/Sc/M	6	$\sigma_e^2 + 2\sigma_{S/C/Sc/M}^2 + 20\sigma_{C/Sc/M}^2$
S/C/Sc/M	108	$\sigma_e^2 + 2\sigma_{S/C/Sc/M}^2$
Exam (E)	1	$\sigma_e^2 + \sigma_{SE/C/Sc/M}^2 + 10\sigma_{CE/Sc/M}^2 + 20\sigma_{ScE/M}^2 + 120\theta_E^2$
ME	1	$\sigma_e^2 + \sigma_{SE/C/Sc/M}^2 + 10\sigma_{CE/Sc/M}^2 + 20\sigma_{ScE/M}^2 + 60\theta_{ME}^2$
Sc × E/M	4	$\sigma_e^2 + \sigma_{SE/C/Sc/M}^2 + 10\sigma_{CE/Sc/M}^2 + 20_{Sc \times E/M}^2$
CE/Sc/M	6	$\sigma_e^2 + \sigma_{SB/C/Sc/M}^2 + 10\sigma_{CE/Sc/M}^2$
SE/C/Sc/M	108	$\sigma_e^2 + \sigma_{SE/C/Sc/M}^2$

9.4

SV	df	EMS
Between G	7	
Orientation (O)	1	$\sigma_e^2 + 6\sigma_{G/O}^2 + 24\theta_O^2$
G/O	6	$\sigma_e^2 + 6\sigma_{G/O}^2$
Within G	40	
Between Problems (P)	5	
Stress (St)	1	$\sigma_e^2 + 3\sigma_{GO/St}^2 + 24\theta_{St}^2$
P/St	4	$\sigma_e^2 + \sigma_{GP/St}^2 + 8\theta_{P/St}^2$
O × St	1	$\sigma_e^2 + 3\sigma_{GO/St}^2 + 12\theta_{O \times St}^2$
OP/St	4	$\sigma_e^2 + \sigma_{GP/St}^2 + 8\theta_{OP/St}^2$
G × St/O	6	$\sigma_e^2 + \sigma_{G \times St/O}^2$
GP/St × O	24	$\sigma_e^2 + \sigma_{GP/St \times O}^2$

Note: P is treated here as a fixed-effect variable. In an experiment of the sort described, the problems might well be selected at random from a larger pool of items. In that case, the *EMS* are changed and certain problems arise in finding error terms. The issue will be considered in Chapter 11.

9.5

SV	df	EMS
Between S	71	
Between G	35	
Probability (P)	2	$\sigma_e^2 + 16\sigma_{G/P}^2 + 192\theta_P^2$
G/P	33	$\sigma_e^2 + 16\sigma_{G/P}^2$
Within G	36	
Role (R)	1	$\sigma_e^2 + 8\sigma_{GR/P}^2 + 288\theta_P^2$
PR	2	$\sigma_e^2 + 8\sigma_{GR/P}^2 + 96\theta_{PR}^2$
GR/P	33	$\sigma_e^2 + 8\sigma_{GR/P}^2$
Within S	504	
Trial Type (T)	1	$\sigma_e^2 + 8\sigma_{GT/P}^2 + 288\theta_T^2$
PT	2	$\sigma_e^2 + 8\sigma_{GT/P}^2 + 96\theta_{PT}^2$
GT/P	33	$\sigma_e^2 + 8\sigma_{GT/P}^2$
RT	1	$\sigma_e^2 + 4\sigma_{GRT/P}^1 + 144\theta_{RT}^2$
PRT	2	$\sigma_e^2 + 4\sigma_{GRT/P} + 8\theta_{PRT}^2$
GRT/P	33	$\sigma_e^2 + 4\sigma_{GRT/P}$
Blocks (B)	3	$\sigma_e^2 + 4\sigma_{GB/P}^2 + 144\theta_B^2$
PB	6	$\sigma_a^2 + 4\sigma_{GB/P}^2 + 48\theta_{PB}^2$
GB/P	99	$\sigma_e^2 + 4\sigma_{GB/P}^2$
RB	3	$\sigma_e^2 + 2\sigma_{GRB/P}^2 + 72\theta_{RB}^2$
PRB	6	$\sigma_e^2 + 2\sigma_{GRB/P}^2 + 24\theta_{PRB}^2$
GRB/P	99	$\sigma_e^2 + 2\sigma_{GRB/P}^2$
TB	3	$\sigma_e^2 + 2\sigma_{GTB/P}^2 + 72\theta_{TB}^2$
PTB	6	$\sigma_e^2 + 2\sigma_{GTB/P}^2 + 24\theta_{PTB}^2$
GTB/P	99	$\sigma_e^2 + 2\sigma_{GTB/P}^2$
RTB	3	$\sigma_e^2 + \sigma_{GRTB/P}^2 + 36\theta_{RTB}^2$
PRTB	6	$\sigma_e^2 + \sigma_{G_{RTB/P}} + 120\theta_{PRTB}^2$
GRTB/P	99	$\sigma_e^2 + \sigma_{GRTB/P}^2$

9.6

SV	df
D	2
Sc/D	27
I	1
DI	2
Sc × I/D	27
S/Sc × I/D	540

We now can test hypotheses about variability due to schools (*Sc*) and their interaction with *I*. Furthermore, if these sources do contribute variability the earlier pooled error term yielded a biased *F* test.

CHAPTER 10

10.1

SV	df
Between S	35
Sequences (R)	3
Instruction (I)	2
RI	6
S/RI	24
Within S	108
Location (L)	3
Column (C)	3
IL	6
IC	6
Between-cells res.	6
I × B. cells res.	12
Within-cells res.	72

10.2

SV	df
Between S	59
Alcohol (A)	2
S/A	57
Within S	180
Mode of presentation (M)	1
Tasks (T)	1
M × T	1
Columns (C)	3
AM	2
AT	2
AMT	2
AC	6
Residual	162

10.3

SV	df
Between cells	24
S	4
C	4
Lists (*L*)	4
Between-cells res.	12
Within cells	150
Position (*P*)	6
LP	24
CP	24
SP	24
Residual	72

10.4

SV	df
Between S	47
Between G	15
Sequences (*R*)	3
G/R	12
S/G/R	32
Within S	144
C	3
Problems (*P*)	3
Between-cells res.	6
Within-cells res.	132
GP/R(GC/R)	36
SP/GR(SC/GR)	96

10.5 (a) 9×9 Latin square with each subject tested under all combinations of two of the variables (e.g., payoff and probability) and 27 subjects at each level of initial stake. We could use nine different squares or replicate the same square nine times. This approach affords precise tests of two of the three main effects (we would have less precision in testing the between-subjects effect) and of all interactions of interest. However, it requires nine measurements/subject.

(b) 3×3 Latin square with each subject tested under all levels of one variable and nine subjects at each combinations of levels of the other two variables. The design saves measurement/subject but yields less precise tests of the two between-S variables and their interaction. Nonadditivity may also be more of a problem in the smaller square.

(c) No repeated measurements. Using the Latin square principle, we could have nine cells with nine subjects in each. But all interaction information is lost and, if the three factors interact, F tests against W. cells res. will be positively biased while those against B. cells res. will be negatively biased.

(d) Complete factorial design, 27 cells with three subjects in each. There are 54 error *df*, a reasonable number, no need to hold subjects for several measures, and no additivity problems. On the other hand, (a) and (b), will generally be more efficient.

10.6
$$SS_A = SS_{(E_1+E_3)-(E_2+E_4)}$$

$$SS_{BC} = SS_{(E_1+E_2)-(E_3+E_4)}$$

$$SS_{ABC} = SS_{(E_1+E_4)-(E_2+E_3)}$$

CHAPTER 11

11.2

SV	df	EMS
A	2	$\sigma_e^2 + 50\sigma_{AB}^2 + 200\sigma_A^2$
B	3	$\sigma_e^2 + 50\sigma_{AB}^2 + 150\sigma_B^2$
C	4	$\sigma_e^2 + 40\sigma_{AC}^2 + 30\sigma_{BC}^2 + 10\sigma_{ABC}^2 + 120\theta_C^2$
AB	6	$\sigma_e^2 + 50\sigma_{AB}^2$
AC	8	$\sigma_e^2 + 10\sigma_{ABC}^2 + 40\sigma_{AC}^2$
BC	12	$\sigma_e^2 + 10\sigma_{ABC}^2 + 30\sigma_{BC}^2$
ABC	24	$\sigma_e^2 + 10\sigma_{ABC}^2$
Within	540	σ_e^2

From this table and the *MS* values, we obtain

SV	Estimate
A	$0; MS_A < MS_{AB}$
B	$\dfrac{MS_B - MS_{AB}}{150} = .053$
C	$\dfrac{MS_C - (MS_{AC} + MS_{BC} - MS_{ABC})}{120} = .045$
AB	$\dfrac{MS_{AB} - MS_W}{50} = .08$
AC	$\dfrac{MS_{AC} - MS_{ABC}}{40} = .0025$
BC	$\dfrac{MS_{BC} - MS_{ABC}}{30} = .033$
ABC	$\dfrac{MS_{ABC} - MS_W}{10} = .15$
Within	2.0

Note that in testing *C*, we have

$$df = \frac{(MS_{AC} + MS_{BC} - MS_{ABC})^2}{\dfrac{(MS_{AC})^2}{8} + \dfrac{(MS_{BC})^2}{12} + \dfrac{(MS_{ABC})^2}{24}} \approx 6$$

11.3 (a) No; the expectations are

$$E(MS_{Sh}) = \sigma_e^2 + 4\sigma_{S \times Sh}^2 + 32\theta_{Sh}^2$$

$$E(MS_R) = \sigma_e^2 + 3\sigma_{S \times R}^2 + 24\theta_R^2$$

Even if $\theta_{Sh}^2 = \theta_R^2$, differences in MS would be obtained due to the interaction components and the θ^2 coefficients.

(b) No; the F ratios are distributed on different *dfs*.

(c) Point estimates of θ^2 or of ω^2 (proportions of total variance) would be most sensible.

11.4 The total variance is

$$\sigma_Y^2 = \sigma_S^2 + \delta_C^2 + \delta_A^2 + \sigma_e^2$$

We desire

$$\hat{\omega}_A^2 = \frac{\hat{\delta}_A^2}{\hat{\sigma}_Y^2} = \left(\frac{a-1}{a}\right) \hat{\theta}_A^2 / \hat{\sigma}_Y^2$$

We have

$$\hat{\theta}_A^2 = \frac{MS_A - MS_{B.\text{cells res.}}}{a}$$

$$\hat{\theta}_C^2 = \frac{MS_C - MS_{B.\text{cells res.}}}{a}$$

$$\hat{\sigma}_S^2 = \frac{MS_S - MS_{B.\text{cells res.}}}{a}$$

$$\hat{\sigma}_e^2 = MS_{B.\text{cells res.}}$$

Substituting in the expression for $\hat{\omega}_A^2$, and dividing numerator and denominator by $MS_{B.\text{cells res.}}$, we have

$$\hat{\omega}_A^2 = \frac{F_A - 1}{(a/a - 1)F_S + F_A + F_C + a - 2}$$

CHAPTER 12

12.1

	SS_y	SS_x	SP	$SS_{y'}$
A	328.56	49.88	-128.02	434.75
B	258.73	44.83	107.69	129.89
AB	348.08	79.21	166.04	145.56
S/AB	586.66	343.83	203.77	465.89

SV	df	MS	F
A	1	434.75	17.73
B	1	129.89	5.30
AB	1	145.56	5.94
S/AB	19	24.52	

12.2

	SS_y	SS_x	SP	$SS_{y'}$
A	262.89	2.00	14.33	268.82
C	6.22	6.67	−0.67	15.14
S	81.56	0.67	7.33	51.96
Res	14.89	4.67	7.67	2.29

SV	df	MS	F
A	2	134.41	58.60
C	2	7.57	3.30
S	2	25.98	11.32
Res	1	2.29	

12.3 The point is that we have only one covariate measure for each subject. Since there is no within-subject variability on X, covariance is meaningless for the Within-S SV. The Between-S SV can be adjusted. The simplest approach is to sum the four scores for each subject and then carry out the usual one-factor analysis of covariance.

12.4 Define a new covariate, Z:

$$Z = X^2$$

Now Y is a linear function of Z and the usual covariance may be carried out. Of course, this assumes the usual covariance model; for example, that the regression of Y on Z is linear and the same for all treatment populations.

CHAPTER 13

13.1 (a) *Scheffé*. Let $EW = .05$. Then the criterion F is $4\,F_{.05;4,70} = 10.00$. That is, in order to obtain significance,

$$\frac{\hat{\psi}^2/\frac{2}{15}}{297.44} > 10$$

Or

$$\hat{\psi} > 8.96$$

Then, A differs significantly from D and E.

13.2 (a)
$$SS_{\hat{\psi}} = \frac{[(3)(5) + (-2)(5) + (-2)(7) + (-2)(9) + (3)(2)]^2}{30/6}$$

$$= \frac{(21)^2}{5} = 88.2$$

$$F = \frac{88.2}{4} = 22.05$$

$$F_{.05;1,25} = 4.24$$

The result is significant. Note that we have a single a priori hypothesis.
(b) *Scheffé*. Let the weights be $-\frac{1}{2}, +\frac{1}{3}, +\frac{1}{3}, +\frac{1}{3}, -\frac{1}{2}$ so that the results are on the original scale. Then the limits are

$$3.5 \pm \sqrt{(4)(2.76)(4)(\tfrac{5}{6})(6)} = 3.5 \pm 2.5$$

Tukey. The limits are

$$3.5 \pm \sqrt{(\tfrac{4}{6})(4.16)} = 3.5 \pm 3.4$$

13.3 (a) $$SS_{\hat{\psi}} = \frac{[13 + 22 - (2)(11)]^2}{(4)(6)} = 7.04$$

(b) $SS_{\hat{\psi}} = \dfrac{[5 + 11 - (2)(5)]^2}{(2)(6)} + \dfrac{[8 + 11 - (2)(6)]^2}{(2)(6)} - 7.04 = .04$

The error term for both (a) and (b) is

$$\frac{[4 + 5 - (2)(3)]^2}{6} + \cdots + \frac{[3 + 7 - (2)(4)]^2}{6}$$

$$- \frac{[5 + 11 - (2)(5)]^2}{(2)(6)} - \frac{[8 + 11 - (2)(6)]^2}{(2)(6)}$$

$$= 7.83 - 7.08 = .75$$

13.4 If we assign the weights $+1, -1$ to B_1 and B_2 and $+2, -1, -1$ to the A levels, we have part of AB on 1 df. Since $0, +1, -1$ gives an orthogonal A contrast, the solution follows.

$$SS_{\hat{\psi}_1} = \frac{\begin{aligned}&[(1)(2)(21) + (1)(-1)(16) + (1)(-1)(9) + (-1)(2)(4)\\ &\qquad + (-1)(-1)(12) + (-1)(-1)(7)]^2\end{aligned}}{(5)(12)}$$

$$SS_{\hat{\psi}_2} = \frac{[(1)(1)(16) + (1)(-1)(9) + (-1)(1)(12) + (-1)(-1)(7)]^2}{(5)(4)}$$

$$SS_{AB} = SS_{\hat{\psi}_1} + SS_{\hat{\psi}_2}$$

13.5 *Tukey*. The criterion value of q is

$$q_{.05;5,12} = 4.51$$

A difference is significant if

$$\hat{\psi} > (4.51)\sqrt{\tfrac{10}{5}} = 6.36$$

D_1 differs from D_3 and D_4 and D_2 differs from D_4.
Dunnett. The criterion value of d is

$$d_{.05;5,12} = 2.88$$

A difference is significant if

$$\hat{\psi} > (2.88)\sqrt{(10)(\tfrac{2}{5})} = 5.76$$

C differs from D_1 and D_4.

The Dunnett test has more power for contrasts involving C (the confidence interval would also be narrower). The Dunnett test is adjusting EW over a set of four comparisons ($D_j - C$) while the Tukey test must maintain the EW over a considerably larger set. Consequently, the Dunnett test has a larger EC, and more power, for those few comparisons within its range of consideration. The situation is much like that of two- versus one-tailed tests; by reducing the range of application, we increase power to assess certain differences while losing information about others completely.

13.6 The six schools form a 3×2 array:

	Public			*Private*	
East	*Midwest*	*West*	*East*	*Midwest*	*West*
Penn State	Minnesota	California	Yale	Northwestern	Stanford

Thus, the overall design is a $3 \times 3 \times 2$.
Let Public, Private $= A_1, A_2$. East, Midwest, West $= B_1, B_2, B_3$.
Clinicians, Experimentalists, Social—C_1, C_2, C_3 and let

$$T_{jkm} = \text{total for } A_j B_k C_m$$

$$T_j = \text{total for } A_j = \sum_k \sum_m T_{jkm}$$

$$T_{jk} = \text{total for } A_j B_k = \sum_m T_{jkm}$$

etc.

H_1: $(2)(T_{.2.}) - T_{.1.} - T_{.3.}$
H_2: $[(2)(T_{12.}) - T_{11.} - T_{13.}] - [(2)(T_{22.}) - T_{21.} - T_{23.}]$
H_3: $(2)[(2)(T_{.21}) - T_{.11} - T_{.31}] + (-1)[(2)(T_{.22}) - T_{.12} - T_{.32}] + (-1)[(2)(T_{.23}) - T_{.13} - T_{.33}]$
H_4: $(5)(T_{121}) - T_{111} - T_{131} - T_{211} - T_{221} - T_{231}$
H_5: $(2)[(5)(T_{121}) - T_{111} - T_{131} - T_{211} - T_{221} - T_{231}] + (-1)[(5)(T_{122}) - (T_{112}) - T_{132} - T_{222} - T_{232}] + (-1)[(5)(T_{123}) - T_{113} - T_{213} - T_{223} - T_{233}]$

CHAPTER 14

14.1 A (lin) $SS = \dfrac{[(-3)(39) - 25 + 17 + (3)(41)]^2}{(10)(20)}$

$= 0.02$

A (quad) $SS = \dfrac{(39 - 25 - 17 + 41)^2}{(10)(4)}$

$= 36.1$

$$A\text{(cub)} \qquad SS = \frac{[-39 + 3(25) + (-3)(17) + 41]^2}{(10)(20)}$$

$$= 2.42$$

$$A\text{(lin)} \times B \qquad SS = \frac{[(-3)(15) - 7 + 5 + (3)(15)]^2 + [(-3)(24) - 18 + 12 + (3)(26)]^2}{(5)(20)}$$

$$- SS_{A\text{(lin)}} = .02$$

$$A\text{(quad)} \times B \qquad SS = \frac{(15 - 7 - 5 + 15)^2 + (24 - 18 - 12 + 26)^2}{(5)(4)}$$

$$- SS_{A\text{(quad)}} = .1$$

$$A\text{(cub)} \times B \qquad SS = \frac{[-15 + (3)(7) - (3)(5) + 15]^2 + [-24 + (3)(18) - (3)(12) + 26]^2}{(5)(20)}$$

$$- SS_{A\text{(cub)}} = 1.94$$

14.2 Whether there is a sexual content main effect will depend upon whether or not high-guilt responses decline at the same rate as low-guilt responses increase as a function of sexual content. There should clearly be a guilt main effect, since the two groups have equal response frequency for low sexual content but the low group has a higher frequency at higher-content levels. There should be an interaction, since the two curves diverge. In particular, $G \times S\text{(lin)}$ should be significant. The residual, $G \times S\text{(curv)}$, should also be tested.

14.3 (a)
$$SS_{P\text{(lin)}} = \frac{(30)^2}{(15)(20)} = 3.0$$

$$SS_{P\text{(lin)} \times T} = \frac{(17)^2 + (9)^2 + (4)^2}{(5)(20)} - 3.0 = .86$$

(b)
$$SS_{\hat{\psi}} = \frac{[(-2)(-17) - 9 - 4]^2}{(5)(20)(6)} = .735$$

14.4 A_1 apparently exhibits a linear and cubic trend; A_2 a quadratic trend; A_3 a linear trend. Assuming that the groups start at the same level, we would expect an A main effect, A_2 performing less well over the five tests (T). There might also be a T main effect; averaging two linear components $(A_1$ and $A_3)$ with a zero linear component should yield $T\text{(lin)}$, on similar grounds we could look for $T\text{(quad)}$ and $T\text{(cub)}$. Most clearly, the hypothesis implies $A \times T\text{(lin)}$, $A \times T\text{(quad)}$, and $A \times T\text{(cub)}$.

14.5 H_1: Test the simple effect of D for blank interval data.
H_2: Discarding the blank interval data, test for $D\text{(lin)}$ and $D\text{(quad)}$.
H_3: Test $I \times D\text{(lin)}$.

14.6   Method: Linear, branching, reading $= A_1, A_2, A_3$
Material: Math, English, Social Studies $= B_1, B_2, B_3$.
Time/Day: 15, 30, 45 $= C_1, C_2, C_3$.

H_1: $2T_{.1.} - T_{.2.} - T_{.3.}$
H_2: $T_{1..} + T_{2..} - T_{3..}$
H_3: $T(\text{lin}) = T_{..1} - T_{..3}$
        $T(\text{quad}) = T_{..1} - 2T_{..2} + T_{..3}$
H_4: $(2T_{11.} - T_{12.} - T_{13.}) + (2T_{21.} - T_{22.} - T_{23.})$
$$-2(T_{31.} - T_{32.} - T_{33.})$$
H_5: $(2T_{11.} - T_{12.} - T_{13.}) - 2T_{21.} - T_{22.} - T_{23.})$
H_6: $(2T_{.11} - T_{.21} - T_{.31}) - 2T_{.13} - T_{.23} - T_{.33})$

14.7   (a)
$$b_0' = \frac{30}{3} = 10$$

$$b_1' = \frac{\sum_i \sum_{ij} \bar{Y}_j}{\sum_i \zeta_{ij}'^2} = 3$$

$$b_2' = -5$$

$$\bar{Y}_{ij} = 10 + 3\xi_{1j}' - 5\xi_{2j}'$$

To check:

$$\bar{Y}_1 = 10 + 3(-1) - 5(+1) = 2$$
$$\bar{Y}_2 = 10 + 3(0) - 5(-2) = 20$$
$$\bar{Y}_3 = 10 + 3(+1) - 5(+1) = 8$$

(b) $\xi_{11} = \alpha_1$ $\xi_{12} = \alpha_2$
    $\xi_{21} = \alpha_1 + 1$ $\xi_{22} = \alpha_2 + \beta_2 + 1$
    $\xi_{31} = \alpha_1 + 2$ $\xi_{32} = \alpha_2 + 2\beta_2 + 4$

$0 = 3\alpha_1 + 3$ $0 = 3\alpha_2 + 3\beta_2 + 5$
$\alpha_1 = -1$ $0 = -\alpha_2 + (0)(\alpha_2 + \beta_2 + 1)$
$$+(1)(\alpha_2 + 2\beta_2 + 4)$$
$$\beta_2 = -2\alpha_2 = \tfrac{1}{3}$$

(c) $\bar{Y}_j = 10 + 3(\alpha_1 + X) - (5)(3)(\alpha_2 + X\beta_2 + X^2)$
$$= 2 + 33X = 15X^2$$

Note: We multiply $(\alpha_2 + \beta_2 + X^2)$ by (3) to obtain the integer ξ_{j2}.
To check:

$$\bar{Y}_1 = 2 + 33(0) - 15(0) = 2$$
$$\bar{Y}_2 = 2 + 33(1) - 15(1) = 20$$
$$\bar{Y}_3 = 2 + 33(2) - 15(4) = 8$$

Appendix Tables

427

TABLE A-1
Random Numbers

Line\Col.	(1)	(2)	(3)	(4)	(5)	(6)	(7)	(8)	(9)	(10)	(11)	(12)	(13)	(14)
1	10480	15011	01536	02011	81647	91646	69179	14194	62590	36207	20969	99570	91291	90700
2	22368	46573	25595	85393	30995	89198	27982	53402	93965	34095	52666	19174	39615	99505
3	24130	48360	22527	97265	76393	64809	15179	24830	49340	32081	30680	19655	63348	58629
4	42167	93093	06243	61680	07856	16376	39440	53537	71341	57004	00849	74917	97758	16379
5	37570	39975	81837	16656	06121	91782	60468	81305	49684	60672	14110	06927	01263	54613
6	77921	06907	11008	42751	27756	53498	18602	70659	90655	15053	21916	81825	44394	42880
7	99562	72905	56420	69994	98872	31016	71194	18738	44013	48840	63213	21069	10634	12952
8	96301	91977	05463	07972	18876	20922	94595	56869	69014	60045	18425	84903	42508	32307
9	89579	14342	63661	10281	17453	18103	57740	84378	25331	12566	58678	44947	05585	56941
10	85475	36857	53342	53988	53060	59533	38867	62300	08158	17983	16439	11458	18593	64952
11	28918	69578	88231	33276	70997	79936	56865	05859	90106	31595	01547	85590	91610	78188
12	63553	40961	48235	03427	49626	69445	18663	72695	52180	20847	12234	90511	33703	90322
13	09429	93969	52636	92737	88974	33488	36320	17617	30015	08272	84115	27756	30613	74952
14	10365	61129	87529	85689	48237	52267	67689	93394	01511	26358	85104	20285	29975	89868
15	07119	97336	71048	08178	77233	13916	47564	81056	97735	85977	29372	74461	28551	90707
16	51085	12765	51821	51259	77452	16308	60756	92144	49442	53900	70960	63990	75601	40719
17	02368	21382	52404	60268	89368	19885	55322	44819	01188	65255	64835	44919	05944	55157
18	01011	54092	33362	94904	31273	04146	18594	29852	71585	85030	51132	01915	92747	64951
19	52162	53916	46369	58586	23216	14513	83149	98736	23495	64350	94738	17752	35156	35749
20	07056	97628	33787	09998	42698	06691	76988	13602	51851	46104	88916	19509	25625	58104
21	48663	91245	85828	14346	09172	30168	90229	04734	59193	22178	30421	61666	99904	32812
22	54164	58492	22421	74103	47070	25306	76468	26384	58151	06646	21524	15227	96909	44592
23	32639	32363	05597	24200	13363	38005	94342	28728	35806	06912	17012	64161	18296	22851
24	29334	27001	87637	87308	58731	00256	45834	15398	46557	41135	10367	07684	36188	18510
25	02488	33062	28834	07351	19731	92420	60952	61280	50001	67658	32586	86679	50720	94953
26	81525	72295	04839	96423	24878	82651	66566	14778	76797	14780	13300	87074	79666	95725
27	29676	20591	68086	26432	46901	20849	89768	81536	86645	12659	92259	57102	80428	25280
28	00742	57392	39064	66432	84673	40027	32832	61362	98947	96067	64760	64584	96096	98253
29	05366	04213	25669	26422	44407	44048	37937	63904	45766	66134	75470	66520	34693	90449
30	91921	26418	64117	94305	26766	25940	39972	22209	71500	64568	91402	42416	07844	69618

TABLE A-1 (*Continued*)

Line\Col.	(1)	(2)	(3)	(4)	(5)	(6)	(7)	(8)	(9)	(10)	(11)	(12)	(13)	(14)
31	00582	04711	87917	77341	42206	35126	74087	99547	81817	42607	43808	76655	62028	76630
32	00725	69884	62797	56170	86324	88072	76222	36086	84637	93161	76038	65855	77919	88006
33	69011	65795	95876	55293	18988	27354	26575	08625	40801	59920	29841	80150	12777	48501
34	25976	57948	29888	88604	67917	48708	18912	82271	65424	69774	33611	54262	85963	03547
35	09763	83473	73577	12908	30883	18317	28290	35797	05998	41688	34952	37888	38917	88050
36	91567	42595	27958	30134	04024	86385	29880	99730	55536	84855	29080	09250	79656	73211
37	17955	56349	90999	49127	20044	59931	06115	20542	18059	02008	73708	83517	36103	42791
38	46503	18584	18845	49618	02304	51038	20655	58727	28168	15475	56942	53389	20562	87338
39	92157	89634	94824	78171	84610	82834	09922	25417	44137	48413	25555	21246	35509	20468
40	14577	62765	35605	81263	39667	47358	56873	56307	61607	49518	89656	20103	77490	18062
41	98427	07523	33362	64270	01638	92477	66969	98420	04880	45585	46565	04102	46880	45709
42	34914	63976	88720	82765	34476	17032	87589	40836	32427	70002	70663	88863	77775	69348
43	70060	28277	39475	46473	23219	53416	94970	25832	69975	94884	19661	72828	00102	66794
44	53976	54914	06990	67245	68350	82948	11398	42878	80287	88267	47363	46634	06541	97809
45	76072	29515	40980	07391	58745	25774	22987	80059	39911	96189	41151	14222	60697	59583
46	90725	52210	83974	29992	65831	38857	50490	83765	55657	14361	31720	57375	56228	41546
47	64364	67412	33339	31926	14883	24413	59744	92351	97473	89286	35931	04110	23726	51900
48	08962	00358	31662	25388	61642	34072	81249	35648	56891	69352	48373	45578	78547	81788
49	95012	68379	93526	70765	10592	04542	76463	54328	02349	17247	28865	14777	62730	92277
50	15664	10493	20492	38391	91132	21999	59516	81652	27195	48223	46751	22923	32261	85653
51	16408	81899	04153	53381	79401	21438	83035	92350	36693	31238	59649	91754	72772	02338
52	18629	81953	05520	91962	04739	13092	97662	24822	94730	06496	35090	04822	86774	98289
53	73115	35101	47498	87637	99016	71060	88824	71013	18735	20286	23153	72924	35165	43040
54	57491	16703	23167	49323	45021	33132	12544	41035	80780	45393	44812	12515	98931	91202
55	30405	83946	23792	14422	15059	45799	22716	19792	09983	74353	68668	30429	70735	25499
56	16631	35006	85900	98275	32388	52390	16815	69298	82732	38480	73817	32523	41961	44437
57	96773	20206	42559	78985	05300	22164	24369	54224	35083	19687	11052	91491	60383	19746
58	38935	64202	14349	82674	66523	44133	00697	35552	35970	19124	63318	29686	03387	59846
59	31624	76384	17403	53363	44167	64486	64758	75366	76554	31601	12614	33072	60332	92325
60	78919	19474	23632	27889	47914	02584	37680	20801	72152	39339	34806	08930	85001	87820

Abridged from "Table of 105,000 Random Decimal Digits," Statement 4914, Bureau of Transport Economics and Statistics, Interstate Commerce Commission, 1949,

Unit Normal Distribution

$$[P(z \leq z_{1-\alpha}) = 1 - \alpha]$$

$1 - \alpha$	$z_{1-\alpha}$	$1 - \alpha$	$z_{1-\alpha}$	$1 - \alpha$	$z_{1-\alpha}$
.50	0.00	.75	0.67	.950	1.645
.51	0.03	.76	0.71	.955	1.695
.52	0.05	.77	0.74	.960	1.751
.53	0.08	.78	0.77	.965	1.812
.54	0.10	.79	0.81	.970	1.881
.55	0.13	.80	0.84	.975	1.960
.56	0.15	.81	0.88	.980	2.054
.57	0.18	.82	0.92	.985	2.170
.58	0.20	.83	0.95	.990	2.326
.59	0.23	.84	0.99	.995	2.576
.60	0.25	.85	1.04	.996	2.652
.61	0.28	.86	1.08	.997	2.748
.62	0.30	.87	1.13	.998	2.878
.63	0.33	.88	1.17	.999	3.090
.64	0.36	.89	1.23		
.65	0.39	.90	1.28	.9995	3.291
.66	0.41	.91	1.34	.99995	3.891
.67	0.44	.92	1.41		
.68	0.47	.93	1.48	.999995	4.417
.69	0.50	.94	1.55		
				.9999995	5.327
.70	0.52				
.71	0.55				
.72	0.58				
.73	0.61				
.74	0.64				

TABLE A-3
Distribution of t

df	Probability												
	.9	.8	.7	.6	.5	.4	.3	.2	.1	.05	.02	.01	.001
1	.158	.325	.510	.727	1.000	1.376	1.963	3.078	6.314	12.706	31.821	63.657	636.619
2	.142	.289	.445	.617	.816	1.061	1.386	1.886	2.920	4.303	6.965	9.925	31.598
3	.137	.277	.424	.584	.765	.978	1.250	1.638	2.353	3.182	4.541	5.841	12.924
4	.134	.271	.414	.569	.741	.941	1.190	1.533	2.132	2.776	3.747	4.604	8.610
5	.132	.267	.408	.559	.727	.920	1.156	1.476	2.015	2.571	3.365	4.032	6.869
6	.131	.265	.404	.553	.718	.906	1.134	1.440	1.943	2.447	3.143	3.707	5.959
7	.130	.263	.402	.549	.711	.896	1.119	1.415	1.895	2.365	2.998	3.499	5.408
8	.130	.262	.399	.546	.706	.889	1.108	1.397	1.860	2.306	2.896	3.355	5.041
9	.129	.261	.398	.543	.703	.883	1.100	1.383	1.833	2.262	2.821	3.250	4.781
10	.129	.260	.397	.542	.700	.879	1.093	1.372	1.812	2.228	2.764	3.169	4.587
11	.129	.260	.396	.540	.697	.876	1.088	1.363	1.796	2.201	2.718	3.106	4.437
12	.128	.259	.395	.539	.695	.873	1.083	1.356	1.782	2.179	2.681	3.055	4.318
13	.128	.259	.394	.538	.694	.870	1.079	1.350	1.771	2.160	2.650	3.012	4.221
14	.128	.258	.393	.537	.692	.868	1.076	1.345	1.761	2.145	2.624	2.977	4.140
15	.128	.258	.393	.536	.691	.866	1.074	1.341	1.753	2.131	2.602	2.947	4.073
16	.128	.258	.392	.535	.690	.865	1.071	1.337	1.746	2.120	2.583	2.921	4.015
17	.128	.257	.392	.534	.689	.863	1.069	1.333	1.740	2.110	2.567	2.898	3.965
18	.127	.257	.392	.534	.688	.862	1.067	1.330	1.734	2.101	2.552	2.878	3.922
19	.127	.257	.391	.533	.688	.861	1.066	1.328	1.729	2.093	2.539	2.861	3.883
20	.127	.257	.391	.533	.687	.860	1.064	1.325	1.725	2.086	2.528	2.845	3.850

df													
21	.127	.257	.391	.532	.686	.859	1.063	1.323	1.721	2.080	2.518	2.831	3.819
22	.127	.256	.390	.532	.686	.858	1.061	1.321	1.717	2.074	2.508	2.819	3.792
23	.127	.256	.390	.532	.685	.858	1.060	1.319	1.714	2.069	2.500	2.807	3.767
24	.127	.256	.390	.531	.685	.857	1.059	1.318	1.711	2.064	2.492	2.797	3.745
25	.127	.256	.390	.531	.684	.856	1.058	1.316	1.708	2.060	2.485	2.787	3.725
26	.127	.256	.390	.531	.684	.856	1.058	1.315	1.706	2.056	2.479	2.779	3.707
27	.127	.256	.389	.531	.684	.855	1.057	1.314	1.703	2.052	2.473	2.771	3.690
28	.127	.256	.389	.530	.683	.855	1.056	1.313	1.701	2.048	2.467	2.763	3.674
29	.127	.256	.389	.530	.683	.854	1.055	1.311	1.699	2.045	2.462	2.756	3.659
30	.127	.256	.389	.530	.683	.854	1.055	1.310	1.697	2.042	2.457	2.750	3.646
40	.126	.255	.388	.529	.681	.851	1.050	1.303	1.684	2.021	2.423	2.704	3.551
60	.126	.254	.387	.527	.679	.848	1.046	1.296	1.671	2.000	2.390	2.660	3.460
120	.126	.254	.386	.526	.677	.845	1.041	1.289	1.658	1.980	2.358	2.617	3.373
∞	.126	.253	.385	.524	.674	.842	1.036	1.282	1.645	1.960	2.326	2.576	3.291

This table is adapted from Table III of R. A. Fisher and F. Yates, *Statistical Tables for Biological, Agricultural and Medical Research*, Oliver and Boyd Ltd., Edinburgh, by permission of the authors and publishers. Note that the probabilities are two-tailed; $\alpha = P(|t| \geq t_\alpha)$.

TABLE A-4
Distribution of χ^2

df	Probability .99	.98	.95	.90	.80	.70	.50	.30	.20	.10	.05	.02	.01	.001
1	$.0^3157$	$.0^3628$	.00393	.0158	.0642	.148	.455	1.074	1.642	2.706	3.841	5.412	6.635	10.827
2	.0201	.0404	.103	.211	.446	.713	1.386	2.408	3.219	4.605	5.991	7.824	9.210	13.815
3	.115	.185	.352	.584	1.005	1.424	2.366	3.665	4.642	6.251	7.815	9.837	11.345	16.266
4	.297	.429	.711	1.064	1.649	2.195	3.357	4.878	5.989	7.779	9.488	11.668	13.277	18.467
5	.554	.752	1.145	1.610	2.343	3.000	4.351	6.064	7.289	9.236	11.070	13.388	15.086	20.515
6	.872	1.134	1.635	2.204	3.070	3.828	5.348	7.231	8.558	10.645	12.592	15.033	16.812	22.457
7	1.239	1.564	2.167	2.833	3.822	4.671	6.346	8.383	9.803	12.017	14.067	16.622	18.475	24.322
8	1.646	2.032	2.733	3.490	4.594	5.527	7.344	9.524	11.030	13.362	15.507	18.168	20.090	26.125
9	2.088	2.532	3.325	4.168	5.380	6.393	8.343	10.656	12.242	14.684	16.919	19.679	21.666	27.877
10	2.558	3.059	3.940	4.865	6.179	7.267	9.342	11.781	13.442	15.987	18.307	21.161	23.209	29.588
11	3.053	3.609	4.575	5.578	6.989	8.148	10.341	12.899	14.631	17.275	19.675	22.618	24.725	31.264
12	3.571	4.178	5.226	6.304	7.807	9.034	11.340	14.011	15.812	18.549	21.026	24.054	26.217	32.909
13	4.107	4.765	5.892	7.042	8.634	9.926	12.340	15.119	16.985	19.812	22.362	25.472	27.688	34.528
14	4.660	5.368	6.571	7.790	9.467	10.821	13.339	16.222	18.151	21.064	23.685	26.873	29.141	36.123
15	5.229	5.985	7.261	8.547	10.307	11.721	14.339	17.322	19.311	22.307	24.996	28.259	30.578	37.697
16	5.812	6.614	7.962	9.312	11.152	12.624	15.338	18.418	20.465	23.542	26.296	29.633	32.000	39.252
17	6.408	7.255	8.672	10.085	12.002	13.531	16.338	19.511	21.615	24.769	27.587	30.995	33.409	40.790
18	7.015	7.906	9.390	10.865	12.857	14.440	17.338	20.601	22.760	25.989	28.869	32.346	34.805	42.312
19	7.633	8.567	10.117	11.651	13.716	15.352	18.338	21.689	23.900	27.204	30.144	33.687	36.191	43.820
20	8.260	9.237	10.851	12.443	14.578	16.266	19.337	22.775	25.038	28.412	31.410	35.020	37.566	45.315
21	8.897	9.915	11.591	13.240	15.445	17.182	20.337	23.858	26.171	29.615	32.671	36.343	38.932	46.797
22	9.542	10.600	12.338	14.041	16.314	18.101	21.337	24.939	27.301	30.813	33.924	37.659	40.289	48.268
23	10.196	11.293	13.091	14.848	17.187	19.021	22.337	26.018	28.429	32.007	35.172	38.968	41.638	49.728
24	10.856	11.992	13.848	15.659	18.062	19.943	23.337	27.096	29.553	33.196	36.415	40.270	42.980	51.179
25	11.524	12.697	14.611	16.473	18.940	20.867	24.337	28.172	30.675	34.382	37.652	41.566	44.314	52.620

df													
26	12.198	13.409	15.379	17.292	19.820	21.792	25.336	29.246	31.795	35.563	38.885	42.856	45.642	54.052
27	12.879	14.125	16.151	18.114	20.703	22.719	26.336	30.319	32.912	36.741	40.113	44.140	46.963	55.476
28	13.565	14.847	16.928	18.939	21.588	23.647	27.336	31.391	34.027	37.916	41.337	45.419	48.278	56.893
29	14.256	15.574	17.708	19.768	22.475	24.577	28.336	32.461	35.139	39.087	42.557	46.693	49.588	58.302
30	14.953	16.306	18.493	20.599	23.364	25.508	29.336	33.530	36.250	40.256	43.773	47.962	50.892	59.703
32	16.362	17.783	20.072	22.271	25.148	27.373	31.336	35.665	38.466	42.585	46.194	50.487	53.486	62.487
34	17.789	19.275	21.664	23.952	26.938	29.242	33.336	37.795	40.676	44.903	48.602	52.995	56.061	65.247
36	19.233	20.783	23.269	25.643	28.735	31.115	35.336	39.922	42.879	47.212	50.999	55.489	58.619	67.985
38	20.691	22.304	24.884	27.343	30.537	32.992	37.335	42.045	45.076	49.513	53.384	57.969	61.162	70.703
40	22.164	23.838	26.509	29.051	32.345	34.872	39.335	44.165	47.269	51.805	55.759	60.436	63.691	73.402
42	23.650	25.383	28.144	30.765	34.157	36.755	41.335	46.282	49.456	54.090	58.124	62.892	66.206	76.084
44	25.148	26.939	29.787	32.487	35.974	38.641	43.335	48.396	51.639	56.369	60.481	65.337	68.710	78.750
46	26.657	28.504	31.439	34.215	37.795	40.529	45.335	50.507	53.818	58.641	62.830	67.771	71.201	81.400
48	28.177	30.080	33.098	35.949	39.621	42.420	47.335	52.616	55.993	60.907	65.171	70.197	73.683	84.037
50	29.707	31.664	34.764	37.689	41.449	44.313	49.335	54.723	58.164	63.167	67.505	72.613	76.154	86.661
52	31.246	33.256	36.437	39.433	43.281	46.209	51.335	56.827	60.332	65.422	69.832	75.021	78.616	89.272
54	32.793	34.856	38.116	41.183	45.117	48.106	53.335	58.930	62.496	67.673	72.153	77.422	81.069	91.872
56	34.350	36.464	39.801	42.937	46.955	50.005	55.335	61.031	64.658	69.919	74.468	79.815	83.513	94.461
58	35.913	38.078	41.492	44.696	48.797	51.906	57.335	63.129	66.816	72.160	76.778	82.201	85.950	97.039
60	37.485	39.699	43.188	46.459	50.641	53.809	59.335	65.227	68.972	74.397	79.082	84.580	88.379	99.607
62	39.063	41.327	44.889	48.226	52.487	55.714	61.335	67.322	71.125	76.630	81.381	86.953	90.802	102.166
64	40.649	42.960	46.595	49.996	54.336	57.620	63.335	69.416	73.276	78.860	83.675	89.320	93.217	104.716
66	42.240	44.599	48.305	51.770	56.188	59.527	65.335	71.508	75.424	81.085	85.965	91.681	95.626	107.258
68	43.838	46.244	50.020	53.548	58.042	61.436	67.335	73.600	77.571	83.308	88.250	94.037	98.028	109.791
70	45.442	47.893	51.739	55.329	59.898	63.346	69.334	75.689	79.715	85.527	90.531	96.388	100.425	112.317

For odd values of df between 30 and 70, the means of the tabular values for $df - 1$ and $df + 1$ may be taken. For larger values of df, the expression $\sqrt{2\chi^2} - \sqrt{2df - 1}$ may be used as a normal deviate with unit variance, remembering that the probability for χ^2 corresponds with that of a single tail of the normal curve.

This table is adapted from R. A. Fisher and F. Yates, *Statistical Tables for Biological, Agricultural and Medical Research*, Oliver and Boyd Ltd., Edinburgh, by permission of the authors and publishers.

435

TABLE A–5 Per cent points in the distribution of F

df_2		df_1 1	2	3	4	5	6	8	12	24	∞
1	0.1%	405284	500000	540379	562500	576405	585937	598144	610667	623497	63661
	0.5%	16211	20000	21615	22500	23056	23437	23925	24426	24940	2546
	1 %	4052	4999	5403	5625	5764	5859	5981	6106	6234	636
	2.5%	647.79	799.50	864.16	899.58	921.85	937.11	956.66	976.71	997.25	1018.3
	5 %	161.45	199.50	215.71	224.58	230.16	233.99	238.88	243.91	249.05	254.3
	10 %	39.86	49.50	53.59	55.83	57.24	58.20	59.44	60.70	62.00	63.3
	20 %	9.47	12.00	13.06	13.73	14.01	14.26	14.59	14.90	15.24	15.5
	25 %	5.83	7.50	8.20	8.58	8.82	8.98	9.19	9.41	9.63	9.8
2	0.1	998.5	999.0	999.2	999.2	999.3	999.3	999.4	999.4	999.5	999.5
	0.5	198.50	199.00	199.17	199.25	199.30	199.33	199.37	199.42	199.46	199.5
	1	98.49	99.00	99.17	99.25	99.30	99.33	99.36	99.42	99.46	99.5
	2.5	38.51	39.00	39.17	39.25	39.30	39.33	39.37	39.42	39.46	39.5
	5	18.51	19.00	19.16	19.25	19.30	19.33	19.37	19.41	19.45	19.5
	10	8.53	9.00	9.16	9.24	9.29	9.33	9.37	9.41	9.45	9.4
	20	3.56	4.00	4.16	4.24	4.28	4.32	4.36	4.40	4.44	4.4
	25	2.56	3.00	3.15	3.23	3.28	3.31	3.35	3.39	3.44	3.4
3	0.1	167.5	148.5	141.1	137.1	134.6	132.8	130.6	128.3	125.9	123.5
	0.5	55.55	49.80	47.47	46.20	45.39	44.84	44.13	43.39	42.62	41.8
	1	34.12	30.81	29.46	28.71	28.24	27.91	27.49	27.05	26.60	26.1
	2.5	17.44	16.04	15.44	15.10	14.89	14.74	14.54	14.34	14.12	13.9
	5	10.13	9.55	9.28	9.12	9.01	8.94	8.84	8.74	8.64	8.5
	10	5.54	5.46	5.39	5.34	5.31	5.28	5.25	5.22	5.18	5.1
	20	2.68	2.89	2.94	2.96	2.97	2.97	2.98	2.98	2.98	2.9
	25	2.02	2.28	2.36	2.39	2.41	2.42	2.44	2.45	2.46	2.4
4	0.1	74.14	61.25	56.18	53.44	51.71	50.53	49.00	47.41	45.77	44.0
	0.5	31.33	26.28	24.26	23.16	22.46	21.98	21.35	20.71	20.03	19.3
	1	21.20	18.00	16.69	15.98	15.52	15.21	14.80	14.37	13.93	13.4
	2.5	12.22	10.65	9.98	9.60	9.36	9.20	8.98	8.75	8.51	8.2
	5	7.71	6.94	6.59	6.39	6.26	6.16	6.04	5.91	5.77	5.6
	10	4.54	4.32	4.19	4.11	4.05	4.01	3.95	3.90	3.83	3.7
	20	2.35	2.47	2.48	2.48	2.48	2.47	2.47	2.46	2.44	2.4
	25	1.81	2.00	2.05	2.06	2.07	2.08	2.08	2.08	2.08	2.0
5	0.1	47.04	36.61	33.20	31.09	29.75	28.84	27.64	26.42	25.14	23.7
	0.5	22.79	18.31	16.53	15.56	14.94	14.51	13.96	13.38	12.78	12.1
	1	16.26	13.27	12.06	11.39	10.97	10.67	10.29	9.89	9.47	9.0
	2.5	10.01	8.43	7.76	7.39	7.15	6.98	6.76	6.52	6.28	6.0
	5	6.61	5.79	5.41	5.19	5.05	4.95	4.82	4.68	4.53	4.3
	10	4.06	3.78	3.62	3.52	3.45	3.40	3.34	3.27	3.19	3.1
	20	2.18	2.26	2.25	2.24	2.23	2.22	2.20	2.18	2.16	2.1
	25	1.70	1.85	1.89	1.89	1.89	1.89	1.89	1.89	1.88	1.8
6	0.1	35.51	27.00	23.70	21.90	20.81	20.03	19.03	17.99	16.89	15.7
	0.5	18.64	14.54	12.92	12.03	11.46	11.07	10.57	10.03	9.47	8.8
	1	13.74	10.92	9.78	9.15	8.75	8.47	8.10	7.72	7.31	6.8
	2.5	8.81	7.26	6.60	6.23	5.99	5.82	5.60	5.37	5.12	4.8
	5	5.99	5.14	4.76	4.53	4.39	4.28	4.15	4.00	3.84	3.6
	10	3.78	3.46	3.29	3.18	3.11	3.05	2.98	2.90	2.82	2.7
	20	2.07	2.13	2.11	2.09	2.08	2.06	2.04	2.02	1.99	1.9
	25	1.62	1.76	1.78	1.79	1.79	1.78	1.78	1.77	1.75	1.7

This table is abridged from Table 5 of R. A. Fisher and F. Yates, *Statistical Tables for Biological, Agricultural and Medical Research*, Oliver and Boyd Ltd., Edinburgh, by permission of the authors and publishers. The 0.5%, 2.5%, and 25% points are reprinted by permission from "Tables of Percentage Points of the Inverted Beta (F) Distribution," *Biometrika*, 33: 73–88 (April 1943).

df_1 df_2		1	2	3	4	5	6	8	12	24	∞
7	0.1%	29.22	21.69	18.77	17.19	16.21	15.52	14.63	13.71	12.73	11.69
	0.5%	16.24	12.40	10.88	10.05	9.52	9.16	8.68	8.18	7.65	7.08
	1 %	12.25	9.55	8.45	7.85	7.46	7.19	6.84	6.47	6.07	5.65
	2.5%	8.07	6.54	5.89	5.52	5.29	5.12	4.90	4.67	4.42	4.14
	5 %	5.59	4.74	4.35	4.12	3.97	3.87	3.73	3.57	3.41	3.23
	10 %	3.59	3.26	3.07	2.96	2.88	2.83	2.75	2.67	2.58	2.47
	20 %	2.00	2.04	2.02	1.99	1.97	1.96	1.93	1.91	1.87	1.83
	25 %	1.57	1.70	1.72	1.72	1.71	1.71	1.70	1.68	1.67	1.65
8	0.1	25.42	18.49	15.83	14.39	13.49	12.86	12.04	11.19	10.30	9.34
	0.5	14.69	11.04	9.60	8.81	8.30	7.95	7.50	7.01	6.50	5.95
	1	11.26	8.65	7.59	7.01	6.63	6.37	6.03	5.67	5.28	4.86
	2.5	7.57	6.06	5.42	5.05	4.82	4.65	4.43	4.20	3.95	3.67
	5	5.32	4.46	4.07	3.84	3.69	3.58	3.44	3.28	3.12	2.93
	10	3.46	3.11	2.92	2.81	2.73	2.67	2.59	2.50	2.40	2.29
	20	1.95	1.98	1.95	1.92	1.90	1.88	1.86	1.83	1.79	1.74
	25	1.53	1.66	1.67	1.55	1.66	1.65	1.64	1.62	1.60	1.58
9	0.1	22.86	16.39	13.90	12.56	11.71	11.13	10.37	9.57	8.72	7.81
	0.5	13.61	10.11	8.72	7.96	7.47	7.13	6.69	6.23	5.73	5.19
	1	10.56	8.02	6.99	6.42	6.06	5.80	5.47	5.11	4.73	4.31
	2.5	7.21	5.71	5.08	4.72	4.48	4.32	4.10	3.87	3.61	3.33
	5	5.12	4.26	3.86	3.63	3.48	3.37	3.23	3.07	2.90	2.71
	10	3.36	3.01	2.81	2.69	2.61	2.55	2.47	2.38	2.28	2.16
	20	1.91	1.94	1.90	1.87	1.85	1.83	1.80	1.76	1.72	1.67
	25	1.51	1.62	1.63	1.63	1.62	1.61	1.60	1.58	1.56	1.53
10	0.1	21.04	14.91	12.55	11.28	10.48	9.92	9.20	8.45	7.64	6.76
	0.5	12.83	9.43	8.08	7.34	6.87	6.54	6.12	5.66	5.17	4.64
	1	10.04	7.56	6.55	5.99	5.64	5.39	5.06	4.71	4.33	3.91
	2.5	6.94	5.46	4.83	4.47	4.24	4.07	3.85	3.62	3.37	3.08
	5	4.96	4.10	3.71	3.48	3.33	3.22	3.07	2.91	2.74	2.54
	10	3.28	2.92	2.73	2.61	2.52	2.46	2.38	2.28	2.18	2.06
	20	1.88	1.90	1.86	1.83	1.80	1.78	1.75	1.72	1.67	1.62
	25	1.49	1.60	1.60	1.60	1.59	1.58	1.56	1.54	1.52	1.48
11	0.1	19.69	13.81	11.56	10.35	9.58	9.05	8.35	7.63	6.85	6.00
	0.5	12.23	8.91	7.60	6.88	6.42	6.10	5.68	5.24	4.76	4.23
	1	9.65	7.20	6.22	5.67	5.32	5.07	4.74	4.40	4.02	3.60
	2.5	6.72	5.26	4.63	4.28	4.04	3.88	3.66	3.43	3.17	2.88
	5	4.84	3.98	3.59	3.36	3.20	3.09	2.95	2.79	2.61	2.40
	10	3.23	2.86	2.66	2.54	2.45	2.39	2.30	2.21	2.10	1.97
	20	1.86	1.87	1.83	1.80	1.77	1.75	1.72	1.68	1.63	1.57
	25	1.46	1.58	1.58	1.58	1.56	1.55	1.54	1.51	1.49	1.45
12	0.1	18.64	12.97	10.80	9.63	8.89	8.38	7.71	7.00	6.25	5.42
	0.5	11.75	8.51	7.23	6.52	6.07	5.76	5.35	4.91	4.43	3.90
	1	9.33	6.93	5.95	5.41	5.06	4.82	4.50	4.16	3.78	3.36
	2.5	6.55	5.10	4.47	4.12	3.89	3.73	3.51	3.28	3.02	2.72
	5	4.75	3.88	3.49	3.26	3.11	3.00	2.85	2.69	2.50	2.30
	10	3.18	2.81	2.61	2.48	2.39	2.33	2.24	2.15	2.04	1.90
	20	1.84	1.85	1.80	1.77	1.74	1.72	1.69	1.65	1.60	1.54
	25	1.46	1.56	1.56	1.55	1.54	1.53	1.51	1.49	1.46	1.42

df_2	df_1	1	2	3	4	5	6	8	12	24	∞
13	0.1%	17.81	12.31	10.21	9.07	8.35	7.86	7.21	6.52	5.78	4.97
	0.5%	11.37	8.19	6.93	6.23	5.79	5.48	5.08	4.64	4.17	3.65
	1 %	9.07	6.70	5.74	5.20	4.86	4.62	4.30	3.96	3.59	3.16
	2.5%	6.41	4.97	4.35	4.00	3.77	3.60	3.39	3.15	2.89	2.60
	5 %	4.67	3.80	3.41	3.18	3.02	2.92	2.77	2.60	2.42	2.21
	10 %	3.14	2.76	2.56	2.43	2.35	2.28	2.20	2.10	1.98	1.85
	20 %	1.82	1.83	1.78	1.75	1.72	1.69	1.66	1.62	1.57	1.51
	25 %	1.45	1.55	1.55	1.53	1.52	1.51	1.49	1.47	1.44	1.40
14	0.1	17.14	11.78	9.73	8.62	7.92	7.43	6.80	6.13	5.41	4.60
	0.5	11.06	7.92	6.68	6.00	5.56	5.26	4.86	4.43	3.96	3.44
	1	8.86	6.51	5.56	5.03	4.69	4.46	4.14	3.80	3.43	3.00
	2.5	6.30	4.86	4.24	3.89	3.66	3.50	3.29	3.05	2.79	2.49
	5	4.60	3.74	3.34	3.11	2.96	2.85	2.70	2.53	2.35	2.13
	10	3.10	2.73	2.52	2.39	2.31	2.24	2.15	2.05	1.94	1.80
	20	1.81	1.81	1.76	1.73	1.70	1.67	1.64	1.60	1.55	1.48
	25	1.44	1.53	1.53	1.52	1.51	1.50	1.48	1.45	1.42	1.38
15	0.1	16.59	11.34	9.34	8.25	7.57	7.09	6.47	5.81	5.10	4.31
	0.5	10.80	7.70	6.48	5.80	5.37	5.07	4.67	4.25	3.79	3.26
	1	8.68	6.36	5.42	4.89	4.56	4.32	4.00	3.67	3.29	2.87
	2.5	6.20	4.77	4.15	3.80	3.58	3.41	3.20	2.96	2.70	2.40
	5	4.54	3.68	3.29	3.06	2.90	2.79	2.64	2.48	2.29	2.07
	10	3.07	2.70	2.49	2.36	2.27	2.21	2.12	2.02	1.90	1.76
	20	1.80	1.79	1.75	1.71	1.68	1.66	1.62	1.58	1.53	1.46
	25	1.43	1.52	1.52	1.51	1.49	1.48	1.46	1.44	1.41	1.36
16	0.1	16.12	10.97	9.00	7.94	7.27	6.81	6.19	5.55	4.85	4.06
	0.5	10.58	7.51	6.30	5.64	5.21	4.91	4.52	4.10	3.64	3.11
	1	8.53	6.23	5.29	4.77	4.44	4.20	3.89	3.55	3.18	2.75
	2.5	6.12	4.69	4.08	3.73	3.50	3.34	3.12	2.89	2.63	2.32
	5	4.49	3.63	3.24	3.01	2.85	2.74	2.59	2.42	2.24	2.01
	10	3.05	2.67	2.46	2.33	2.24	2.18	2.09	1.99	1.87	1.72
	20	1.79	1.78	1.74	1.70	1.67	1.64	1.61	1.56	1.51	1.43
	25	1.42	1.51	1.51	1.50	1.48	1.47	1.45	1.43	1.39	1.34
17	0.1	15.72	10.66	8.73	7.68	7.02	6.56	5.96	5.32	4.63	3.85
	0.5	10.38	7.35	6.16	5.50	5.07	4.78	4.39	3.97	3.51	2.98
	1	8.40	6.11	5.18	4.67	4.34	4.10	3.79	3.45	3.08	2.65
	2.5	6.04	4.62	4.01	3.66	3.44	3.28	3.06	2.82	2.56	2.25
	5	4.45	3.59	3.20	2.96	2.81	2.70	2.55	2.38	2.19	1.96
	10	3.03	2.64	2.44	2.31	2.22	2.15	2.06	1.96	1.84	1.69
	20	1.78	1.77	1.72	1.68	1.65	1.63	1.59	1.55	1.49	1.42
	25	1.42	1.51	1.51	1.49	1.47	1.46	1.44	1.41	1.38	1.33
18	0.1	15.38	10.39	8.49	7.46	6.81	6.35	5.76	5.13	4.45	3.67
	0.5	10.22	7.21	6.03	5.37	4.96	4.66	4.28	3.86	3.40	2.87
	1	8.28	6.01	5.09	4.58	4.25	4.01	3.71	3.37	3.00	2.57
	2.5	5.98	4.56	3.95	3.61	3.38	3.22	3.01	2.77	2.50	2.19
	5	4.41	3.55	3.16	2.93	2.77	2.66	2.51	2.34	2.15	1.92
	10	3.01	2.62	2.42	2.29	2.20	2.13	2.04	1.93	1.81	1.66
	52	1.77	1.76	1.71	1.67	1.64	1.62	1.58	1.53	1.48	1.40
	20	1.41	1.50	1.49	1.48	1.46	1.45	1.43	1.40	1.37	1.32

df_2	df_1	1	2	3	4	5	6	8	12	24	∞
19	0.1 %	15.08	10.16	8.28	7.26	6.61	6.18	5.59	4.97	4.29	3.52
	0.5 %	10.07	7.09	5.92	5.27	4.85	4.56	4.18	3.76	3.31	2.78
	1 %	8.18	5.93	5.01	4.50	4.17	3.94	3.63	3.30	2.92	2.49
	2.5 %	5.92	4.51	3.90	3.56	3.33	3.17	2.96	2.72	2.45	2.13
	5 %	4.38	3.52	3.13	2.90	2.74	2.63	2.48	2.31	2.11	1.88
	10 %	2.99	2.61	2.40	2.27	2.18	2.11	2.02	1.91	1.79	1.63
	20 %	1.76	1.75	1.70	1.66	1.63	1.61	1.57	1.52	1.46	1.39
	25 %	1.41	1.50	1.49	1.48	1.46	1.44	1.42	1.40	1.36	1.31
20	0.1	14.82	9.95	8.10	7.10	6.46	6.02	5.44	4.82	4.15	3.38
	0.5	9.94	6.99	5.82	5.17	4.76	4.47	4.09	3.68	3.22	2.69
	1	8.10	5.85	4.94	4.43	4.10	3.87	3.56	3.23	2.86	2.42
	2.5	5.87	4.46	3.86	3.51	3.29	3.13	2.91	2.68	2.41	2.09
	5	4.35	3.49	3.10	2.87	2.71	2.60	2.45	2.28	2.08	1.84
	10	2.97	2.59	2.38	2.25	2.16	2.09	2.00	1.89	1.77	1.61
	20	1.76	1.75	1.70	1.65	1.62	1.60	1.56	1.51	1.45	1.37
	25	1.40	1.49	1.48	1.47	1.45	1.44	1.42	1.39	1.35	1.29
21	0.1	14.59	9.77	7.94	6.95	6.32	5.88	5.31	4.70	4.03	3.26
	0.5	9.83	6.89	5.73	5.09	4.68	4.39	4.01	3.60	3.15	2.61
	1	8.02	5.78	4.87	4.37	4.04	3.81	3.51	3.17	2.80	2.36
	2.5	5.83	4.42	3.82	3.48	3.25	3.09	2.87	2.64	2.37	2.04
	5	4.32	3.47	3.07	2.84	2.68	2.57	2.42	2.25	2.05	1.81
	10	2.96	2.57	2.36	2.23	2.14	2.08	1.98	1.88	1.75	1.59
	20	1.75	1.74	1.69	1.65	1.61	1.59	1.55	1.50	1.44	1.36
	25	1.40	1.49	1.48	1.46	1.44	1.43	1.41	1.38	1.34	1.29
22	0.1	14.38	9.61	7.80	6.81	6.19	5.76	5.19	4.58	3.92	3.15
	0.5	9.73	6.81	5.65	5.02	4.61	4.32	3.94	3.54	3.08	2.55
	1	7.94	5.72	4.82	4.31	3.99	3.76	3.45	3.12	2.75	2.31
	2.5	5.79	4.38	3.78	3.44	3.22	3.05	2.84	2.60	2.33	2.00
	5	4.30	3.44	3.05	2.82	2.66	2.55	2.40	2.23	2.03	1.78
	10	2.95	2.56	2.35	2.22	2.13	2.06	1.97	1.86	1.73	1.57
	20	1.75	1.73	1.68	1.64	1.61	1.58	1.54	1.49	1.43	1.35
	25	1.40	1.48	1.47	1.46	1.44	1.42	1.40	1.37	1.33	1.28
23	0.1	14.19	9.47	7.67	6.69	6.08	5.65	5.09	4.48	3.82	3.05
	0.5	9.63	6.73	5.58	4.95	4.54	4.26	3.88	3.47	3.02	2.48
	1	7.88	5.66	4.76	4.26	3.94	3.71	3.41	3.07	2.70	2.26
	2.5	5.75	4.35	3.75	3.41	3.18	3.02	2.81	2.57	2.30	1.97
	5	4.28	3.42	3.03	2.80	2.64	2.53	2.38	2.20	2.00	1.76
	10	2.94	2.55	2.34	2.21	2.11	2.05	1.95	1.84	1.72	1.55
	20	1.74	1.73	1.68	1.63	1.60	1.57	1.53	1.49	1.42	1.34
	25	1.39	1.47	1.47	1.45	1.43	1.41	1.40	1.37	1.33	1.27
24	0.1	14.03	9.34	7.55	6.59	5.98	5.55	4.99	4.39	3.74	2.97
	0.5	9.55	6.66	5.52	4.89	4.49	4.20	3.83	3.42	2.97	2.43
	1	7.82	5.61	4.72	4.22	3.90	3.67	3.36	3.03	2.66	2.21
	2.5	5.72	4.32	3.72	3.38	3.15	2.99	2.78	2.54	2.27	1.94
	5	4.26	3.40	3.01	2.78	2.62	2.51	2.36	2.18	1.98	1.73
	10	2.93	2.54	2.33	2.19	2.10	2.04	1.94	1.83	1.70	1.53
	20	1.74	1.72	1.67	1.63	1.59	1.57	1.53	1.48	1.42	1.33
	25	1.39	1.47	1.46	1.44	1.43	1.41	1.39	1.36	1.32	1.26

df_2	df_1	1	2	3	4	5	6	8	12	24	∞
25	0.1%	13.88	9.22	7.45	6.49	5.88	5.46	4.91	4.31	3.66	2.89
	0.5%	9.48	6.60	5.46	4.84	4.43	4.15	3.78	3.37	2.92	2.38
	1 %	7.77	5.57	4.68	4.18	3.86	3.63	3.32	2.99	2.62	2.17
	2.5%	5.69	4.29	3.69	3.35	3.13	2.97	2.75	2.51	2.24	1.91
	5 %	4.24	3.38	2.99	2.76	2.60	2.49	2.34	2.16	1.96	1.71
	10 %	2.92	2.53	2.32	2.18	2.09	2.02	1.93	1.82	1.69	1.52
	20 %	1.73	1.72	1.66	1.62	1.59	1.56	1.52	1.47	1.41	1.32
	25 %	1.39	1.47	1.46	1.44	1.42	1.41	1.39	1.36	1.32	1.25
26	0.1	13.74	9.12	7.36	6.41	5.80	5.38	4.83	4.24	3.59	2.82
	0.5	9.41	6.54	5.41	4.79	4.38	4.10	3.73	3.33	2.87	2.33
	1	7.72	5.53	4.64	4.14	3.82	3.59	3.29	2.96	2.58	2.13
	2.5	5.66	4.27	3.67	3.33	3.10	2.94	2.73	2.49	2.22	1.88
	5	4.22	3.37	2.98	2.74	2.59	2.47	2.32	2.15	1.95	1.69
	10	2.91	2.52	2.31	2.17	2.08	2.01	1.92	1.81	1.68	1.50
	20	1.73	1.71	1.66	1.62	1.58	1.56	1.52	1.47	1.40	1.31
	25	1.38	1.46	1.45	1.44	1.42	1.41	1.38	1.35	1.31	1.25
27	0.1	13.61	9.02	7.27	6.33	5.73	5.31	4.76	4.17	3.52	2.75
	0.5	9.34	6.49	5.36	4.74	4.34	4.06	3.69	3.28	2.83	2.29
	1	7.68	5.49	4.60	4.11	3.78	3.56	3.26	2.93	2.55	2.10
	2.5	5.63	4.24	3.65	3.31	3.08	2.92	2.71	2.47	2.19	1.85
	5	4.21	3.35	2.96	2.73	2.57	2.46	2.30	2.13	1.93	1.67
	10	2.90	2.51	2.30	2.17	2.07	2.00	1.91	1.80	1.67	1.49
	20	1.73	1.71	1.66	1.61	1.58	1.55	1.51	1.46	1.40	1.30
	25	1.38	1.46	1.45	1.43	1.42	1.40	1.38	1.35	1.31	1.24
28	0.1	13.50	8.93	7.19	6.25	5.66	5.24	4.69	4.11	3.46	2.70
	0.5	9.28	6.44	5.32	4.70	4.30	4.02	3.65	3.25	2.79	2.25
	1	7.64	5.45	4.57	4.07	3.75	3.53	3.23	2.90	2.52	2.06
	2.5	5.61	4.22	3.63	3.29	2.06	2.90	2.69	2.45	2.17	1.83
	5	4.20	3.34	2.95	2.71	2.56	2.44	2.29	2.12	1.91	1.65
	10	2.89	2.50	2.29	2.16	2.06	2.00	1.90	1.79	1.66	1.48
	20	1.72	1.71	1.65	1.61	1.57	1.55	1.51	1.46	1.39	1.30
	25	1.38	1.46	1.45	1.43	1.41	1.40	1.38	1.34	1.30	1.24
29	0.1	13.39	8.85	7.12	6.19	5.59	5.18	4.64	4.05	3.41	2.64
	0.5	9.23	6.40	5.28	4.66	4.26	3.98	3.61	3.21	2.76	2.21
	1	7.60	5.42	4.54	4.04	3.73	3.50	3.20	2.87	2.49	2.03
	2.5	5.59	4.20	3.61	3.27	3.04	2.88	2.67	2.43	2.15	1.81
	5	4.18	3.33	2.93	2.70	2.54	2.43	2.28	2.10	1.90	1.64
	10	2.89	2.50	2.28	2.15	2.06	1.99	1.89	1.78	1.65	1.47
	20	1.72	1.70	1.65	1.60	1.57	1.54	1.50	1.45	1.39	1.29
	25	1.38	1.45	1.45	1.43	1.41	1.40	1.37	1.34	1.30	1.23
30	0.1	13.29	8.77	7.05	6.12	5.53	5.12	4.58	4.00	3.36	2.59
	0.5	9.18	6.35	5.24	4.62	4.23	3.95	3.58	3.18	2.73	2.18
	1	7.56	5.39	4.51	4.02	3.70	3.47	3.17	2.84	2.47	2.01
	2.5	5.57	4.18	3.59	3.25	3.03	2.87	2.65	2.41	2.14	1.79
	5	4.17	3.32	2.92	2.69	2.53	2.42	2.27	2.09	1.89	1.62
	10	2.88	2.49	2.28	2.14	2.05	1.98	1.88	1.77	1.64	1.46
	20	1.72	1.70	1.64	1.60	1.57	1.54	1.50	1.45	1.38	1.28
	25	1.38	1.45	1.44	1.42	1.41	1.39	1.37	1.34	1.29	1.23

df_2 \ df_1		1	2	3	4	5	6	8	12	24	∞
40	0.1%	12.61	8.25	6.60	5.70	5.13	4.73	4.21	3.64	3.01	2.23
	0.5%	8.83	6.07	4.98	4.37	3.99	3.71	3.35	2.95	2.50	1.93
	1 %	7.31	5.18	4.31	3.83	3.51	3.29	2.99	2.66	2.29	1.80
	2.5%	5.42	4.05	3.46	3.13	2.90	2.74	2.53	2.29	2.01	1.64
	5 %	4.08	3.23	2.84	2.61	2.45	2.34	2.18	2.00	1.79	1.51
	10 %	2.84	2.44	2.23	2.09	2.00	1.93	1.83	1.71	1.57	1.38
	20 %	1.70	1.68	1.62	1.57	1.54	1.51	1.47	1.41	1.34	1.24
	25 %	1.36	1.44	1.42	1.41	1.39	1.37	1.35	1.31	1.27	1.19
60	0.1	11.97	7.76	6.17	5.31	4.76	4.37	3.87	3.31	2.69	1.90
	0.5	8.49	5.80	4.73	4.14	3.76	3.49	3.13	2.74	2.29	1.69
	1	7.08	4.98	4.13	3.65	3.34	3.12	2.82	2.50	2.12	1.60
	2.5	5.29	3.93	3.34	3.01	2.79	2.63	2.41	2.17	1.88	1.48
	5	4.00	3.15	2.76	2.52	2.37	2.25	2.10	1.92	1.70	1.39
	10	2.79	2.39	2.18	2.04	1.95	1.87	1.77	1.66	1.51	1.29
	20	1.68	1.65	1.59	1.55	1.51	1.48	1.44	1.38	1.31	1.18
	25	1.35	1.42	1.41	1.39	1.37	1.35	1.32	1.29	1.24	1.15
120	0.1	11.38	7.31	5.79	4.95	4.42	4.04	3.55	3.02	2.40	1.56
	0.5	8.18	5.54	4.50	3.92	3.55	3.28	2.93	2.54	2.09	1.43
	1	6.85	4.79	3.95	3.48	3.17	2.96	2.66	2.34	1.95	1.38
	2.5	5.15	3.80	3.23	2.89	2.67	2.52	2.30	2.05	1.76	1.31
	5	3.92	3.07	2.68	2.45	2.29	2.17	2.02	1.83	1.61	1.25
	10	2.75	2.35	2.13	1.99	1.90	1.82	1.72	1.60	1.45	1.19
	20	1.66	1.63	1.57	1.52	1.48	1.45	1.41	1.35	1.27	1.12
	25	1.34	1.40	1.39	1.37	1.35	1.33	1.30	1.26	1.21	1.10
∞	0.1	10.83	6.91	5.42	4.62	4.10	3.74	3.27	2.74	2.13	1.00
	0.5	7.88	5.30	4.28	3.72	3.35	3.09	2.74	2.36	1.90	1.00
	1	6.64	4.60	3.78	3.32	3.02	2.80	2.51	2.18	1.79	1.00
	2.5	5.02	3.69	3.12	2.79	2.57	2.41	2.19	1.94	1.64	1.00
	5	3.84	2.99	2.60	2.37	2.21	2.09	1.94	1.75	1.52	1.00
	.10	2.71	2.30	2.08	1.94	1.85	1.77	1.67	1.55	1.38	1.00
	20	1.64	1.61	1.55	1.50	1.46	1.43	1.38	1.32	1.23	1.00
	25	1.32	1.39	1.37	1.35	1.33	1.37	1.28	1.24	1.18	1.00

TABLE A-6
Distribution of $F_{\max}$ Statistic

df for s_X^{2j}	$1 - \alpha$	a = number of variances								
		2	3	4	5	6	7	8	9	10
4	.95	9.60	15.5	20.6	25.2	29.5	33.6	37.5	41.4	44.6
	.99	23.2	37.	49.	59.	69.	79.	89.	97.	106.
5	.95	7.15	10.8	13.7	16.3	18.7	20.8	22.9	24.7	26.5
	.99	14.9	22.	28.	33.	38.	42.	46.	50.	54.
6	.95	5.82	8.38	10.4	12.1	13.7	15.0	16.3	17.5	18.6
	.99	11.1	15.5	19.1	22.	25.	27.	30.	32.	34.
7	.95	4.99	6.94	8.44	9.70	10.8	11.8	12.7	13.5	14.3
	.99	8.89	12.1	14.5	16.5	18.4	20.	22.	23.	24.
8	.95	4.43	6.00	7.18	8.12	9.03	9.78	10.5	11.1	11.7
	.99	7.50	9.9	11.7	13.2	14.5	15.8	16.9	17.9	18.9
9	.95	4.03	5.34	6.31	7.11	7.80	8.41	8.95	9.45	9.91
	.99	6.54	8.5	9.9	11.1	12.1	13.1	13.9	14.7	15.3
10	.95	3.72	4.85	5.67	6.34	6.92	7.42	7.87	8.28	8.66
	.99	5.85	7.4	8.6	9.6	10.4	11.1	11.8	12.4	12.9
12	.95	3.28	4.16	4.79	5.30	5.72	6.09	6.42	6.72	7.00
	.99	4.91	6.1	6.9	7.6	8.2	8.7	9.1	9.5	9.9
15	.95	2.86	3.54	4.01	4.37	4.68	4.95	5.19	5.40	5.59
	.99	4.07	4.9	5.5	6.0	6.4	6.7	7.1	7.3	7.5
20	.95	2.46	2.95	3.29	3.54	3.76	3.94	4.10	4.24	4.37
	.99	3.32	3.8	4.3	4.6	4.9	5.1	5.3	5.5	5.6
30	.95	2.07	2.40	2.61	2.78	2.91	3.02	3.12	3.21	3.29
	.99	2.63	3.0	3.3	3.4	3.6	3.7	3.8	3.9	4.0
60	.95	1.67	1.85	1.96	2.04	2.11	2.17	2.22	2.26	2.30
	.99	1.96	2.2	2.3	2.4	2.4	2.5	2.5	2.6	2.6
∞	.95	1.00	1.00	1.00	1.00	1.00	1.00	1.00	1.00	1.00
	.99	1.00	1.00	1.00	1.00	1.00	1.00	1.00	1.00	1.00

This table is abridged from Table 31 in *Biometrika Tables for Statisticians*, Vol. 1, 2d ed. (New York: Cambridge, 1958). Edited by E. S. Pearson and H. O. Hartley. Reproduced with the permission of E. S. Pearson and the trustees of *Biometrika*.

TABLE A-7

Critical Values for Cochran's Test for Homogeneity of Variance

$$C = (\text{largest } s^2)/(\textstyle\sum s_j^2)$$

df for s_j^2	$1-\alpha$	a = number of variances										
		2	3	4	5	6	7	8	9	10	15	20
1	.95	.9985	.9669	.9065	.8412	.7808	.7271	.6798	.6385	.6020	.4709	.3894
	.99	.9999	.9933	.9676	.9279	.8828	.8376	.7945	.7544	.7175	.5747	.4799
2	.95	.9750	.8709	.7679	.6838	.6161	.5612	.5157	.4775	.4450	.3346	.2705
	.99	.9950	.9423	.8643	.7885	.7218	.6644	.6152	.5727	.5358	.4069	.3297
3	.95	.9392	.7977	.6841	.5981	.5321	.4800	.4377	.4027	.3733	.2758	.2205
	.99	.9794	.8831	.7814	.6957	.6258	.5685	.5209	.4810	.4469	.3317	.2654
4	.95	.9057	.7457	.6287	.5441	.4803	.4307	.3910	.3584	.3311	.2419	.1921
	.99	.9586	.8335	.7212	.6329	.5635	.5080	.4627	.4251	.3934	.2882	.2288
5	.95	.8772	.7071	.5895	.5065	.4447	.3974	.3595	.3286	.3029	.2195	.1735
	.99	.9373	.7933	.6761	.5875	.5195	.4659	.4226	.3870	.3572	.2593	.2048
6	.95	.8534	.6771	.5598	.4783	.4184	.3726	.3362	.3067	.2823	.2034	.1602
	.99	.9172	.7606	.6410	.5531	.4866	.4347	.3932	.3592	.3308	.2386	.1877
7	.95	.8332	.6530	.5365	.4564	.3980	.3535	.3185	.2901	.2666	.1911	.1501
	.99	.8988	.7335	.6129	.5259	.4608	.4105	.3704	.3378	.3106	.2228	.1748
8	.95	.8159	.6333	.5175	.4387	.3817	.3384	.3043	.2768	.2541	.1815	.1422
	.99	.8823	.7107	.5897	.5037	.4401	.3911	.3522	.3207	.2945	.2104	.1646
9	.95	.8010	.6167	.5017	.4241	.3682	.3259	.2926	.2659	.2439	.1736	.1357
	.99	.8674	.6912	.5702	.4854	.4229	.3751	.3373	.3067	.2813	.2002	.1567
16	.95	.7341	.5466	.4366	.3645	.3135	.2756	.2462	.2226	.2032	.1429	.1108
	.99	.7949	.6059	.4884	.4094	.3529	.3105	.2779	.2514	.2297	.1612	.1248
36	.95	.6602	.4748	.3720	.3066	.2612	.2278	.2022	.1820	.1655	.1144	.0879
	.99	.7067	.5153	.4057	.3351	.2858	.2494	.2214	.1992	.1811	.1251	.0960
144	.95	.5813	.4031	.3093	.2513	.2119	.1833	.1616	.1446	.1308	.0889	.0675
	.99	.6062	.4230	.3251	.2644	.2229	.1929	.1700	.1521	.1376	.0934	.0709

Adapted from C. Eisenhart, M. W. Hastay, and W. A. Wallis, eds., *Techniques of Statistical Analysis*, chap. 15 (New York: McGraw-Hill, 1947).

Power Function for Analysis of Variance Test

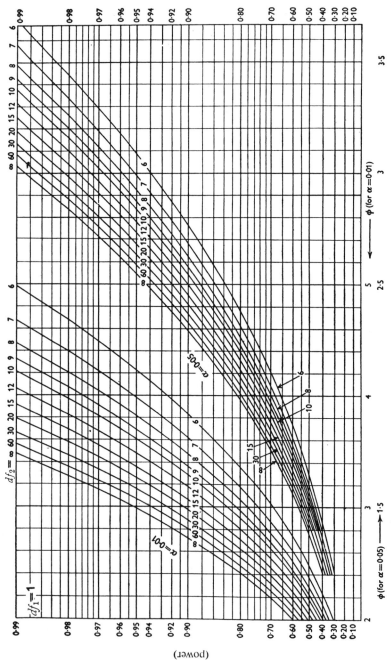

Adapted from E. S. Pearson and H. O. Hartley, "Charts of the Power Function for Analysis of Variance Tests Derived From the Non-Central F-Distribution," *Biometrika*, 38:112-130 (1951), with the permission of E. S. Pearson and the trustees of *Biometrika*.

TABLE A-8 (*Continued*)

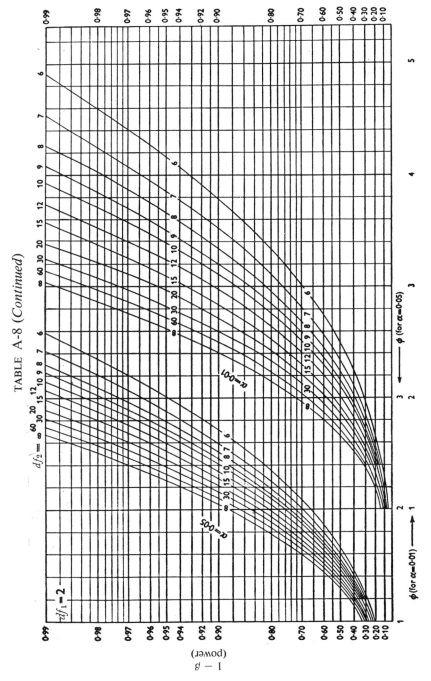

445

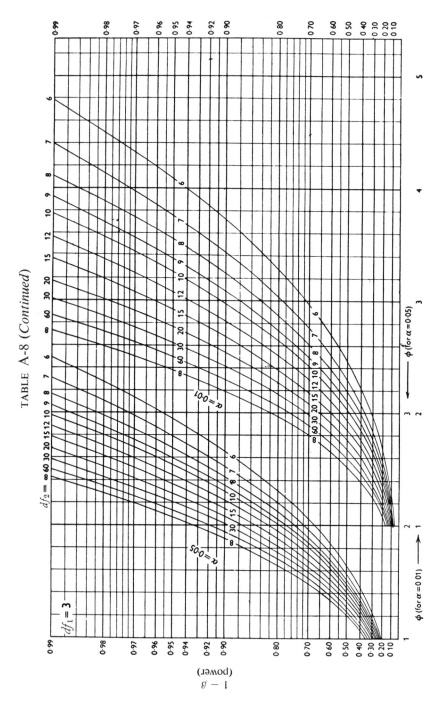

TABLE A-8 (*Continued*)

447

TABLE A-8 (Continued)

TABLE A-8 (*Continued*)

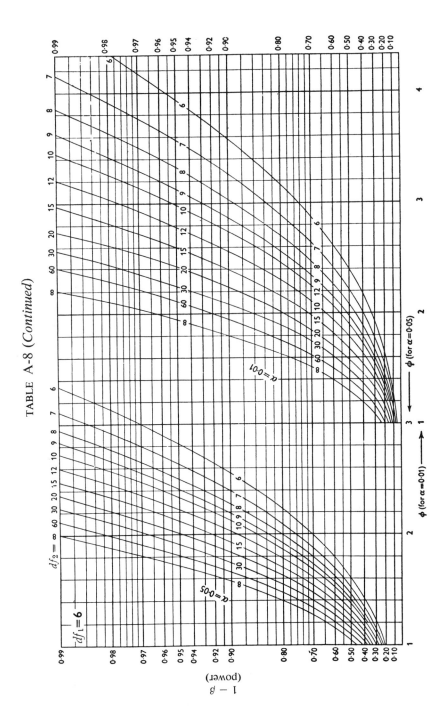

449

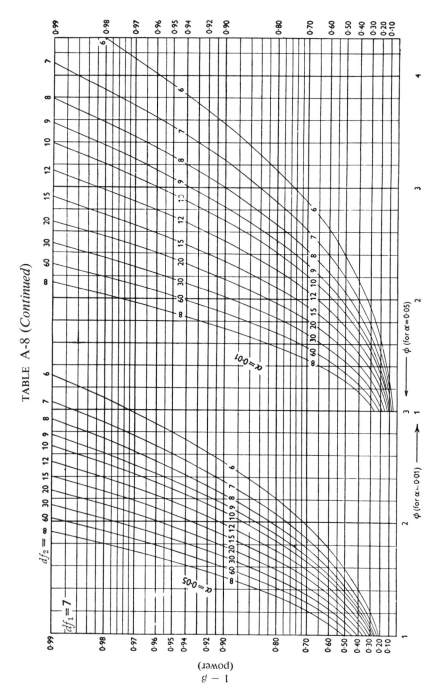

TABLE A-8 (Continued)

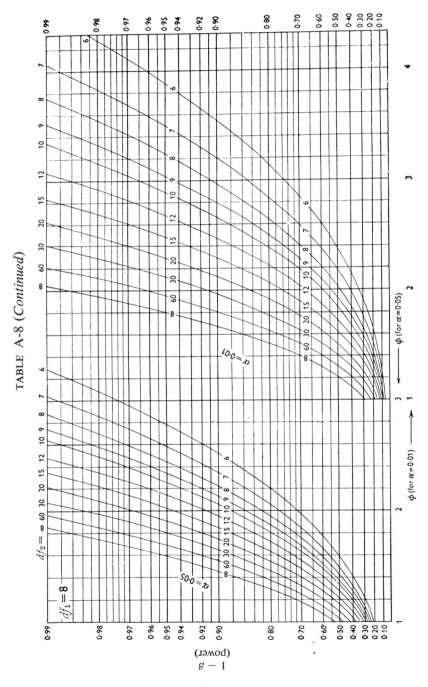

451

TABLE A-9
Distribution of the Studentized Range Statistic

df for $s_{\bar{X}}$	$1-\alpha$	Number of ordered means													
		2	3	4	5	6	7	8	9	10	11	12	13	14	15
1	.95	18.0	27.0	32.8	37.1	40.4	43.1	45.4	47.4	49.1	50.6	52.0	53.2	54.3	55.4
	.99	90.0	135	164	186	202	216	227	237	246	253	260	266	272	277
2	.95	6.09	8.3	9.8	10.9	11.7	12.4	13.0	13.5	14.0	14.4	14.7	15.1	15.4	15.7
	.99	14.0	19.0	22.3	24.7	26.6	28.2	29.5	30.7	31.7	32.6	33.4	34.1	34.8	35.4
3	.95	4.50	5.91	6.82	7.50	8.04	8.48	8.85	9.18	9.46	9.72	9.95	10.2	10.4	10.5
	.99	8.26	10.6	12.2	13.3	14.2	15.0	15.6	16.2	16.7	17.1	17.5	17.9	18.2	18.5
4	.95	3.93	5.04	5.76	6.29	6.71	7.05	7.35	7.60	7.83	8.03	8.21	8.37	8.52	8.66
	.99	6.51	8.12	9.17	9.96	10.6	11.1	11.5	11.9	12.3	12.6	12.8	13.1	13.3	13.5
5	.95	3.64	4.60	5.22	5.67	6.03	6.33	6.58	6.80	6.99	7.17	7.32	7.47	7.60	7.72
	.99	5.70	6.97	7.80	8.42	8.91	9.32	9.67	9.97	10.2	10.5	10.7	10.9	11.1	11.2
6	.95	3.46	4.34	4.90	5.31	5.63	5.89	6.12	6.32	6.49	6.65	6.79	6.92	7.03	7.14
	.99	5.24	6.33	7.03	7.56	7.97	8.32	8.61	8.87	9.10	9.30	9.49	9.65	9.81	9.95
7	.95	3.34	4.16	4.69	5.06	5.36	5.61	5.82	6.00	6.16	6.30	6.43	6.55	6.66	6.76
	.99	4.95	5.92	6.54	7.01	7.37	7.68	7.94	8.17	8.37	8.55	8.71	8.86	9.00	9.12
8	.95	3.26	4.04	4.53	4.89	5.17	5.40	5.60	5.77	5.92	6.05	6.18	6.29	6.39	6.48
	.99	4.74	5.63	6.20	6.63	6.96	7.24	7.47	7.68	7.87	8.03	8.18	8.31	8.44	8.55
9	.95	3.20	3.95	4.42	4.76	5.02	5.24	5.43	5.60	5.74	5.87	5.98	6.09	6.19	6.28
	.99	4.60	5.43	5.96	6.35	6.66	6.91	7.13	7.32	7.49	7.65	7.78	7.91	8.03	8.13
10	.95	3.15	3.88	4.33	4.65	4.91	5.12	5.30	5.46	5.60	5.72	5.83	5.93	6.03	6.11
	.99	4.48	5.27	5.77	6.14	6.43	6.67	6.87	7.05	7.21	7.36	7.48	7.60	7.71	7.81
11	.95	3.11	3.82	4.26	4.57	4.82	5.03	5.20	5.35	5.49	5.61	5.71	5.81	5.90	5.99
	.99	4.39	5.14	5.62	5.97	6.25	6.48	6.67	6.84	6.99	7.13	7.26	7.36	7.46	7.56

12	.95	3.08	3.77	4.20	4.51	4.75	4.95	5.12	5.27	5.40	5.51	5.62	5.71	5.80	5.88
	.99	4.32	5.04	5.50	5.84	6.10	6.32	6.51	6.67	6.81	6.94	7.06	7.17	7.26	7.36
13	.95	3.06	3.73	4.15	4.45	4.69	4.88	5.05	5.19	5.32	5.43	5.53	5.63	5.71	5.79
	.99	4.26	4.96	5.40	5.73	5.98	6.19	6.37	6.53	6.67	6.79	6.90	7.01	7.10	7.19
14	.95	3.03	3.70	4.11	4.41	4.64	4.83	4.99	5.13	5.25	5.36	5.46	5.55	5.64	5.72
	.99	4.21	4.89	5.32	5.63	5.88	6.08	6.26	6.41	6.54	6.66	6.77	6.87	6.96	7.05
16	.95	3.00	3.65	4.05	4.33	4.56	4.74	4.90	5.03	5.15	5.26	5.35	5.44	5.52	5.59
	.99	4.13	4.78	5.19	5.49	5.72	5.92	6.08	6.22	6.35	6.46	6.56	6.66	6.74	6.82
18	.95	2.97	3.61	4.00	4.28	4.49	4.67	4.82	4.96	5.07	5.17	5.27	5.35	5.43	5.50
	.99	4.07	4.70	5.09	5.38	5.60	5.79	5.94	6.08	6.20	6.31	6.41	6.50	6.58	6.65
20	.95	2.95	3.58	3.96	4.23	4.45	4.62	4.77	4.90	5.01	5.11	5.20	5.28	5.36	5.43
	.99	4.02	4.64	5.02	5.29	5.51	5.69	5.84	5.97	6.09	6.19	6.29	6.37	6.45	6.52
24	.95	2.92	3.53	3.90	4.17	4.37	4.54	4.68	4.81	4.92	5.01	5.10	5.18	5.25	5.32
	.99	3.96	4.54	4.91	5.17	5.37	5.54	5.69	5.81	5.92	6.02	6.11	6.19	6.26	6.33
30	.95	2.89	3.49	3.84	4.10	4.30	4.46	4.60	4.72	4.83	4.92	5.00	5.08	5.15	5.21
	.99	3.89	4.45	4.80	5.05	5.24	5.40	5.54	5.56	5.76	5.85	5.93	6.01	6.08	6.14
40	.95	2.86	3.44	3.79	4.04	4.23	4.39	4.52	4.63	4.74	4.82	4.91	4.98	5.05	5.11
	.99	3.82	4.37	4.70	4.93	5.11	5.27	5.39	5.50	5.60	5.69	5.77	5.84	5.90	5.96
60	.95	2.83	3.40	3.74	3.98	4.16	4.31	4.44	4.55	4.65	4.73	4.81	4.88	4.94	5.00
	.99	3.76	4.28	4.60	4.82	4.99	5.13	5.25	5.36	5.45	5.53	5.60	5.67	5.73	5.79
120	.95	2.80	3.36	3.69	3.92	4.10	4.24	4.36	4.48	4.56	4.64	4.72	4.78	4.84	4.90
	.99	3.70	4.20	4.50	4.71	4.87	5.01	5.12	5.21	5.30	5.38	5.44	5.51	5.56	5.61
∞	.95	2.77	3.31	3.63	3.86	4.03	4.17	4.29	4.39	4.47	4.55	4.62	4.68	4.74	4.80
	.99	3.64	4.12	4.40	4.60	4.76	4.88	4.99	5.08	5.16	5.23	5.29	5.35	5.40	5.45

This table is abridged from Table II.2 in *The Probability Integrals of the Range and of the Studentized Range*, prepared by H. Leon Harter, Donald S. Clemm, and Eugene H. Guthrie. These tables are published in WADC tech. Rep. 58–484, vol. 2, 1959, Wright Air Development Center, and are reproduced with the permission of the authors. Note that the tabled values are two-tailed.

Distribution of d Statistic in Comparing Treatment Means With a Control

df for MS_{error}	$1 - \alpha$	Number of means (including control)								
		2	3	4	5	6	7	8	9	10
6	.95	1.94	2.34	2.56	2.71	2.83	2.92	3.00	3.07	3.12
	.975	2.45	2.86	3.18	3.41	3.60	3.75	3.88	4.00	4.11
	.99	3.14	3.61	3.88	4.07	4.21	4.33	4.43	4.51	4.59
	.995	3.71	4.22	4.60	4.88	5.11	5.30	5.47	5.61	5.74
7	.95	1.89	2.27	2.48	2.62	2.73	2.82	2.89	2.95	3.01
	.975	2.36	2.75	3.04	3.24	3.41	3.54	3.66	3.76	3.86
	.99	3.00	3.42	3.66	3.83	3.96	4.07	4.15	4.23	4.30
	.995	3.50	3.95	4.28	4.52	4.17	4.87	5.01	5.13	5.24
8	.95	1.86	2.22	2.42	2.55	2.66	2.74	2.81	2.87	2.92
	.975	2.31	2.67	2.94	3.13	3.28	3.40	3.51	3.60	3.68
	.99	2.90	3.29	3.51	3.67	3.79	3.88	3.96	4.03	4.09
	.995	3.36	3.77	4.06	4.27	4.44	4.58	4.70	4.81	4.90
9	.95	1.83	2.18	2.37	2.50	2.60	2.68	2.75	2.81	2.86
	.975	2.26	2.61	2.86	3.04	3.18	3.29	3.39	3.48	3.55
	.99	2.82	3.19	3.40	3.55	3.66	3.75	3.82	3.89	3.94
	.995	3.25	3.63	3.90	4.09	4.24	4.37	4.48	4.57	4.65
10	.95	1.81	2.15	2.34	2.47	2.56	2.64	2.70	2.76	2.81
	.975	2.23	2.57	2.81	2.97	3.11	3.21	3.31	3.39	3.46
	.99	2.76	3.11	3.31	3.45	3.56	3.64	3.71	3.78	3.83
	.995	3.17	3.53	3.78	3.95	4.10	4.21	4.31	4.40	4.47
11	.95	1.80	2.13	2.31	2.44	2.53	2.60	2.67	2.72	2.77
	.975	2.20	2.53	2.76	2.92	3.05	3.15	3.24	3.31	3.38
	.99	2.72	3.06	3.25	3.38	3.48	3.56	3.63	3.69	3.74
	.995	3.11	3.45	3.68	3.85	3.98	4.09	4.18	4.26	4.33
12	.95	1.78	2.11	2.29	2.41	2.50	2.58	2.64	2.69	2.74
	.975	2.18	2.50	2.72	2.88	3.00	3.10	3.18	3.25	3.32
	.99	2.68	3.01	3.19	3.32	3.42	3.50	3.56	3.62	3.67
	.995	3.05	3.39	3.61	3.76	3.89	3.99	4.08	4.15	4.22
13	.95	1.77	2.09	2.27	2.39	2.48	2.55	2.61	2.66	2.71
	.975	2.16	2.48	2.69	2.84	2.96	3.06	3.14	3.21	3.27
	.99	2.65	2.97	3.15	3.27	3.37	3.44	3.51	3.56	3.61
	.995	3.01	3.33	3.54	3.69	3.81	3.91	3.99	4.06	4.13
14	.95	1.76	2.08	2.25	2.37	2.46	2.53	2.59	2.64	2.69
	.975	2.14	2.46	2.67	2.81	2.93	3.02	3.10	3.17	3.23
	.99	2.62	2.94	3.11	3.23	3.32	3.40	3.46	3.51	3.56
	.995	2.98	3.29	3.49	3.64	3.75	3.84	3.92	3.99	4.05

This table is adapted from "A Multiple Comparison Procedure for Comparing Several Treatments With a Control," *Journal of the American Statistical Association*, 50: 1096–1121 (1955), with the permission of the author, C. W. Dunnett, and the editor. Note that the tabled values are one-tailed.

df for MS_{error}	$1 - \alpha$	Number of means (including control)								
		2	3	4	5	6	7	8	9	10
16	.95	1.75	2.06	2.23	2.34	2.43	2.50	2.56	2.61	2.65
	.975	2.12	2.42	2.63	2.77	2.88	2.96	3.04	3.10	3.16
	.99	2.58	2.88	3.05	3.17	3.26	3.33	3.39	3.44	3.48
	.995	2.92	3.22	3.41	3.55	3.65	3.74	3.82	3.88	3.93
18	.95	1.73	2.04	2.21	2.32	2.41	2.48	2.53	2.58	2.62
	.975	2.10	2.40	2.59	2.73	2.84	2.92	2.99	3.05	3.11
	.99	2.55	2.84	3.01	3.12	3.21	3.27	3.33	3.38	3.42
	.995	2.88	3.17	3.35	3.48	3.58	3.67	3.74	3.80	3.85
20	.95	1.72	2.03	2.19	2.30	2.39	2.46	2.51	2.56	2.60
	.975	2.09	2.38	2.57	2.70	2.81	2.89	2.96	3.02	3.07
	.99	2.53	2.81	2.97	3.08	3.17	3.23	3.29	3.34	3.38
	.995	2.85	3.13	3.31	3.43	3.53	3.61	3.67	3.73	3.78
24	.95	1.71	2.01	2.17	2.28	2.36	2.43	2.48	2.53	2.57
	.975	2.06	2.35	2.53	2.66	2.76	2.84	2.91	2.96	3.01
	.99	2.49	2.77	2.92	3.03	3.11	3.17	3.22	3.27	3.31
	.995	2.80	3.07	3.24	3.36	3.45	3.52	3.58	3.64	3.69
30	.95	1.70	1.99	2.15	2.25	2.33	2.40	2.45	2.50	2.54
	.975	2.04	2.32	2.50	2.62	2.72	2.79	2.86	2.91	2.96
	.99	2.46	2.72	2.87	2.97	3.05	3.11	3.16	3.21	3.24
	.995	2.75	3.01	3.17	3.28	3.37	3.44	3.50	3.55	3.59
40	.95	1.68	1.97	2.13	2.23	2.31	2.37	2.42	2.47	2.51
	.975	2.02	2.29	2.47	2.58	2.67	2.75	2.81	2.86	2.90
	.99	2.42	2.68	2.82	2.92	2.99	3.05	3.10	3.14	3.18
	.995	2.70	2.95	3.10	3.21	3.29	3.36	3.41	3.46	3.50
60	.95	1.67	1.95	2.10	2.21	2.28	2.35	2.39	2.44	2.48
	.975	2.00	2.27	2.43	2.55	2.63	2.70	2.76	2.81	2.85
	.99	2.39	2.64	2.78	2.87	2.94	3.00	3.04	3.08	3.12
	.995	2.66	2.90	3.04	3.14	3.22	3.28	3.33	3.38	3.42
120	.95	1.66	1.93	2.08	2.18	2.26	2.32	2.37	2.41	2.45
	.975	1.98	2.24	2.40	2.51	2.59	2.66	2.71	2.76	2.80
	.99	2.36	2.60	2.73	2.82	2.89	2.94	2.99	3.03	3.06
	.995	2.62	2.84	2.98	3.08	3.15	3.21	3.25	3.30	3.33
∞	.95	1.64	1.92	2.06	2.16	2.23	2.29	2.34	2.38	2.42
	.975	1.96	2.21	2.37	2.47	2.55	2.62	2.67	2.71	2.75
	.99	2.33	2.56	2.68	2.77	2.84	2.89	2.93	2.97	3.00
	.995	2.58	2.79	2.92	3.01	3.08	3.14	3.18	3.22	3.25

TABLE A-11

Coefficients of Orthogonal Polynomials

a	Polynomial	$X = 1$	2	3	4	5	6	7	8	9	10	$\Sigma \xi'^2$	λ
3	Linear	−1	0	1								2	1
	Quadratic	1	−2	1								6	3
	Linear	−3	−1	1	3							20	2
4	Quadratic	1	−1	−1	1							4	1
	Cubic	−1	3	−3	1							20	10⅓
	Linear	−2	−1	0	1	2						10	1
5	Quadratic	2	−1	−2	−1	2						14	1
	Cubic	−1	2	0	−2	1						10	5⁄6
	Quartic	1	−4	6	−4	1						70	35⁄12
	Linear	−5	−3	−1	1	3	5					70	2
6	Quadratic	5	−1	−4	−4	−1	5					84	3⁄2
	Cubic	−5	7	4	−4	−7	5					180	5⁄3
	Quartic	1	−3	2	2	−3	1					28	7⁄12
	Linear	−3	−2	−1	0	1	2	3				28	1
7	Quadratic	5	0	−3	−4	−3	0	5				84	1
	Cubic	−1	1	1	0	−1	−1	1				6	1⁄6
	Quartic	3	−7	1	6	1	−7	3				154	7⁄12
	Linear	−7	−5	−3	−1	1	3	5	7			168	2
	Quadratic	7	1	−3	−5	−5	−3	1	7			168	1
8	Cubic	−7	5	7	3	−3	−7	−5	7			264	2⁄3
	Quartic	7	−13	−3	9	9	−3	−13	7			616	7⁄12
	Quintic	−7	23	−17	−15	15	17	−23	7			2184	7⁄10
	Linear	−4	−3	−2	−1	0	1	2	3	4		60	1
	Quadratic	28	7	−8	−17	−20	−17	−8	7	28		2772	3
9	Cubic	−14	7	13	9	0	−9	−13	−7	14		990	5⁄6
	Quartic	14	−21	−11	9	18	9	−11	−21	14		2002	7⁄12
	Quintic	−4	11	−4	−9	0	9	4	−11	4		468	3⁄20
	Linear	−9	−7	−5	−3	−1	1	3	5	7	9	330	2
	Quadratic	6	2	−1	−3	−4	−4	−3	−1	2	6	132	1⁄2
10	Cubic	−42	14	35	31	12	−12	−31	−35	−14	42	8580	5⁄3
	Quartic	18	−22	−17	3	18	18	3	−17	−22	18	2860	5⁄12
	Quintic	−6	14	−1	−11	−6	6	11	1	−14	6	780	1⁄10

This table is adapted with permission from B. J. Winer, *Statistical Principles in Experimental Design* (New York: McGraw Hill, 1962).

456

K	EW	5	7	10	12	15	20	24	30	40	60	120
							df_{error}					
2	0.01	4.7742	4.0265	3.5815	3.4284	3.2864	3.1531	3.0920	3.0297	2.9701	2.9134	2.8618
	0.05	3.1614	2.8438	2.6349	2.5615	2.4898	2.4238	2.3921	2.3600	2.3274	2.2986	2.2735
	0.10	2.5689	2.3640	2.2269	2.1793	2.1306	2.0854	2.0624	2.0438	2.0203	2.0014	1.9822
3	0.01	5.2476	4.3537	3.8292	3.6474	3.4819	3.3296	3.2591	3.1869	3.1218	3.0581	2.9961
	0.05	3.5319	3.1281	2.8705	2.7793	2.6957	2.6122	2.5736	2.5372	2.4974	2.4628	2.4306
	0.10	2.9101	2.6421	2.4686	2.4040	2.3438	2.2850	2.2567	2.2312	2.2054	2.1793	2.1562
4	0.01	5.6099	4.5909	4.0027	3.8073	3.6245	3.4561	3.3759	3.3009	3.2278	3.1569	3.0901
	0.05	3.8073	3.3375	3.0383	2.9362	2.8371	2.7448	2.6993	2.6565	2.6168	2.5763	2.5391
	0.10	3.1614	2.8438	2.6349	2.5615	2.4898	2.4238	2.3921	2.3600	2.3274	2.2986	2.2735
5	0.01	5.8999	4.7842	4.1432	3.9291	3.7346	3.5505	3.4656	3.3851	3.3072	3.2320	3.1597
	0.05	4.0324	3.5007	3.1719	3.0571	2.9459	2.8471	2.7964	2.7492	2.7054	2.6610	2.6204
	0.10	3.3688	2.9981	2.7621	2.6833	2.6021	2.5278	2.4921	2.4560	2.4218	2.3921	2.3620
6	0.01	6.1337	4.9459	4.2581	4.0294	3.8198	3.6302	3.5405	3.4561	3.3723	3.2918	3.2177
	0.05	4.2185	3.6356	3.2755	3.1554	3.0352	2.9272	2.8755	2.8253	2.7742	2.7274	2.6851
	0.10	3.5319	3.1281	2.8705	2.7793	2.6957	2.6122	2.5736	2.5371	2.4974	2.4628	2.4306
7	0.01	6.3552	5.0779	4.3571	4.1158	3.8971	3.6965	3.6022	3.5143	3.4271	3.3439	3.2650
	0.05	4.3848	3.7531	3.3688	3.2374	3.1112	2.9965	2.9402	2.8887	2.8362	2.7858	2.7405
	0.10	3.6813	3.2374	2.9572	2.8626	2.7708	2.6851	2.6412	2.6030	2.5610	2.5249	2.4883
8	0.01	6.5364	5.2043	4.4431	4.1918	3.9608	3.7537	3.6572	3.5614	3.4733	3.3864	3.3081
	0.05	4.5218	3.8571	3.4499	3.3088	3.1795	3.0571	2.9973	2.9427	2.8871	2.8337	2.7862
	0.10	3.8073	3.3375	3.0383	2.9362	2.8371	2.7448	2.6993	2.6565	2.6168	2.5763	2.5391
9	0.01	6.7128	5.3120	4.5209	4.2581	4.0217	3.8058	3.7022	3.6068	3.5145	3.4275	3.3444
	0.05	4.6507	3.9469	3.5156	3.3723	3.2359	3.1074	3.0446	2.9877	2.9297	2.8780	2.8253
	0.10	3.9306	3.4228	3.1074	2.9985	2.8928	2.7986	2.7522	2.7050	2.6610	2.6204	2.5833
10	0.01	6.8707	5.4043	4.5870	4.3193	4.0706	3.8486	3.7466	3.6458	3.5505	3.4613	3.3741
	0.05	4.7742	4.0265	3.5815	3.4284	3.2864	3.1531	3.0920	3.0297	2.9701	2.9134	2.8618
	0.10	4.0324	3.5007	3.1719	3.0571	2.9459	2.8471	2.7964	2.7491	2.7054	2.6610	2.6204
15	0.01	7.4932	5.7980	4.8537	4.5495	4.2733	4.0228	3.9072	3.7970	3.6895	3.5912	3.4966
	0.05	5.2476	4.3537	3.8292	3.6474	3.4819	3.3296	3.2591	3.1869	3.1218	3.0581	2.9961
	0.10	4.4592	3.8058	3.4092	3.2755	3.1474	3.0258	2.9691	2.9144	2.8618	2.8113	2.7632
20	0.01	7.9786	6.0791	5.0508	4.7151	4.4148	4.1468	4.0191	3.9024	3.7901	3.6804	3.5798
	0.05	5.6099	4.5909	4.0027	3.8073	3.6245	3.4561	3.3759	3.3009	3.2278	3.1569	3.0901
	0.10	4.7742	4.0265	3.5815	3.4284	3.2864	3.1531	3.0920	3.0297	2.9701	2.9134	2.8618
25	0.01	8.3687	6.3076	5.2005	4.8445	4.5265	4.2418	4.1080	3.9837	3.8650	3.7501	3.6443
	0.05	5.8999	4.7842	4.1432	3.9291	3.7346	3.5505	3.4656	3.3851	3.3072	3.2320	3.1597
	0.10	5.0272	4.2074	3.7186	3.5505	3.3961	3.2527	3.1832	3.1170	3.0542	2.9951	2.9373
30	0.01	8.6867	6.5030	5.3321	4.9540	4.6266	4.3176	4.1808	4.0505	3.9243	3.8058	3.6955
	0.05	6.1337	4.9459	4.2581	4.0294	3.8198	3.6302	3.5405	3.4561	3.3723	3.2918	3.2177
	0.10	5.2476	4.3537	3.8292	3.6474	3.4819	3.3296	3.2591	3.1869	3.1218	3.0581	2.9961

We wish to express our gratitude to Miss Jane Perlmutter who wrote the program to obtain these values and to the University of Massachusetts Computing Center for providing computer time. Note that the tabled values are two-tailed.

K	EW	5	7	10	12	15	df_{error} 20	24	30	40	60	120
35	0.01	8.9736	6.6660	5.4380	5.0449	4.6969	4.3869	4.2418	4.1044	3.9753	3.8554	3.7386
	0.05	6.3552	5.0779	4.3571	4.1158	3.8971	3.6965	3.6022	3.5143	3.4271	3.3439	3.2650
	0.10	5.4351	4.4845	3.9238	3.7330	3.5585	3.3996	3.3220	3.2490	3.1776	3.1112	3.0469
40	0.01	9.2387	6.8185	5.5351	5.1270	4.7654	4.4417	4.2951	4.1540	4.0191	3.8948	3.7763
	0.05	6.5364	5.2043	4.4431	4.1918	3.9608	3.7537	3.6572	3.5614	3.4733	3.3864	3.3081
	0.10	5.6099	4.5909	4.0027	3.8073	3.6245	3.4561	3.3759	3.3009	3.2278	3.1569	3.0901
45	0.01	9.4650	6.9484	5.6199	5.1984	4.8237	4.4926	4.3408	4.1955	4.0594	3.9314	3.8102
	0.05	6.7128	5.3120	4.5209	4.2581	4.0217	3.8058	3.7022	3.6068	3.5145	3.4275	3.3444
	0.10	5.7578	4.6955	4.0780	3.8687	3.6804	3.5049	3.4236	3.3444	3.2693	3.1968	3.1289
50	0.01	9.6736	7.0653	5.6930	5.2646	4.8780	4.5388	4.3801	4.2330	4.0944	3.9626	3.8384
	0.05	6.8707	5.4043	4.5870	4.3193	4.0706	3.8486	3.7466	3.6458	3.5505	3.4613	3.3741
	0.10	5.8999	4.7842	4.1432	3.9291	3.7346	3.5505	3.4656	3.3851	3.3072	3.2320	3.1597
100	0.01	11.1879	7.8828	6.2124	5.6930	5.2398	4.8368	4.6563	4.4810	4.3202	4.1688	4.0255
	0.05	7.9786	6.0791	5.0508	4.7151	4.4148	4.1468	4.0191	3.9024	3.7901	3.6804	3.5798
	0.10	6.8707	5.4043	4.5870	4.3193	4.0706	3.8486	3.7466	3.6458	3.5505	3.4613	3.3741
250	0.01	8.3687	6.3076	5.2005	4.8445	4.5265	4.2418	4.1080	3.9837	3.8650	3.7501	3.6443
	0.05	5.8999	4.7842	4.1432	3.9291	3.7346	3.5505	3.4656	3.3851	3.3072	3.2320	3.1597
	0.10	5.0272	4.2074	3.7186	3.5505	3.3961	3.2527	3.1832	3.1170	3.0542	2.9951	2.9373
500	0.01	15.5370	10.1031	7.5249	6.7622	6.1111	5.5438	5.2902	5.0539	4.8349	4.6307	4.4434
	0.05	11.1879	7.8828	6.2124	5.6930	5.2398	4.8368	4.6563	4.4810	4.3202	4.1688	4.0255
	0.10	9.6736	7.0653	5.6930	5.2646	4.8790	4.5388	4.2801	4.2330	4.0944	3.9626	3.8384
1000	0.01	17.8990	11.2212	8.1545	7.2604	6.5025	5.8543	5.5648	5.2990	5.0516	4.8253	4.6131
	0.05	12.8956	8.7857	6.7581	6.1400	5.6073	5.1393	4.9284	4.7278	4.5429	4.3716	4.2088
	0.10	11.1879	7.8828	6.2124	5.6930	5.2398	4.8368	4.6563	4.4810	4.3202	4.1688	4.0255
2000	0.01	20.5717	12.4338	8.8132	7.7805	6.9053	6.1718	5.8463	5.5438	5.2682	5.0145	4.7816
	0.05	14.8505	9.7656	7.3344	6.6076	5.9864	5.4455	5.2005	4.9735	4.7631	4.5675	4.3865
	0.10	12.8956	8.7857	6.7581	6.1400	5.6073	5.1393	4.9284	4.7278	4.5429	4.3716	4.2088
3000	0.01	22.3259	13.2039	9.2161	8.0944	7.1505	6.3595	6.0093	5.6899	5.3970	5.1248	4.8786
	0.05	16.1299	10.3796	7.6890	6.8918	6.2094	5.6239	5.3619	5.1193	4.8933	4.6832	4.4883
	0.10	14.0089	9.3485	7.0916	6.4130	5.8288	5.3174	5.0865	4.8733	4.6743	4.4867	4.3137
4000	0.01	23.6703	13.7837	9.5144	8.3241	7.3242	6.4917	6.1264	5.7903	5.4847	5.2037	4.9468
	0.05	17.0898	10.8498	7.9486	7.0969	6.3723	5.7516	5.4762	5.2220	4.9829	4.7631	4.5593
	0.10	14.8505	9.7656	7.3344	6.6076	5.9864	5.4455	5.2005	4.9735	4.7631	4.5675	4.3865
5000	0.01	24.7926	14.2488	9.7525	8.5026	7.4675	6.5984	6.2169	5.8702	5.5543	5.2637	4.9986
	0.05	17.8990	11.2212	8.1545	7.2604	6.5025	5.8543	5.5648	5.2990	5.0516	4.8253	4.6131
	0.10	15.5370	10.1031	7.5249	6.7622	6.1111	5.5438	5.2902	5.0539	4.8349	4.6307	4.4434

Index